Novel Techniques in Sensory Characterization and Consumer Profiling

Novel Techniques in Sensory Characterization and Consumer Profiling

Edited by
Paula Varela
Gastón Ares

CRC Press
Taylor & Francis Group
Boca Raton London New York

CRC Press is an imprint of the
Taylor & Francis Group, an **informa** business

CRC Press
Taylor & Francis Group
6000 Broken Sound Parkway NW, Suite 300
Boca Raton, FL 33487-2742

First issued in paperback 2016

© 2014 by Taylor & Francis Group, LLC
CRC Press is an imprint of Taylor & Francis Group, an Informa business

No claim to original U.S. Government works

Version Date: 20140320

ISBN 13: 978-1-138-03427-3 (pbk)
ISBN 13: 978-1-4665-6629-3 (hbk)

Visit the Taylor & Francis Web site at
http://www.taylorandfrancis.com

and the CRC Press Web site at
http://www.crcpress.com

Contents

Preface

Sensory characterization is one of the most powerful, sophisticated, and extensively applied tools in sensory science, in both academia and industry. It aims at providing a complete description of the sensory characteristics of products. Sensory characterization is extensively applied in the industry for the development and marketing of new products, the reformulation of existing products, the optimization of manufacturing processes, the monitoring of sensory characteristics of the products available in the market, the implementation of sensory quality assurance programs, the establishment of relationships between sensory and instrumental methods, and for estimating sensory shelf life.

Descriptive analysis techniques, such as QDA® and Spectrum®, applied with trained assessor panels have been the most common methodologies for this purpose for the last 50 years. However, due to the cost and time needed for their application, several alternative methods have been recently developed. These methods do not require training; can be performed by trained or semitrained assessors, or even naive consumers; and have been reported to be a reliable option when quick information about the sensory characteristics of a set of products is needed. The application of these novel methodologies for sensory characterization with consumers allows to better understand their perception of products, providing a description based on consumers' perception and vocabulary. Novel methodologies for sensory characterization have been rapidly gaining popularity and have become one of the most active and dynamic areas of research in sensory and consumer science in the last five years. This type of methodology opens new opportunities for those companies that cannot afford training and maintaining a trained sensory panel, or when quick information about the sensory characteristics of products is needed.

However, one of the main challenges that many researchers face in the application of novel methodologies is that information on how to implement them appears in a large number of articles published in different scientific journals. In this context, the aim of this book is to provide a comprehensive overview of classical and novel alternative methodologies characterization of food and nonfood products. The most common novel methodologies for sensory characterization are described and accompanied by detailed information for their implementation, discussion of examples of applications, and case studies. Data analysis of the majority of the methodologies is implemented in the statistical free software R, which makes the book useful for people unfamiliar with complex statistical software.

We hope that this book provides the reader a complete, actual, and critical view of new trends in sensory characterization and that it encourages the application of novel methodologies for sensory characterization. Additional material is available from the CRC Web site: http://www. crcpress.com/product/isbn/9781466566293.

Paula Varela
Gastón Ares

Acknowledgments

I would like to express my gratitude to Dr. Susana Fiszman. I would not have been able to pursue this project without her support.

Special thanks to all the collaborators who joined the project for their hard work.

Thanks to my family and friends for their unconditional support and for teaching me that every dream is possible with effort and dedication.

Quique and Gael, you are the pillars of my life, thanks for your infinite love.

Paula Varela

I am extremely grateful to my colleagues from Universidad de la República, Ana Giménez, Leticia Vidal, and Lucía Antúnez, for all their work and support. I would not have been able to edit this book without you. I would also like to thank all the authors for trusting us and joining the project.

I would also like to thank my family and friends for their continuous love and support. Thanks for giving me the strength to achieve all my dreams.

Gastón Ares

Editors

Paula Varela graduated as food engineer from Universidad de la República (Uruguay) and earned her PhD in food science and technology from Universidad Politécnica de Valencia (Spain). She has wide experience both in academic and industrial research in sensory and consumer science, having worked in collaboration with various research groups from Europe, Asia, and South America. She recently joined Nofima (Norway) as senior scientist. Dr. Varela has published more than 60 SCI papers and various book chapters, and has made several contributions to international symposiums. Also, she has taught graduate, undergraduate, and professional courses, and she has supervised various master and doctoral theses in sensory and consumer science. Dr. Varela collaborates as reviewer in various journals in this field and is a member of the editorial board of *Food Research International*. In the last years, her research has focused on the exploration of new methodologies to further understanding consumer perception, in particular sensory descriptive techniques with the use of consumers and the influence of nonsensory parameters in consumer food choice.

Gastón Ares is a food engineer. He received his PhD in chemistry, with a focus on sensory and consumer science, from Universidad de la República (Uruguay) in 2009. He has worked as professor and researcher in the Food Science and Technology Department of the Chemistry Faculty at the same university since 2005. He has authored more than 80 articles in international refereed journals and numerous presentations in scientific meetings. He was awarded the 2007 Rose Marie Pangborn Sensory Science Scholarship, granted to PhD students in sensory science worldwide. In 2011, he won the Food Quality and Preference Award for a young researcher for his contributions to sensory and consumer science. He is a member of the editorial boards of both the *Journal of Sensory Studies* and *Food Quality and Preference*, as well as an associate editor of *Food Research International*.

Contributors

Hervé Abdi
School of Behavioral and Brain
  Sciences
The University of Texas at Dallas
Richardson, Texas

Gastón Ares
Departamento de Ciencia y
  Tecnología de Alimentos
Facultad de Química
Universidad de la República
Montevideo, Uruguay

Rafael Silva Cadena
Departamento de Ciencia y
  tecnología de Alimentos
Facultad de Química
Universidad de la República
Montevideo, Uruguay

John C. Castura
Compusense Inc.
Guelph, Ontario, Canada

Sylvie Chollet
Groupe ISA
Institut Supérieur d'Agriculture
Food Institute
Catholic University
Lille, France

Christian Dehlholm
Department of Food Technology
Danish Technological Institute
Aarhus, Denmark

Julien Delarue
Laboratoire de Perception
  Sensorielle et Sensométrie
Ingénierie Procédés Aliments
AgroParisTech
Massy, France

Mara V. Galmarini
Facultad de Ciencias Agrarias,
  Pontificia Universidad Católica
  Argentina
and
CONICET - Consejo Nacional de
  Investigaciones Científicas y
  Técnicas
Buenos Aires, Argentina

Hildegarde Heymann
Department of Viticulture and
  Enology
University of California at Davis
Davis, California

Helene Hopfer
Department of Viticulture and
  Enology
University of California at Davis
Davis, California

Ellena S. King
Department of Viticulture and
  Enology
University of California at Davis
Davis, California

Sébastien Lê
Applied Mathematics Laboratory
Agrocampus Ouest
Rennes, France

Michael Meyners
Procter & Gamble Service GmbH
Schwalbach, Germany

Richard Popper
Peryam & Kroll Research
    Corporation
Chicago, Illinois

Pieter H. Punter
OP&P Product Research
Utrecht, the Netherlands

Ronan Symoneaux
Groupe Ecole Supérieure
    d'Agriculture
Unité de Recherche GRAPPE
Angers, France

Paula Tarancón
Instituto de Agroquímica y
    Tecnología de Alimentos (CSIC)
Valencia, Spain

Amparo Tárrega
Instituto de Agroquímica y
    Tecnología de Alimentos (CSIC)
Valencia, Spain

Eric Teillet
Sensostat
Centre des Sciences du Goût et de
    l'Alimentation
Dijon, France

Dominique Valentin
AgroSup Dijon
Université de Bourgogne
Dijon, France

Paula Varela
Instituto de Agroquímica y
    Tecnología de Alimentos (CSIC)
Valencia, Spain

and

Nofima AS
Ås, Norway

Leticia Vidal
Departamento de Ciencia y
    tecnología de Alimentos
Facultad de Química
Universidad de la República
Montevideo, Uruguay

Thierry Worch
QI Statistics
Berkshire, United Kingdom

1 Introduction

Paula Varela and Gastón Ares

CONTENTS

1.1 INTRODUCTION

This book tackles what has become a hot topic in sensory and consumer science in the last 10 years: sensory description by nontrained assessors. Sensory description methods—also known as sensory characterization or product profiling—are nowadays used as never before in the food industry with the use of consumers, and the line between sensory and consumer science is becoming blurred. The hypothesis that consumers are capable of accurately describing products from a sensory point of view is becoming day by day more accepted within the sensory science community. Furthermore, the realization that this kind of approach gives quick and flexible answers to the changing needs of the industry has increased interest in consumer product profiling tools. Industries as varied as food and beverages, cosmetic, personal care, sound, fabrics, or automotive are more and more interested in these kinds of methodologies.

Sensory characterization provides a representation of the qualitative and quantitative aspects of human perception, enabling measurement of the sensory reaction to the stimuli resulting from the use of a product and allowing correlations to other parameters (Murray et al. 2001; Stone and Sidel 2004; Lawless and Heymann 2010; Moussaoui and Varela 2010; Varela and Ares 2012). It is the most potent and frequently used instrument in sensory science. Sensory descriptive analysis acts as a bridge between different areas of research, product development, and consumer science, providing a link between the products' characteristics and consumer perception. Its use has steadily grown in the end of the twentieth century and the beginning of the twenty-first century. Figure 1.1 clearly shows the rise in the number of publications featuring sensory characterization since the 1960s.

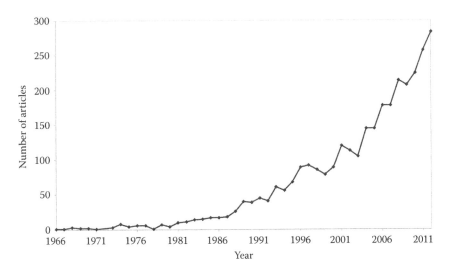

FIGURE 1.1 Number of publications featuring the words *sensory* and *characterization* from 1966 to 2012. (Extracted from Scopus.)

Describing the sensory characteristics of products has been common practice in the industry since long ago, guiding product development to match consumers' needs or getting closer to a gold standard product; to identify changes in the products as a result of changing ingredients or processes; to estimate shelf life, for quality control; and to correlate with physical or chemical measurements. Most importantly, sensory characterization allows informed business decisions. In academic research, it has been also a very important resource, helping to explain how changes in composition or structural and microstructural features determine different sensory characteristics, enabling the establishment of correlations with analytical measurements and allowing to better understand the mechanisms underlying sensory perception (Gacula 1997; Szczesniak, 2002; Stone and Sidel 2004; Moussaoui and Varela 2010; Varela and Ares 2012).

In classic sensory science, sensory analytical tests have been performed by highly trained assessors, while consumers have been considered mainly for qualitative and quantitative affective testing purposes (Stone and Sidel 2004). Therefore, sensory characterization has been traditionally performed with highly trained assessors.

There are various methods for describing the sensory characteristics of products, and there is plenty of literature extensively reviewing them in the last years. From the moment Cairncross and Sjöstrom presented the flavor profile method in 1950 to these days (Cairncross and Sjöstrom 1950), many ideas, practical aspects, and modifications have led to the most currently used technique, generic descriptive analysis, which comprises a combination of the basic elements of QDA™ and Spectrum™ (Lawless and Heymann 2010). Chapter 2 of this book is devoted to this topic, presenting the history behind descriptive analysis and detailing the most common current techniques with a fresh, up-to-date view.

Descriptive panels are highly specialized instruments that provide very detailed, robust, consistent, and reproducible results, which are stable in time and within a certain sensory space (Stone and Sidel 2004). Creating and maintaining a well-trained, calibrated sensory panel can be quite expensive and time consuming, though. It requires the selection of assessors with sensory abilities above the average, their training as individuals and as a group, the close control of their performance, and the maintenance of the panel throughout time in terms of performance and motivation. The high economic and time-consuming aspect of having a trained descriptive panel can be an issue both in industry and academic environments. Small companies cannot afford having them, big industries often have too many products that make having descriptive panels unmanageable or too expensive, and short-term projects or lack of funding sometimes put difficulties in academia to set up and maintain them. It was natural then that sensory science would flow toward more flexible and rapid sensory tools that would give extra agility to sensory characterization, both in terms of timing and training requirements (Varela and Ares 2012).

The development of free choice profiling (FCP) and repertory grid (RG) methods in the 1980s (Williams and Arnold 1985; Thomson and McEwan 1988) was a turning point, as they opened the door to the use of nontrained assessors for sensory description. Since then, descriptive analysis spiraled to what it is today, with a vast array of methods that vary in approach and outcome, with different degrees of difficulty and that can be used with panels varying in number of people and degree of training. FCP and RG are thoroughly reviewed in Chapter 6 of this book. With the emergence of these two methods, semitrained assessors and consumers started to be used for product sensory characterization, with the added realization that by allowing panelists to select their own attributes, it was possible to identify characteristics, which may not have been considered using the classic approach, as well as economizing time and resources without having to train a panel. In parallel, the advancements in multivariate statistics also played an important role in the advent of most of the novel consumer profiling techniques; the development of generalized Procrustes analysis (GPA) as statistical tool (Gower 1975) made it possible to analyze data coming from sets that differed in the number of attributes evaluated by each assessor and also having differences in the use of the scale. This statistical technique allowed the analysis of FCP data and a few years after, analysis of flash profiling data sets. Chapter 3 presents the theory and application of some of the most useful multivariate statistical tools applied in the analysis of data sets coming from consumer profiling methods.

The development of descriptive techniques continued since that time to our days. The emerging profiling methods, alternative to conventional descriptive analysis, can be performed with semitrained assessors (trained in sensory recognition and characterization but not in the specific category of products or in scaling) or with the use of nontrained people, obtaining sensory maps very close to those provided classic descriptive analysis with highly trained panels (Varela and Ares 2012). The present book details those new tools: flash profiling (Dairou and Sieffermann 2002) in Chapter 7, sorting (Schiffman et al. 1981; Lawless et al. 1995) in Chapter 8, projective mapping or Napping® (Risvik et al. 1994; Pagès 2005) in Chapter 9, and other techniques less frequently used for sensory profiling like the evaluation of open-ended questions (ten Kleij and Musters 2003) in Chapter 12 or still in early development as polarized sensory positioning (PSP) (Teillet et al. 2010; de Saldamando et al. 2013) in Chapter 10. Also, Chapter 11 is centered in a very consumer-friendly technique, the use of check-all-that-applies (CATA) questions (Adams et al. 2007).

Generally speaking, novel methodologies for sensory characterization or consumer profiling techniques are based on different approaches. There are methods based on the evaluation of individual attributes, as commonly done in conventional profiling: FCP, CATA, and flash profiling. Other methods are based on the evaluation of global differences, as sorting and Napping®. Other alternatives are the comparison with product references (PSP) and methods based on global evaluation or description of individual products, like open-ended questions. Each approach would be better suited for different particular applications, which this book aims to shed light upon. There are many studies comparing the outcomes of these methods with classic profiling tools and have generally reported a good correspondence with product mapping coming from consumers' assessment (Risvik et al. 1994; Bárcenas et al. 2004; Ares et al. 2010; Moussaoui and Varela 2010; Bruzzone et al. 2012).

On a slightly different perspective, within sensory profiling techniques, there are some that not only describe the product but also aim at determining the ideal level of the product attributes. Those techniques take advantage of the benefit of having consumers as sensory assessors, by also taking into account preferences together with attribute levels. In this line, Chapter 4 reviews the ideal profile method (Worch 2012) and Chapter 5 the use of just-about-right scales in consumer profiling (Moskowitz 1972).

All the aforementioned techniques give *static* evaluations of the sensory perceptions, where attributes are regarded as single events and their temporal aspects are not considered. The use of dynamic sensory analysis methods such as temporal dominance of sensations (TDSs) (Labbe et al. 2009), reviewed in Chapter 13, has proved useful for studying the temporal dimension of sensory perception in the mouth. In TDS, assessors evaluate which sensation is dominant over time and can also score its intensity. This approach makes it possible to monitor the behavior of the food piece when it is broken down and physically transformed in the mouth, as well as the release of flavors and aromas (Albert et al. 2012; Bruzzone et al. 2013; Laguna et al. 2013).

Conventional descriptive analysis, however, has not been substituted by novel profiling methods, and it is not expected to happen in the near future, as classic descriptive tools perform better in some cases, when a very detailed sensory description is required and also because they are more stable. That means more robust and consistent throughout time, when there is a need to compare samples from different lots or different moments in the development process and also within a sensory space, when comparing various sample sets with few samples in common. Furthermore, novel descriptive tools emerged, in fact, from a necessity to gather product descriptions directly from consumers, apart from gaining rapidity and flexibility, as complementary tools to sensory and consumer science (Moussaoui and Varela 2010; Varela and Ares 2012).

The final part of this book, Chapter 14, provides a critical comparison, discussing advantages and disadvantages of all the exposed methodologies as well as general recommendations for their application; added to this, a good part of the chapter presents an overview of the remaining challenges. There is still a good deal of research to be done in this field, particularly in terms of the practical aspects of the implementation of some of the methods, as well as regarding their statistical robustness and the validity of the results obtained with them. Panel performance and repeatability are two angles of these novel techniques that are currently much discussed in the sensory and consumer community and that need to be further investigated.

This book aims at providing a complete description of the novel methodologies developed in the last few years. Each method is accompanied by detailed information for the practical implementation, discussion of examples of applications, and case studies. Also, most chapters present instructions on how to analyze the data coming from the methodologies, starting from the raw data, presenting how to build the data tables, and explaining the analysis procedure with the use of the statistical free software R, including the corresponding script, which hopefully would make the book very useful for users that are nonfamiliar with complex statistical software.

ACKNOWLEDGMENTS

The authors are grateful to the Spanish Ministry of Science and Innovation for financial support (AGL2009-12785-C02-01) and for the contract awarded to author P. Varela (Juan de la Cierva Program). They would also like to thank the Comisión Sectorial de Investigación Científica (CSIC), Universidad de la República for financial support.

REFERENCES

Adams, J., Williams, A., Lancaster, B., and Foley, M. 2007. Advantages and uses of check-all-that-apply response compared to traditional scaling of attributes for salty snacks. In *Seventh Pangborn Sensory Science Symposium.* Minneapolis, MN, August 12–16, 2007.

Albert, A., Salvador, A., Schlich, P., and Fiszman, S.M. 2012. Comparison between temporal dominance of sensations (TDS) and key-attribute sensory profiling for evaluating solid food with contrasting textural layers: Fish sticks. *Food Quality and Preference* 24: 111–118.

Ares, G., Barreiro, C., Deliza, R., Giménez, A., and Gámbaro, A. 2010. Application of a check-all-that-apply question to the development of chocolate milk desserts. *Journal of Sensory Studies* 25: 67–86.

Bárcenas, P., Pérez Elortondo, F.J., and Albisu, M. 2004. Projective mapping in sensory analysis of ewes milk cheeses: A study on consumers and trained panel performance. *Food Research International* 37: 723–729.

Bruzzone, F., Ares, G., and Giménez, A. 2012. Consumers' texture perception of milk desserts II—Comparison with trained assessors' data. *Journal of Texture Studies* 43: 214–226.

Bruzzone, F., Ares, G., and Giménez, A. 2013. Temporal aspects of yoghurt texture perception. *International Dairy Journal* 29: 124–134.

Cairncross, S.E. and Sjöstrom, L.B. 1950. Flavor profiles: A new approach to flavor problems. *Food Technology* 4: 308–311.

Dairou, V. and Sieffermann, J.-M. 2002. A comparison of 14 jams characterized by conventional profile and a quick original method, flash profile. *Journal of Food Science* 67: 826–834.

de Saldamando, L., Delgado, P., Herencia, P., Giménez, A., and Ares, G. 2013. Polarized sensory positioning: Do conclusions depend on the poles? *Food Quality and Preference* 29: 25–32.

Gacula, M.C. 1997. *Descriptive Sensory Analysis in Practice.* Trumbull, CT: Food and Nutrition Press.

Gower, J.C. 1975. Generalised Procrustes analysis. *Psychometrika* 40: 33–50.

Labbe, D., Schlich, P., Pineau, N., Gilbert, F., and Martin, N. 2009. Temporal dominance of sensations and sensory profiling: A comparative study. *Food Quality and Preference* 20: 216–221.

Laguna, L., Varela, P., Salvador, A., and Fiszman, S.M. 2013 A new sensory tool to analyze the oral trajectory of biscuits with different fat and fiber content. *Food Research International* 51: 544–553.

Lawless, H.T. and Heymann, H. 2010. *Sensory Evaluation of Food. Principles and Practices*, 2nd edn., pp. 227–253. New York: Springer.

Lawless, H.T., Sheng, N., and Knoops, S.S.C.P. 1995. Multidimensional scaling of sorting data applied to cheese perception. *Food Quality and Preference* 6: 91–98.

Moskowitz, H.R. 1972. Subjective ideals and sensory optimization in evaluating perceptual dimensions of food. *Journal of Applied Psychology* 56: 60–66.

Moussaoui, K.A. and Varela, P. 2010. Exploring consumer product profiling techniques and their linkage to a quantitative descriptive analysis. *Food Quality and Preference* 21: 1088–1099.

Murray, J.M., Delahunty, C.M., and Baxter, I.A. 2001. Descriptive sensory analysis: Past, present and future. *Food Research International* 34: 461–471.

Pagès, J. 2005. Collection and analysis of perceived product inter-distances using multiple factor analysis: Application to the study of 10 white wines from the Loire Valley. *Food Quality and Preference* 16: 642–649.

Risvik, E., McEvan, J.A., Colwill, J.S., Rogers, R., and Lyon, D.H. 1994. Projective mapping: A tool for sensory analysis and consumer research. *Food Quality and Preference* 5: 263–269.

Schiffman, S.S., Reynolds, M.L., and Young, F.W. 1981. *Introduction to Multidimensional Scaling.* New York: Academic Press.

Stone, H. and Sidel, J.L. 2004. *Sensory Evaluation Practices.* London, U.K.: Elsevier Academic Press.

Szczesniak, A.S. 2002. Texture is a sensory property. *Food Quality and Preference* 13: 215–225.

Teillet, E., Schlich, P., Urbano, C., Cordelle, S., and Guichard, E. 2010. Sensory methodologies and the taste of water. *Food Quality and Preference* 21: 967–976.

ten Kleij, F. and Musters, P.A.D. 2003. Text analysis of open-ended survey responses: A complementary method to preference mapping. *Food Quality and Preference* 14: 43–52.

Thomson, D. and McEwan, J. 1988. An application of the repertory grid method to investigate consumer perceptions of foods. *Appetite* 10: 181–193.

Varela, P. and Ares, G. 2012. Sensory profiling, the blurred line between sensory and consumer science. A review of novel methods for product characterization. *Food Research International* 48: 893–908.

Williams, A. and Arnold, G. 1985. A comparison of the aromas of 6 coffees characterized by conventional profiling, free-choice profiling and similarity scaling methods. *Journal of the Science of Food and Agriculture* 36: 204–214.

Worch, T. 2012. The ideal profile analysis: From the validation to the statistical analysis of ideal profile data. PhD dissertation. http://www.opp.nl/uk/. Accessed June 10, 2013.

2 Classical Descriptive Analysis

Hildegarde Heymann, Ellena S. King, and Helene Hopfer

CONTENTS

2.1 INTRODUCTION AND HISTORY OF DESCRIPTIVE ANALYSIS

Classical or generic descriptive analysis (DA) is the gold standard technique in sensory science (Lawless and Heymann 2010). The method allows the experimenter to describe all the sensory attributes associated with a product and sensory differences among products. The technique is used extensively, particularly in the food, beverage, and personal care industries, as can be seen from a few examples published in 2012 on food products (Alasalvar et al. 2012; Cakir et al. 2012; Elmaci and Onogur 2012; Paulsen et al. 2012;

Zeppa et al. 2012), beverages (Garcia-Carpintero et al. 2012; Keenan et al. 2012; Ng et al. 2012; Parker et al. 2012; Sokolowsky and Fischer 2012), and other consumer products (Bacci et al. 2012; Verriele et al. 2012).

The current technique of DA originates from three different methods: Flavor Profile (FP®), Quantitative Descriptive Analysis (QDA®), and the Spectrum Method®. FP® was invented by Jean Caul and coworkers (Cairncross and Sjostrom 1950; Sjostrom et al. 1957) when they evaluated the effect of monosodium glutamate on food flavor. In this method, a group of panelists and the panel leader describe the products by consensus using agreed upon terminology and a nonnumerical scale. In the early 1970s, Herbert Stone, Joel Sidel, and others (Stone et al. 1974) created QDA®, which changed the FP® by removing the consensus evaluation of the products and adding a line scale used by each panelist individually, in replicate. This method retained the consensus generation of the attributes but allowed the use of statistical analysis on the data obtained. In the late 1970s, Gail Civille and others (Munoz and Civille 1998) created the Spectrum Method®, which uses absolute scales and attribute lexicons rather than consensus term generation.

In this chapter, we describe the two generic DA techniques that sprang from these predecessors: consensus-trained DA and ballot-trained DA. As will become clear later in this chapter, the major difference between these techniques is in the generation of attributes that the DA panel uses to score perceived intensities of the products. Despite the underlying differences in the training process, it has been shown that the data from different DA panels are very consistent (e.g., Heymann 1994; Lotong et al. 2002; Martin et al. 2000).

2.2 PROCESS OF DESCRIPTIVE ANALYSIS

2.2.1 EXPERIMENTAL DESIGN

As with all studies, the experiment must be designed before the DA can be performed. Since this is not a chapter on experimental design, the following books and chapters would provide an excellent foundation into the design of DA experiments (Gacula et al. 2008; Meullenet et al. 2007, Naes et al. 2010). However, a few key points should always be kept in mind. These are replication, number of panelists, carryover, and number of samples per session. These will be discussed here.

There are scientists who believe that with an extremely well-trained panel, there is no need for replication (Mammasse et al. 2011). However, unless one has spent a great deal of time determining that the panel is truly reproducible (which could take years), it is much better statistically and much faster to add replication in the experimental design. While the original QDA® suggests four to six replications, general agreement among sensory community is that three replications are sufficient and give enough statistical power, if paired with a trained panel of adequate size.

The literature states that the adequate number of panelists is between 8 and 12 (Lawless and Heymann 2010). A recent publication has also affirmed that this seems to be an ideal number (Heymann et al. 2012). However, the number of panelists may be lower when there are large differences among the samples (Mammasse and Schlich 2012), and conversely, if only subtle differences exist among samples, then more panelists would be required.

If samples are likely to cause a carryover effect from one sample to the next, for example, *astringency* in wines or *heat* in products flavored with chilies, then both an adequate rinsing regime and an experimental design that allow the researcher to determine carryover effects are needed. Using a Williams Latin square design or an incomplete block design for carryover effects for the product presentation in the DA is an easy way to evaluate the effects of sample carryover as needed (Ball 1997; Wakeling and MacFie 1995).

The number of samples served within a session is determined by the type of sample and the specific attributes being evaluated. For samples that are evaluated visually or tactilely, but not orally or nasally, panelist fatigue is less likely, and thus, evaluating up to 15 or 20 samples per session is possible. However, if the samples are evaluated for aroma and flavor, then the number of samples per session should be much lower. If the samples are *challenging*, for example, highly astringent wines, spirits, or very spicy salsas, then the number of samples per session would be even fewer. For example, in a study by Cliff and Heymann (1992), panelists evaluated only three samples in a session during an examination of oral pungency. As a general rule of thumb, about six samples per session seems acceptable.

2.2.2 PANELIST SELECTION

First and foremost, the panelists must be motivated and interested in serving on the panel. If this is true, then we have found in over 30 years of training panelists that essentially everyone can be trained and can be a reliable panelist. Secondly, the panelists must be reliable, in that they arrive when they are supposed to and that they follow instructions. Beyond these requirements, we have found that panelist selection is relatively simple. We usually do not do extensive screening, although others encourage this (e.g., Barcenas et al. 2000; Noronha et al. 1995). However, Nachtsheim et al. (2012) found that screening seems to decrease panel performance—this makes some sense, especially if the screening process is onerous and protracted. The panelists may lose interest and motivation before they even start the training for the specific study. On the other hand, screening for competence in a specific task may be important for the outcomes of the study. For instance, one should screen for color blindness, if evaluation of color is part of the DA.

In our laboratory, we have also found that trying to train experts, such as expert wine judges, coffee tasters, or dairy judges, as DA panelists can be very frustrating for both the panelist and the panel leader. It is often easier to train panelists who are novices, as far as the specific product is concerned.

2.2.3   TERM GENERATION AND REFERENCE STANDARDS

The next step after panel selection is term generation. The procedures differ depending on whether the process will involve consensus training or ballot training.

2.2.3.1   Consensus Training

In this process, the panelists are charged with determining, through consensus, the attributes that discriminate among the samples. On the first day, we usually serve them two or three samples from the product set—these samples are chosen to be as different as possible in order for the panelists to feel that they are actually able to do the task at hand. We then ask the panelists, individually and quietly, without conversation, to determine a list of attributes that discriminate among the given samples. They are told that the attribute terms must be actionable in the sense that we can make a reference standard for it. This means that a term such as *green vegetable* would be acceptable but *yummy* would not. In the case of the last descriptor, it must be made clear to the panel that their opinion or preference for the product is not important. Also ambiguous terms like *complexity* should be discouraged since creating reference standards for such a term would likely be impossible.

Once all panelists have assessed the products, we ask each panelist to read the attributes they used. We write all words on a board—grouping words where possible and indicating words that were used multiple times. This process usually takes about 50–60 min.

At the next training session, we give the panelists another subset of samples from the product set (these are frequently more similar to one another than the first subset) and we repeat the process. During this session, we also start showing the panel potential reference standards to anchor the attributes (see Section 2.2.3.3). The process is repeated as many times as is necessary to allow the panel to see all samples in the product set and to ensure that all potential attribute terms have been listed.

Usually, starting in the third training session, the panel leader will work with the panel to determine which of the listed attributes will actually be used in the study. There are usually a number of terms that were used by most, if not all, panelists and these clearly need to be in the final attribute list. There are also frequently a number of terms that are synonymous, and in these cases, it is relatively easy to find a compromise. The more problematic terms are the ones that were used by a few panelists but that do seem to describe specific differences among the samples in the product set. For these terms, the panel leader's negotiation skills become crucial. The trick is to minimize the eventual attribute list, to prevent panelist fatigue, while still covering all the differences among the samples. It is often worth adding one or two additional terms to maintain panel harmony, but care must be taken not to add too many. Frequently, an especially vocal proponent of a specific attribute will be mollified if the panel leader explains that the score sheet will have a line scale labeled "Other" where the panelists can indicate the attribute and then score its intensity. The "Other" attribute is also useful to minimize *dumping*. This occurs when panelists perceive a difference in an attribute but the attribute is not part of the listed terms (Lawless and Heymann 2010).

Once the attribute list has been completed, then the training sessions involve making sure that the entire panel is comfortable with the specific reference standards and, most importantly, that they can identify these standards blind. Once all reference standards have been approved and all panelists can identify all standards blind, then the panelists are shown how to use the computerized data acquisition system (if used) or how to use the score sheet.

Subsequently, they are tested by serving them a subset of the product set, usually in triplicate. These data are evaluated by analysis of variance (ANOVA) and other methods, to determine how consistent and discriminating individual panelists are, as well as the panel as a whole. PanelCheck (http://www.panelcheck.com) and SensoMineR (http://sensominer.free.fr) make it relatively simple and easy to analyze these data. If there are issues, then training continues; however, if all panelists perform to an acceptable standard, then the actual data collection starts.

2.2.3.2 Ballot Training

In a certain sense, consensus training is similar to the panel learning a new language as a child, while ballot training is similar to being taught a language as an adult. In this situation, the panelists do not generate the attributes used to describe differences among the samples but are taught to use an attribute list with reference standards. This attribute list may have been generated as a lexicon for the product category, usually with suggested reference standards included (e.g., Civille and Lyon 1996; Dooley et al. 2009; Lawless et al. 2012; Warmund and Elmore 2008).

Alternatively, the specific attributes (and their reference standards) may have been generated in a consensus training in the first year of a longitudinal study. In this case, the panelists in the second and subsequent years are taught the initial attribute list. It usually takes longer to train a panel using the ballot training method, but there are situations where it may be the only option.

The process for ballot training is similar to consensus training in the sense that in the first session, the panelists are given the two or three most different samples in the product set. They are then asked to use the ballot containing predetermined and defined attributes to describe how the samples differ. During subsequent sessions, this process is repeated until the panel is confident that they understand the attributes, that they can identify the reference standards blind, and that they are consistent in using the attributes. The panel will then be tested in a similar fashion to the consensus-trained panels prior to the actual sample evaluation.

2.2.3.3   Reference Standards

Reference standards have two roles in a DA. First, they anchor the concept assigned to the attributes for the panelists. It is not unusual for two panelists to use different words to describe the same underlying attribute nor is it unusual for two panelists to use the same word to describe different underlying concepts. For example, we had a red wine panel in which most of the panelists said a wine smelled like *Blackberry Jam*, while one panelist insisted it smelled like *Violets*. When the panel leader produced both a *Blackberry Jam* and a *Violet* reference standard, the lone holdout realized that his concept of *Violet* was actually *Blackberry Jam*. We have also had the situation where a number of panelists would agree that a specific sample smelled *Woody*. Yet when the panel leader produced a *Woody* reference standard created by using oak chips in wine, there was intense disagreement. It transpired that for some panelists, *Woody* was actually the aroma associated with the debris found on a forest floor. In this case, the situation was resolved by using one term called *Oak* and another term called *Mushroom*.

Second, reference standards act as translation devices for anyone reading the reports or articles about the study. Lund et al. (2009) used the attribute *Bourbon* to describe differences among Sauvignon blanc wines. On first glance, this term seems nonsensical, until when one realizes that the reference standard used was 1-hexanol, which smells *grassy, chemical*. This is logical since numerous Sauvignon blanc wines are grassy in odor. For this reason, the reference standard recipes should be detailed enough for someone else to recreate them. Tables 2.1 and 2.2 show two reference standard lists—Table 2.1 is inadequate as a translation device and Table 2.2 would be acceptable. Earlier in my career (H. Heymann),

TABLE 2.1
Examples of Reference Standards: Reference Standards for Chocolate Ice Cream Made with Varying Levels of Fat

Attribute	Reference Standard
Color	Light brown to dark brown
Foaminess	Look for bubbly foam
Separation of color	Look for dark and light streaks in melted ice cream
Chocolate	Hershey's™ milk chocolate bar[a] and cocoa used in mix
Cocoa	Cocoa powder and unsweetened chocolate references
Cooked milk aroma	Evaporated milk (Schnucks evaporated milk[b])
Creamy	Combination of thickness and lubricative feeling as ice cream melts—refer to skim milk and cream

Source: Adapted from Prindiville, E.A. et al., *J. Dairy Sci.*, 83, 2216, 2000.

[a] Hershey Foods Corporation, Hershey, PA.

[b] Schnucks Foods, St. Louis, MO.

TABLE 2.2
Examples of Reference Standards: Reference Standards for Sauvignon Blanc Wines

Attribute	Reference Standard
Sweet sweaty passion fruit	2000 ng/L 3-mercaptohexyl acetate (Oxford Chemicals)[a]
Bell pepper (capsicum)	1000 ng/L 2-methoxy-3-isobutylpyrazine (Acros Organics)[a]
Cat pee/boxwood	1000 ng/L 4-mercaptomethyl pentane (Oxford Chemicals)[a]
Passion fruit skin/stalk	2000 ng/L 3-mercaptohexan-1-ol (Interchim)[a]
Bourbon	2400 µg/L hexanol/L (Sigma)[a]
Apple candy	250 mg hexyl acetate/L (Sigma)[a]
Tropical	40 mL Golden Circle Mango juice + 40 mL Golden Circle Golden Pash drink + 200 mL Just Juice Mandarin Passion Fruit juice[b]
Mint	25 mg/L cineole (Sigma)[a]
Fresh asparagus	50 mL steamed asparagus water[b]
Stone fruit	Canned Watties apricot and peach juice soaked in diluted base wine for 30 min (equal parts)[b]

Source: Adapted from Lund, C.M. et al., *Am. J. Viticult. Enol.*, 60, 1, 2009.

[a] Added to diluted base wine (50% Corban Sauvignon blanc and 50% water).

[b] Added equal parts to base wine (Corban Sauvignon blanc).

I published a few papers (e.g., Im et al. 1994; Lin et al. 1998) without reference standards. However, once I realized how important they are to translations, I have tried to make the tables with reference standards as clear as possible.

Reference standards could be made using chemicals, for example, Lund et al. (2009) used 1000 ng/L mercaptomethyl pentane in a diluted base wine (50% Sauvignon blanc wine and 50% water) as a reference standard for cat pee/boxwood. However, in many places, due to environmental health and safety rules, sensory laboratories are not allowed to use chemicals since food and beverages are evaluated in that space. In these cases, it is easier and often the only legal option to use actual products to simulate the required concepts, for example, Robinson et al. (2011) cut 2 cm lengths of leather shoelaces (Kiwi Outdoor shoelaces) into small squares and then soaked them in 50 mL base red wine (Franzia Vintners Select Cabernet Sauvignon) as the reference standard for the *leather* aroma in red wines (Robinson et al. 2011).

As a last resort, the reference standard could be anchored by a verbal definition. This truly should be a rare occurrence since this type of standard is neither good for concept anchoring nor good as a translation device. One of the few times, recently, that we have used verbal descriptors was in a study of chocolate milks, where a few of the milks had a fecal off-odor. The panel leader created a reference standard by scraping fecal matter from the floor of a cow barn. The panelists, after smelling it once, decided that they did not need to smell this reference again, and for the remainder of the study, the attribute was verbally anchored.

Reference standard creation is part science and part art. The most complex part is to determine exactly what the panel means when they say a specific word. For example, in a recent Chardonnay study, the panel wanted an *Apple* reference standard. The panel leader created a number of potential apple standards (Table 2.3) and then asked each panelist to score each standard on a 1–9-point numerical scale in terms of its match to their mental concept of *Apple* as they perceived it in the samples under discussion. A score of 1 was assigned when the reference standard was an exact match and a 9 was assigned when the standard had no relationship to the concept. From this, a median score can be calculated and it is fairly easy to determine which standard should be used. In the case of the Chardonnay panel, Apple 5 was the closest match to the concept of *Apple*. We use this technique for all our reference standards. A similar technique, using an appropriateness scale, has successfully been used by Murray and Delahunty (2000).

Panelists should be able to identify the reference standards blind. This is accomplished by giving them a list of attributes and a set of

TABLE 2.3

Potential Apple Reference Standards for Chardonnay Wines

	Reference Standard	Median[a]
Apple 1	20 g Red Delicious fresh apple, chopped + 25 mL base wine	5
Apple 2	20 g Red Delicious fresh apple, chopped + 25 mL base wine; decanted after 1 h; serve liquid as standard	4
Apple 3	8 g Granny Smith fresh apple, chopped + 25 mL base wine	6.5
Apple 4	8 g Granny Smith fresh apple (in one piece) + 25 mL base wine	4.5
Apple 5	12.5 g Granny Smith fresh apple, chopped + 25 mL wine	1.5[b]
Apple 6	8 g canned Pie Fruit Apples, sliced (Ardmona, Victoria, Australia) + 25 mL base wine	5
Apple 7	10 g canned Granny Smith Apple Slices (WW Select, Woolworths, Australia) + 25 mL base wine	4.5
Apple 8	Orchard Apple Stage 1 Baby Food (only organic, Auckland, New Zealand) + 25 mL wine	4.5
Apple 9	25 mL 100% Granny Smith cold-pressed juice (Preshafruit, Victoria, Australia) + 25 mL base wine	2.5
Base wine	Sunnyvale Dry White Wine, Miranda Wines, Merbein, Victoria, Australia	

[a] Mean score based on 1 = reference standard very similar to the concept of *Apple* in these wines and 9 = reference standard very dissimilar to the concept of *Apple* in these wines.

[b] Standard used in the actual study.

reference standards labeled by three-digit codes. The panelists are asked to match the code of the reference standard to the correct attribute. Once the panelists can identify all standards consistently and accurately, the data collection phase can start. This process is repeated as a reference standard identification test in the booths prior to each evaluation session to ensure that panelists interact with the reference standards. In the case of a computerized data acquisitions system, it is fairly easy to do this and to provide panelists with instant feedback on their accuracy. After the first few sessions, it is extremely rare for panelists to identify the standards incorrectly. If a panelist's performance suddenly drops, it indicates to the panel leader that there may be an issue that needs to be explored.

2.2.4 EVALUATION OF SAMPLES

Once the panel has been trained and tested, then the actual evaluation of the samples can commence. It is usual that this process occurs in individual temperature- and light-controlled booths (Lawless and Heymann 2010), but it is also possible to do the evaluation in a large conference room, as long as the panelists are not within each other's line of sight and there is no discussion or distractions (Snitkjaer et al. 2011).

These days, data acquisition is usually performed through a computerized system (e.g., Compusense, Guelph, Canada; EyeQuestion, Elst, the Netherlands; and FIZZ, Couternon, France), but the use of paper ballots is not unusual. There is an indication that switching from paper ballots to computerized ballots is not detrimental to the data collection (Swaney-Stueve and Heymann 2002), and in some cases, this is helpful, for example, when a computer glitch prevents the use of the computerized acquisition system but the samples have already been prepared.

Panelists must be made to feel welcome and appreciated during the data acquisition phase to ensure continued motivation and interest. It is not unusual to serve them some snacks as a token of appreciation after they complete their sensory sessions. In certain situations, it may also be appropriate to pay panelists.

2.2.5 DATA ANALYSIS

The next chapter in this book is on multivariate data analysis, and thus, we will not provide an in-depth discussion in this chapter. However, it is beneficial to describe the standard sequence in which we start the data analysis process in our laboratory. Assuming that we had a fairly uncomplicated experimental design involving samples, panelists, and replications and that we have no missing values,* we start with a three-way multivariate analysis of variance (MANOVA) with a related series of univariate analyses (ANOVA) of all attributes. In this case, the main effects would be samples, panelists, and replication with the addition of all two-way interactions (panelists by sample, panelist by replication, and sample by replication). The MANOVA tests for the overall significance of all the attributes in the data and the ANOVAs for the individual attributes.

---

* If there are missing values, for example, where a panelist missed a session, that could lead to complications with multivariate data analysis techniques. For this reason, if the number of missing values is less than 10% (and it is usually 2% or less), we usually impute the missing variables by calculating the average of the two (out of three) replication that the panelist actually evaluated. This decreases the overall variability of the data, and thus, the analyst should remove an equivalent number of degrees of freedom from the error or residual term in the MANOVA and the individual ANOVAs (Beale and Little 1975; Little and Rubin 1987).

If the MANOVA is significant with a probability of 5% or less, then we continue and evaluate the significance levels of the individual attributes. If sample is significant, as well as the sample by panelist or the sample by replication interaction terms is, then we need to evaluate the impact of this interaction on the sample effect. The standard in our laboratory is to use the pseudomixed model (Naes and Langsrud 1998) where the F-value for sample is calculated by dividing the mean square (sample) value by either the mean square (sample by panelist) value or the mean square (sample by replication) value. If the calculated sample F-value remains significant, then we assume that the interaction effect is not important and we treat that attribute as significant for the sample effect. If the F-value is not significant, then the interaction has an impact on the sample effect and we treat that attribute as not significant for the sample effect. There are other ways in which these data could be analyzed and we suggest the following references: Lawless (1998), Schlich (1998), and Steinsholt (1998).

For any significant attributes, we would then calculate a mean separation value for the means of the samples for each attribute. We traditionally use Fisher's protected least significant difference (LSD), but these values are somewhat liberal, and if a more conservative value is needed, we would use Tukey's honestly significant difference (HSD). See Gacula et al. (2008) for further discussion on mean separation techniques.

The next step of the data analysis involves a graphical representation of the data. Our preference is the creation of a canonical variate analysis (CVA). To do this, rerun the MANOVA (main effect: wine) since Monrozier and Danzart (2001) have shown that the one-way analysis is more stable in calculating a CVA. We use CVA as a multivariate *mean separation* technique for the MANOVA (Chatfield and Collins 1980). The CVA will separate the mean positions of the samples in a 2D or 3D space. It is possible to calculate the number of significantly discriminating dimensions using Bartlett's test (Bartlett 1947; Chatfield and Collins 1980) as well as the 95% confidence intervals around mean position of each sample (Chatfield and Collins 1980; Owen and Chmielewski 1985). These pieces of information make the CVA more useful than the principal component analysis (PCA). For further discussion on the advantages of CVA over PCA, see Heymann and Noble (1989) and Monrozier and Danzart (2001).

2.3 CASE STUDIES

In this section, we discuss a fairly simple case study involving six commercial Cabernet Sauvignon wines and blends of Cabernet Sauvignon wines. We then discuss a more complex study involving 17 commercial wines from 6 countries. The data sets are available for download from the CRC Web site: http://www.crcpress.com/product/isbn/9781466566293.

TABLE 2.4

Commercial Cabernet Sauvignon Wines and Blends for Case Study 1

Code	Vintage	Blend[a]	Wine Region and/or State	Retail Price (US$)	Alcohol Content (%v/v)
W1	2009	100% CS	Paso Robles, California	6	13.2
W2	2008	88% CS, 10% CF, 2% Merlot	Napa Valley, California	42	14.0
W3	2006	100% CS	Napa Valley, California	68	15.2
W4	2006	100% CS	Napa Valley, California	60	15.2
W5	2007	60% CS, 15% Syrah, 11% Merlot, 10% Petit Verdot, 4% CF	Columbia Valley, Washington State	26	15.5
W6	2008	100% CS	Washington State	50	15.9

[a] CS, Cabernet Sauvignon; CF, Cabernet Franc.

2.3.1 CASE STUDY 1

Six commercial wines (Table 2.4) made with at least 60% Cabernet Sauvignon were evaluated in quadruplicate by 11 trained panelists.* The panel had been trained over five sessions using the consensus method sequence described in Section 2.2.3.1. The panel used 12 attributes (Table 2.5) to describe differences among the wines.

The data were analyzed using R and all R-code is shown in Appendix 2.A. A MANOVA (main effects: panelists, wines, replications, and all two-way interactions; Table 2.6) was followed by a series of ANOVAs (main effects and interactions as in the MANOVA; Figure 2.1) where the pseudo-mixed model was used whenever the wine interactions (wine by panelists and/or wine by replication) were significant. This was the case for HerbalA (herbal aroma), AlcoholA (alcohol aroma), and BurningA (burning aroma), where the former two attributes remained significant after the application of the pseudomixed model and the last one became nonsignificant.

---

* These data are related to, but not part of, the study described in King et al. (in press). The quadruplicate analysis of each sample was an artifact of the specific study and is not the usual way we do replication. Triplicates are more standard.

TABLE 2.5
Attributes and Reference Standards Used for Case Study 1

Attribute	Description	Reference Standard
Aroma (A)		
Fresh fruit	Red apple, banana, orange, peach, pear, pomegranate, grape, mango, citrus	2 pieces red and yellow papaya from canned tropical fruit (Dole), 1/2 cm² piece fresh banana, 1/2 cm² piece fresh apple, and 1/2 cm piece fresh lemon rind
Berry	Blackberry, blueberry, raspberry, strawberry, tart berry, forest fruit	1 fresh strawberry halved, 1 fresh raspberry halved, and 1 fresh blackberry halved
Herbal	Grassy, leafy	1 tsp fresh, cut grass and 1 tsp of green leaves
Barnyard	Brett, band-aid	1 grain 4-ethylphenol
Alcohol		1 tsp Vodka (Sobieski)
Burning	Physical prickling sensation in nose	—
Taste and mouthfeel (T)		
Sourness	Acidity, tart	2 g/L tartaric acid (Fisher Scientific) dissolved in water
Sweetness		15 g/L (D)-fructose (Sigma) dissolved in water
Bitterness		800 mg/L anhydrous caffeine (Sigma) dissolved in water
Alcohol	Warm to hot	150 mL/L Vodka (Sobieski) in water
Viscosity	Thickness of mouthfeel, body of wine, oiliness Low anchor (thin) High anchor (thick)	7 g/L Pectin ex-citrus (Sigma) dissolved in water
Astringency	Dry, tannic, puckering	800 mg/L alum (McCormick) dissolved in water

Source: Adapted from King, E.S. et al., *Am. J. Enol. Viticult.*, 2012.

Fisher's LSD was used as a univariate mean separation technique for all attributes that differed significantly across wines (Table 2.7). A one-way MANOVA with wine as the main effect was followed by a CVA used as a multivariate *mean separations technique* (Figure 2.2). Additionally, a PCA was performed on the covariance matrix of the mean wine values, shown in Figure 2.3.

TABLE 2.6

MANOVA Table Obtained from R (See R-Code in Appendix 2.A) Using a Three-Way Fixed-Effect Model with All Two-Way Interactions for Case Study 1

SoV[a]	Degrees of Freedom (df)	Wilk's Lambda	Approximate F-Value	Numerator df	Denominator df	Probability (P) > F[b]
Wine	5	0.32832	2.9306	60	654.66	2.21e−11*
rep[c]	3	0.61819	2.0203	36	411.42	6.379e−4*
pan[c]	10	0.00104	12.9762	120	1093.52	<2.2e−16*
w/r[c]	15	0.27444	1.0899	180	1335.53	0.2108887
w/p[c]	50	0.01492	1.2056	600	1669.02	2.345e−3*
r/p[c]	30	0.01358	2.0751	360	1591.18	<2.2e−16*
res[c]	150					

[a] SoV, sources of variation.

[b] * indicates 0.05.

[c] rep, replication; pan, panelist; w/r, wine/replication; w/p, wine/panelist; r/p, replication/panelist; res, residuals.

		Fresh Fruit Aroma (FrshFrtA)				Berry Aroma (BerryA)			
SoV[a]	df[a]	SS[a]	MS[a]	F-Value	Pr(>F)	SS[a]	MS[a]	F-Value	Pr(>F)[b]
wine	5	80.06	16.01	3.01	0.01*	86.60	17.32	3.61	0.00*
rep[c]	3	6.44	2.15	0.40	0.75	29.23	9.74	2.03	0.11
pan[c]	10	310.7	31.07	5.84	0.00*	169.6	16.96	3.53	0.00*
w:r[c]	15	33.32	2.22	0.42	0.97	89.56	5.97	1.24	0.24
w:p[c]	50	227.0	4.54	0.85	0.74	290.2	5.80	1.21	0.19
r:p[c]	30	252.5	8.42	1.58	0.04*	350.5	11.68	2.43	0.00*
res[c]	150	798.0	5.32			719.5	4.80		
		Herbal Aroma (HerbalA)				Barnyard Aroma (BrnYrdA)			
wine	5	38.39	7.67	3.86	0.00*	36.82	7.36	2.39	0.04*
Wine[d]	5			2.58	0.037*				
rep[c]	3	1.15	0.38	0.19	0.90	2.58	0.86	0.28	0.84
pan[c]	10	324.4	32.44	16.3	0.00*	455.6	45.56	14.8	0.00*
w:r[c]	15	50.64	3.38	1.70	0.06	34.15	2.277	0.74	0.74
w:p[c]	50	148.6	2.97	1.49	0.03*	212.6	4.25	1.38	0.07
r:p[c]	30	152.8	5.09	2.56	0.00*	94.91	3.16	1.02	0.43
res[c]	150	298.5	1.99			461.5	3.08		
		Alcohol Aroma (AlcoholA)				Burning Aroma (BurningA)			
wine	5	54.36	10.87	6.71	0.00*	18.66	3.73	2.74	0.02*
Wine[e]	5			4.03	0.00*			1.40	0.24
rep[c]	3	14.16	4.72	2.91	0.04*	9.11	3.04	2.23	0.09
pan[c]	10	624.2	62.42	38.5	0.00*	430.5	43.05	31.6	0.00*
w:r[c]	15	37.61	2.51	1.55	0.09	38.14	2.54	1.87	0.03*
w:p[c]	50	134.8	2.70	1.66	0.01*	133.5	2.67	1.96	0.00*
r:p[c]	30	81.47	2.72	1.67	0.02*	59.99	2.00	1.47	0.07
res[c]	150	243.1	1.62			204.3	1.36		
		Sour Taste (SourT)				Sweet Taste (SweetT)			
wine	5	25.13	5.03	1.46	0.20	17.61	3.52	1.08	0.37
rep[c]	3	9.56	3.19	0.92	0.43	1.85	0.62	0.18	0.90
pan[c]	10	422.4	42.24	12.25	0.00*	581.7	58.17	17.81	0.00*
w:r[c]	15	38.48	2.56	0.74	0.73	49.35	3.29	1.01	0.45
w:p[c]	50	156.1	3.12	0.90	0.65	116.7	2.33	0.71	0.91
r:p[c]	30	214.6	7.15	2.07	0.00*	236.9	7.90	2.42	0.00*
res[c]	150	517.2	3.45			490.0			

FIGURE 2.1 ANOVA tables for all attributes evaluated in for Case Study 1. See R-codes in Appendix 2.A.

(continued)

		Bitter Taste (BitterT)				Alcohol Mouthfeel (AlcoholT)			
wine	5	26.17	5.23	1.92	0.09	33.31	6.66	2.99	0.01*
rep[c]	3	36.91	12.30	4.51	0.00*	30.12	10.04	4.51	0.00*
pan[c]	10	464.3	46.43	17.02	0.00*	764.4	76.4	34.32	0.00*
w:r[c]	15	53.22	3.55	1.30	0.21	32.53	2.17	0.97	0.48
w:p[c]	50	95.96	1.92	0.70	0.92	95.05	1.90	0.85	0.74
r:p[c]	30	231.7	7.72	2.83	0.00*	99.66	3.32	1.49	0.06
res[c]	150	409.0	2.73			334.1	2.23		

		Viscous Mouthfeel (ViscosityT)				Astringent Mouthfeel (AstringencyT)			
wine	5	28.36	5.67	2.04	0.07	149.5	29.90	10.01	0.00*
rep[c]	3	9.68	3.23	1.16	0.32	18.60	6.20	2.08	0.11
pan[c]	10	157.9	15.79	5.69	0.00*	410.7	41.07	13.75	0.00*
w:r[c]	15	64.97	4.33	1.56	0.09	34.86	2.32	0.78	0.79
w:p[c]	50	188.0	3.76	1.36	0.08	154.8	3.10	1.04	0.42
r:p[c]	30	257.5	8.58	3.09	0.00*	119.1	3.97	1.33	0.13
res[c]	150	416.1	2.77			447.9	2.99		

[a] SoV, sources of variation; df, degrees of freedom; SS, sums of squares; MS, mean sums of squares; Pr(>F), probability larger than F-value.

[b] * indicates 0.05.

[c] rep, replication; pan, panelist; w:r, wine:replication; w:p, wine:panelist; r:p, replication:panelist; res, residuals.

[d] The shaded wine row lists the pseudomixed model for Herbal Aroma (7.67/2.97 = 2.58) with numerator df = 5 and denominator df = 50.

[e] The shaded wine row lists the pseudomixed model for Alcohol Aroma (10.87/2.70 = 4.03) with numerator df = 5 and denominator df = 50 and the pseudomixed model for Burning Aroma (3.73/2.67 = 1.40) using the w:p interaction (since it was larger than the w:r interaction) with numerator df = 5 and denominator df = 50.

FIGURE 2.1 (continued)   ANOVA tables for all attributes evaluated in for Case Study 1. See R-codes in Appendix 2.A.

The CVA (Figure 2.2) shows that the first two dimensions (both of which were significant) explain a total of 85.5% of the variance ratio in the data space. The 95% confidence ellipse of W1 does not overlap any of the other wines' confidence ellipses. This wine is significantly different from all the other wines, and when we look at the means table (Table 2.7), we find that W1 was higher in Fresh Fruit and berry aromas than all the wines except W4 (for Fresh Fruit). Additionally, W1 had the lowest perceived alcohol aroma and flavor (Table 2.7), and as can be seen in Table 2.4, W1 had the lowest alcohol content as well. According to the CVA (Figure 2.2), W2 is significantly different from W4, W5, and W6 but not from W3. Table 2.7 indicates that W5 and W6 were significantly more astringent in

TABLE 2.7
Means and Fisher's LSD Values for the Significantly
Different Attributes for Wines in Case Study 1

Attribute[a]	W1	W2	W3	W4	W5	W6	LSD[b]
FrshFrtA	3.5 a	2.0 bc	2.3 bc	2.9 ab	1.9 c	2.5 bc	1.0
BerryA	4.3 a	2.5 b	3.3 b	3.0 b	2.7 b	3.0 b	0.9
HerbalA	2.2 a	1.4 bc	0.9 c	1.7 ab	1.5 bc	1.2 bc	0.6
BarnYrdA	2.2 ab	2.4 a	1.8 abc	1.6 bc	1.3 c	1.6 bc	0.7
AlcoholA	1.2 c	2.3 ab	1.8 b	2.3 ab	1.8 b	2.6 a	0.5
AlcoholT	2.2 b	2.8 a	3.1 a	2.8 ab	3.1 a	3.2 a	0.6
AstringencyT	2.1 c	2.6 c	3.4 b	3.4 b	4.2 a	4.1 a	0.7

Note: See R-code in Appendix 2.A.

[a] FrshFrtA, Fresh Fruit aroma; BerryA, berry aroma; HerbalA, herbal aroma; BarnYrdA, barnyard aroma; AlcoholA, alcohol aroma; AlcoholT, alcohol mouthfeel; AstringencyT, astringent mouthfeel.

[b] Means with the same letter within a row are not significantly different.

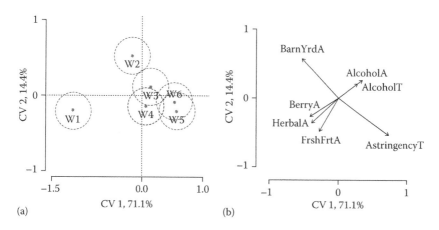

FIGURE 2.2 CVA score (a) and loading (b) plots of the significantly differ-ent sensory attributes for six commercial Cabernet Sauvignon wines and blends evaluated by 11 panelists in quadruplicate in Case Study 1. The circles represent 95% confidence intervals and circles that overlap indicate wines that are not sig-nificantly different.

mouthfeel than any of the other wines and this can also be inferred from their positions on the CVA. In this case study, the CVA (Figure 2.2) and the PCA (Figure 2.3) are very similar but inverted. The PCA explained 83.8% of the variance in the data space, which is slightly less than the first two dimensions of the CVA.

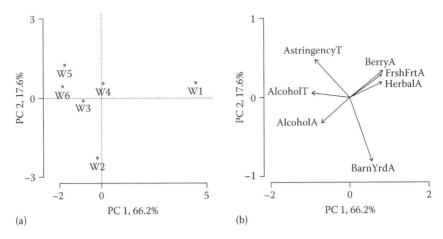

FIGURE 2.3 PCA score (a) and loading (b) plots of the significantly different sensory attributes for six commercial Cabernet Sauvignon wines and blends for six commercial Cabernet Sauvignon wines evaluated by 11 panelists in quadruplicate in Case Study 1.

2.3.2 CASE STUDY 2

In this study, 17 red wines from 6 countries were evaluated by 22 panelists* using 18 attributes (Machado 2009). The wines are listed in Table 2.8 and the attributes and reference standards are shown in Table 2.9.[†] The data were analyzed using R and all R-code is shown in Appendix 2.B. The data analysis process was very similar to Case Study 1. The R-code is shown in Appendix 2.B. The MANOVA was significant for wines (data not shown), and we then did univariate ANOVAs (data not shown) on all the attributes; after evaluating the effects of significant wine panelist interactions on the wine main effect, the following 12 attributes were significantly different at $P < 0.05$ for the following wines: AlcoholA (alcohol aroma), CitrusA (citrus aroma), VeggieA (veggie aroma), CaramelA (caramel aroma), WoodyA (woody aroma), LeatherA (leather aroma), MedicinalA (medicinal aroma), SweetT (sweet taste), BitterT (bitter taste), BodyVisT (viscous mouthfeel), and AStrinT (astringent mouthfeel).

These attributes were used in the CVA (Figure 2.4). The first three dimensions of the CVA were significant and the first two explained 56.3%

---

* This study had an unusually large number of panelists on the panel. The reason was that we were collecting data for a project trying to determine the optimum number of panelists in a descriptive analysis study (Heymann et al. 2012).
† To save space, only the attributes that were significantly different across wines are shown.

TABLE 2.8

Commercial Red Wines from Six Countries Used for Case Study 2

Wine	Varietal Blend	Appellation	Alcohol (%v/v)	Price (US$)
AR1	Malbec	Mendoza, Argentina	13.5	49.99
AR2	65% Malbec, 35% Cabernet Sauvignon	Mendoza, Argentina	14.0	42.99
AU1	Shiraz	Barossa Valley, South Australia	14.5	88.99
AU2	Shiraz	Barossa Valley, South Australia	14.8	65.00
CH1	73% Cabernet Sauvignon, 22%–23% Carmenere, 4%–5% Cabernet Franc	Chile	14.5	64.99
FR1	Bordeaux Blend	Pomerol, Bordeaux	14.0	73.00
FR2	52% Cabernet Sauvignon, 45% Merlot, 3% Petit Verdot	Margaux, Bordeaux	13.5	84.99
FR3	55% Merlot, 45% Cabernet Sauvignon	Pessac Leognan, Bordeaux	13.5	191.99
PT1	Portuguese Blend	Alentejo, Portugal	14.1	16.00
PT2	Portuguese Blend	Dão, Portugal	14.5	16.00
PT3	Portuguese Blend	Douro, Portugal	12.5	16.00
PT4	Portuguese Blend	Alentejo, Portugal	14.0	25.00
PT5	Portuguese Blend	Dão, Portugal	15.4	25.00
PT6	Portuguese Blend (Tinta Roriz, Touriga Nacional, Touriga Franca)	Douro, Portugal	12.2	25.00
PT7	Portuguese Blend	Douro, Portugal	14.0	40.00
US1	Cabernet Sauvignon	Napa Valley, United States	14.9	76.99
US2	Syrah	Napa Valley, United States	14.8	54.99

TABLE 2.9
Attributes and Reference Standards for Significant Attributes Used for Case Study 2

Aroma Attributes	Code	Composition
Alcohol	AlcoholA	15 mL of ethanol solution (25 mL 100% ethanol/475 mL water)
Caramel	CaramelA	1/2 tbs soy sauce + 1/2 tbs molasses + 1/2 tbs butter
Citrus	CitrusA	1 tsp of Bigelow Tea Earl Grey Tea (R.C. Bigelow Fairfield, CT 06825) + 1/2 tsp orange zest + 1/2 tsp lemon zest
Leather	LeatherA	4 × 1/6 in. pieces of Leather Kiwi Outdoor yellow/brown shoelaces
Medicinal	MedicinA	1 drop of 4-ethylphenol solution (1 g/L)
Veggie	VeggieA	10 mL canned Green Bean Brine (Del Monte Fresh cut blue lake French style Green Beans 411 g) + 10 mL canned Asparagus Brine (Raley's fine foods Whole Asparagus Spears 425 g)
Woody	WoodyA	3 chips (evOAK, high vanilla) + 1 drop of Wright's All Natural Hickory Seasoning Liquid Smoke
Taste and mouthfeel attributes		
Astringent	AstrinT	312 mg alum/500 mL water
Bitter taste	BitterT	800 mg caffeine/500 mL water
Sweet taste	SweetT	20 g sucrose/500 mL water
Viscosity/body	BodyVisT	2.5 g pectin/500 mL water
Sour taste	SourT	200 mg citric acid/500 mL water

Source: Adapted from Machado, B., *Revealing the secret preferences for top-rated dry red wines through sensometrics,* MS thesis, University of California at Davis, Davis, CA (Advisor: H. Heymann), 2009.

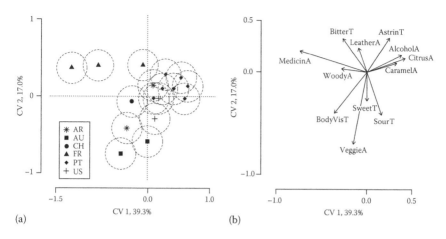

FIGURE 2.4 CVA score (a) and loading (b) plots of the significantly different sensory attributes for 17 commercial red wines from six countries evaluated by 22 panelists in triplicate in Case Study 2. The circles represent 95% confidence intervals and circles that overlap indicate wines that are not significantly different. (See R-codes in Appendix 2.B.)

of the variance ratio. The percentage of the variance ratio explained is much lower than in Case Study 1, but this is not unexpected since there were more wines. The Australian wines were not different from each other, and they differed from all the other wines except an American and an Argentine wine. From the CVA, it would seem that the Australian wines were more vegetative and viscous. The means table (Table 2.10) shows that one of the Australian wines is the most vegetative and that they are both very viscous. The French wines differed from all others except an American and a Portuguese wine. From the CVA, it would seem that the French wines were medicinal and leathery. Table 2.10 indicates that two of the French wines are very high in medicinal and leather intensities. The Portuguese wines clustered together and did not differ from one another.

2.4 CONCLUSIONS

In this chapter, we described classical DA, highlighting the two ways to generate attributes (consensus and ballot training). We also described the intricacies of reference standards. In the final section, we very briefly showed two case studies, with their R-code. The intent was to make sure that the reader has a thorough understanding of the classical DA methodology before plunging into the remainder of the book where more novel techniques in profiling and sensory science will be highlighted.

TABLE 2.10
Means and Fisher's LSD Values for Significantly Different Attributes for Case Study 2

Wine	CitrusA	AlcoholA	VeggieA	CaramelA	WoodyA	LeatherA
AR1	0.7 ef	2.5 d	1.8 b	1.2 ef	3.2 cdef	1.6 cdef
AR2	0.9 ef	2.5 d	0.9 ef	1.6 abcde	2.7 fg	1.3 defgh
AU1	0.7 ef	2.9 bcd	2.6 a	1.2 ef	3.3 bcd	1.9 bc
AU2	0.8 def	3.2 ab	1.5 bcd	1.4 bcde	3.4 bcd	1.1 gh
CH1	0.7 f	3.1 abc	1.6 bc	1.2 ef	3.6 abc	1.6 cde
FR1	0.6 f	2.9 bcd	1.0 ef	1.5 abcde	3.4 bc	1.7 cd
FR2	0.6 f	2.7 cd	1.4 bcde	1.3 cdef	4.0 a	2.1 ab
FR3	0.6 f	2.6 d	1.5 bcd	0.9 f	4.0 a	2.4 a
PT1	0.8 def	2.8 bcd	1.1 def	1.9 ab	2.8 fg	1.5 defg
PT2	1.7 a	2.9 abcd	0.9 ef	1.8 abc	2.8 efg	1.2 fgh
PT3	1.4 abc	2.5 d	1.2 cde	1.3 def	2.4 g	0.9 h
PT4	0.7 f	2.8 bcd	1.4 bcd	1.4 cde	3.4 bcd	1.2 fgh
PT5	1.1 bcd	3.1 abc	1.0 ef	1.7 abcd	3.3 bcde	1.2 efgh
PT6	1.5 ab	2.5 d	0.7 f	1.2 ef	2.9 def	1.0 h
PT7	0.9 def	2.6 d	0.7 f	1.3 def	3.3 bcde	1.3 efgh
US1	0.6 f	3.1 abc	1.6 bc	1.9 a	3.7 ab	1.3 defgh
US2	1.1 cde	3.4 a	1.2 cde	1.3 cdef	2.9 def	1.4 defgh
LSD	0.4	0.5	0.5	0.5	0.5	0.4

Wine	MedicinalA	SweetT	SourT	BitterT	BodyVisT	AstrinT
AR1	1.3 c	1.3 cde	2.7 abcdef	1.6 ef	4.0 bc	3.1 c
AR2	1.1 cd	1.3 cd	2.6 bcdef	2.0 cdef	3.7 c	3.8 cd
AU1	1.2 cd	1.7 b	2.4 ef	1.8 def	4.5 a	3.3 de
AU2	0.9 cde	2.1 a	2.9 ab	1.6 ef	4.5 a	3.3 de
CH1	1.2 c	1.3 cdef	2.8 abcde	2.2 abcd	4.0 bc	4.0 abc
FR1	1.0 cde	0.9 h	2.5 cdef	2.5 ab	3.8 c	4.3 ab
FR2	2.0 b	1.0 fgh	2.5 def	2.7 a	4.0 bc	4.3 abc
FR3	2.9 a	1.0 gh	2.6 bcdef	2.2 abcd	3.9 bc	4.4 a
PT1	0.8 de	1.0 defgh	2.8 abcdef	2.0 bcde	4.1 bc	4.2 abc
PT2	0.6 e	1.2 cdefg	2.6 bcdef	1.7 ef	3.9 bc	4.1 abc
PT3	0.6 e	0.9 h	2.9 abc	1.5 f	2.9 d	3.2 e
PT4	0.7 de	1.0 efgh	3.1 a	2.3 abc	3.8 c	4.4 a
PT5	0.6 e	1.3 cdefg	2.6 bcdef	1.7 ef	4.0 bc	4.2 abc
PT6	0.7 de	0.9 h	2.9 abcd	1.5 f	3.0 d	3.5 de
PT7	0.6 e	1.1 cdefgh	2.4 f	1.7 ef	3.6 c	4.2 abc
US1	0.6 e	1.3 c	2.9 abcde	2.0 cdef	4.2 ab	3.8 bcd
US2	1.1 cd	1.2 cdefg	3.0 a	1.9 cdef	4.3 ab	4.2 abc
LSD	0.4	0.3	0.4	0.5	0.4	0.5

Note: See R-code in Appendix 2.B.

2.A   APPENDIX: R-CODE FOR THE CASE STUDY 1

```
# classical Descriptive Analysis evaluation - example 1
# (c) H. Hopfer, October 2012
# all code comes without any warranty

## --------------read data into R---------------- ##
# in this data set we have 11 judges, 4 replicates,
# 6 wines & 12 sensory attributes

da.d = read.table('data1.csv', sep=',', header=TRUE)
head(da.d)
dim(da.d)

# check your data: 11*4*6 = 264 observations;
# 12 + 3 = 15 columns

# define judge, rep and wine as factor
# columns starting with a letter are automatically set
# as factors by R

for(i in 1:3) {da.d[,i] <- as.factor(da.d[,i])
        print(is.factor(da.d[,i]))}

# combine all attributes and define it as a matrix

da.a = as.matrix(da.d[,-c(1:3)])

# use all columns but the first three (wine, judge, rep)

head(da.a)

## -------------MANOVA-------------- ##
# building a 3-way MANOVA model with all 2-way
# interactions, and run the MANOVA

da.lm = lm(da.a ~ (wine + rep + judge)^2 , data= da.d)
da.maov = manova(da.lm)
summary(da.maov, test='Wilks')

# in MANOVA wine, rep and judge and the interaction
# wine:judge and rep:judge are significant
# at p< 0.05 => continue with individual ANOVA's

## -------------ANOVA--------------- ##
# using the same lm model but now use the ANOVA output

da.aov = aov(da.lm)
aovsum = summary(da.aov)
aovsum

# sign. wine effect at p< 0.05 for: FrshFrtA, BerryA,
# HerbalA, BarnYrdA, AlcoholA, BurningA, AlcoholT,
# AstringencyT
```

```
# sign. wine interaction (p<0.05) with sign. wine
# effect: HerbalA (W:J), AlcoholA (W:J), BurningA
# (W:J,W:R)

## -------------Pseudomixed model-------------- ##
# apply pseudomixed model for attributes with sign.
# wine interactions
# Fnew = (MS_wine)/(MS_wine interaction)
# determine critical F-value for wine effect and W:J
# interaction, df1= df wine and df2 = df W:J

df_W = aovsum[[1]][1,1]
df_WJ = aovsum[[1]][5,1]
newF_crit = qf(0.95, df_W, df_WJ)

# HerbalA => sign. W:J; calculate new F-value and test
# significance for

newF_herbal = aovsum[[3]][1,3]/(aovsum[[3]][5,3])
newF_herbal > newF_crit

# F remains significant at p<0.05 => HerbalA has a
# sign. wine effect

# ditto for AlcA and BurningA

# continue further analyses with only significant
# attributes (significant wine effect)
# create data subset with significant attributes only
# from original data
# sign. attributes are FrshFrtA, BerryA, HerbalA,
# BarnYrdA, AlcA, AlcT, AstrT

da.s = da.d[, c(1:8,13:15)]
head(da.s)
da.s.a = as.matrix(da.s[,-c(1:3)])
head(da.s.a)

## ----------------LSDs---------------- ##
# calculate LSDs or HSDs using the agricolae package
# install the agricolae package first from the CRAN
# and then load it

library(agricolae)

# for LSD use LSD.test, individually for all your
# attributes of interest => see also ?LSD.test

FrshFrtA.lm = lm(FrshFrtA ~ (wine + rep + judge)^2,
  data=da.s)

# build lm models for each attr
```

```
FrshFrtA.LSD = LSD.test(FrshFrtA.lm, trt='wine',
  group = TRUE)

# calculate means and LSD

# ditto for all other significant attributes
# for HSD use HSD.test, analog to LSD.test => see also
# ?HSD.test

## --------------CVA----------------- ##
# run CVA on the significant data set using the
# candisc package
# install candisc package first from the CRAN and load
# it afterwards

library(candisc)

# build MANOVA model with significant attributes only,
# using only the wine effect
# see Monrozier, R.; Danzart, M. (2001) Food Quality
# and Preference 12:393-406

da.s.mlm = lm(da.s.a ~ wine, data = da.s)
da.cva = candisc(da.s.mlm)

# extract CVA output => eigenvalues, variance ratios
# and Bartlett's test for sign. CVs

da.cva

# plots the CVA biplot together with the 95%
# confidence interval circles

plot(da.cva, type = 'n')

## ---------------PCA------------------- ##
# run PCA on data averaged over judges and replicates

# calculate means using G.R. Hirson's mtable function

mtable<- function (x, bycol, firstvarcol){
      #A function to compute a means table for a
      #matrix.
      #x - the data frame with the data
      #bycol - the row or rows used for grouping
      #(usually wine)
      #firstvarcol - the column containing the first
      #variable

      mns<-matrix(nrow=0, ncol=length(levels(as.
      factor(x[,bycol])))
```

```
    for (n in firstvarcol:length(x)){
        m.r<-with(x, tapply(x[,n], x[,bycol], mean))
     mns<-rbind(mns,m.r[])
     }
    mns<-as.data.frame(mns)
    names(mns)<-names(m.r)
    rownames(mns)<-names(x[firstvarcol:length(x)])
    mns<-t(mns)
    return(mns)
    }

da.s.m = mtable(da.s, bycol='wine', firstvarcol=4)

# install SensoMineR package from the CRAN first and
# load it

library(SensoMineR)
da.pca = PCA(da.s.m)
```

2.B   APPENDIX: R-CODE FOR THE CASE STUDY 2

```
# classical Descriptive Analysis evaluation - example 2
# (c) H. Hopfer, October 2012
# all code comes without any warranty

# reading in your DA data
# 22 judges, 3 replicates, 17 wines (2 AR, 2AU, 1 CH,
# 3 FR, 7PT, 2 US), 18 sensory attributes

da.d = read.table('data2.csv', sep=',', header=TRUE)
head(da.d)
levels(da.d$judge)
levels(da.d$Product)

# define judge, rep and wine as factor
# columns starting with a letter are automatically set
# as factors by R

for(i in 1:3) {
      da.d[,i] <- as.factor(da.d[,i])
      print(is.factor(data[,i]))
      }

dim(da.d) # 22*3*17 = 1122 observations; 18 + 3 = 21
  #columns

# combine all attributes and define it as a matrix

da.a = as.matrix(da.d[,-c(1:3)])
# use all columns but the first three (wine, judge, rep)
head(da.a)
```

```
# building a 3-way MANOVA model with all 2-way
# interactions

da.lm = lm(da.a ~ (judge + Product + rep)^2 , data= da.d)

# run the MANOVA

da.maov = manova(da.lm)

summary(da.maov, test='Wilks')  # print MANOVA table

# in MANOVA wine, rep and judge and the interaction
# wine:judge and rep:judge are significant
# at p< 0.05 => continue with individual ANOVA's

# using the same lm model but now use the ANOVA output

da.aov = aov(da.lm)
aovsum = summary(da.aov)
aovsum # print ANOVA tables for each attribute

# sign. W: FloralA, CitrusA, BananaA, RdFruitA, AlcoA,
# VeggieA, HerbA, SpicyA, VanillaA, CaramelA, ChocoA,
# WoodyA, NuttyA, LeatherA, EarthyA, MedicA, SweetT,
# SourT, BitterT, BodyVisT, AstrinT
# sign. W:J: FloralA, CitrusA, BananaA, RdFruitA,
# VeggieA, HerbA, SpicyA, VanillaA, CaramelA, ChocoA,
# NuttyA, LeatherA, MedicA, SourT,
# sign. W:R: BodyVisT, MedicA, ChocoA, CitrusA

# apply pseudomixed model for attributes with sign.
# wine interactions
# Fnew = (MS_wine)/(MS_wine interaction)

# determine critical F-value for wine effect and W:J
# interaction, df1= df wine and df2 = df W:J

df_W = aovsum[[1]][2,1]  # df for wine effect
df_WJ = aovsum[[1]][4,1]
# df for wine:judge interaction
df_WR = aovsum[[1]][6,1]
# df for wine:rep interaction
newF_crit1 = qf(0.95, df_W, df_WJ)  # critical F value
  #for pseudomixed model of wine and W:J
newF_crit2 = qf(0.95, df_W, df_WR)  # critical F value
  #for pseudomixed model of wine and W:R

aovsum[[1]][2,3]/(aovsum[[1]][4,3]) > newF_crit1
  #floral A still sign.
# ditto for all other attributes with significant wine
# and wine-interaction effects
```

```
# extract sign. attributes only
# sign. W: CitrusA, AlcoA, VeggieA, CaramelA, WoodyA,
# LeatherA, EarthyA, MedicA, SweetT, SourT, BitterT,
# BodyVisT, AstrinT

da.s = da.d[,c(1:4,8:10,12,15:21)]
head(da.s)

# calculate means and LSD

library(agricolae)
cit.LSD = LSD.test(lm(CitrusA ~ (judge + Product +
  rep)^2, data=da.s), 'Product', group=TRUE)
cit.LSD

# ditto for all other significant attributes

# run CVA on the significant data set using the
# candisc package
# install candisc package first from the CRAN and load
# it afterwards

library(candisc)

# build MANOVA model with significant attributes only,
# using only the wine effect
# see Monrozier, R.; Danzart, M. (2001) Food Quality
# and Preference 12:393-406

da.s.mlm = lm(as.matrix(da.s[,-c(1:3)]) ~ Product,
  data = da.s)

da.cva = candisc(da.s.mlm)
da.cva  # extract CVA output => eigenvalues, variance
  #ratios and Bartlett's test for sign. CVs
plot(da.cva, type = 'n')  # plots the CVA biplot
  #together with the 95% confidence interval circles

# consult candisc help for further options regarding
# the plot b
```

REFERENCES

Alasalvar, C., Pelvan, E., Bahar, B., Korel, F., and Olmez, H. 2012. Flavour of natural and roasted Turkish hazelnut varieties (*Corylus avellana* L.) by descriptive sensory analysis, electronic nose and chemometrics. *International Journal of Food Science and Technology* 47: 122–131.

Bacci, L., Camilli, F., Drago, S., Magli, M., Vagnoni, E., Mauro, A., and Predieri, S. 2012. Sensory evaluation and instrumental measurements to determine tactile properties of wool fabrics. *Textile Research Journal* 82: 1430–1441.

Ball, R.D. 1997. Incomplete block designs for the minimisation of order and carry-over effects in sensory analysis. *Food Quality and Preference* 8: 111–118.

Barcenas, P., Elortondo, F.J.P., and Albisu, M. 2000. Selection and screening of a descriptive panel for ewes milk cheese sensory profiling. *Journal of Sensory Studies* 15: 79–99.

Bartlett, M.S. 1947. Multivariate analysis (with discussion). *Journal of the Royal Statistical Society Supplement* 9(B): 176–197.

Beale, E.M.L. and Little, R.J.A. 1975. Missing values in multivariate analysis. *Journal of the Royal Statistical Society B* 37: 129–145.

Cairncross, S.E. and Sjostrom, L.B. 1950. Flavor profiles—A new approach to flavor problems. *Food Technology* 4: 308–311.

Cakir, E., Daubert, C.R., Drake, M.A., Vinyard, C.J., Essick, G., and Foegeding, E.A. 2012. The effect of microstructure on the sensory perception and textural characteristics of whey protein/kappa-carrageenan mixed gels. *Food Hydrocolloids* 26: 33–43.

Chatfield, C. and Collins, A.J. 1980. *Introduction to Multivariate Analysis.* London, U.K.: Chapman & Hall.

Civille, G.V. and Lyon, B.G. 1996. *Aroma and Flavor Lexicon for Sensory Evaluation.* West Conshohocken, PA: American Society for Testing and Materials.

Cliff, M.A. and Heymann, H. 1992. Descriptive analysis of oral pungency. *Journal of Sensory Studies* 7: 279–290.

Dooley, L.M., Adhikari, K., and Chambers, E. 2009. A general lexicon for sensory analysis of texture and appearance of lip products. *Journal of Sensory Studies* 24: 581–600.

Elmaci, Y. and Onogur, T.A. 2012. Mandarin peel aroma: Estimation by using headspace/GC/MS and descriptive analysis techniques. *Acta Alimentaria* 41: 131–139.

Gacula, M.C., Singh, J., Bi, J., and Altan, S. 2008. *Statistical Methods in Food and Consumer Research.* New York: Academic Press.

Garcia-Carpintero, E.G., Sanchez-Palomo, E., Gallego, M.A.G., and Gonzalez-Vinas, M.A. 2012. Characterization of impact odorants and sensory profile of Bobal red wines from Spain's La Mancha region. *Flavour and Fragrance Journal* 27: 60–68.

Heymann, H. 1994. A comparison of descriptive analysis of vanilla by two independently trained panels. *Journal of Sensory Studies* 9: 21–32.

Heymann, H., Machado, B., Torri, L., and Robison, A.L. 2012. How many judges should one use for sensory descriptive analysis? *Journal of Sensory Studies* 12: 111–122.

Heymann, H. and Noble, A.C. 1989. Comparison of canonical variate and principal component analyses of wine descriptive analysis data. *Journal of Food Science* 54: 1355–1358.

Im, J.S., Marshall, R.T., and Heymann, H. 1994. Frozen dessert attribute changes with increased amounts of unsaturated fatty-acids. *Journal of Food Science* 59: 1222–1226.

Keenan, D.F., Brunton, N.P., Mitchell, M., Gormley, R., and Butler, F. 2012. Flavour profiling of fresh and processed fruit smoothies by instrumental and sensory analysis. *Food Research International* 45: 17–25.

King, E.S., Dunn, R.L., and Heymann, H. 2012. The influence of alcohol on the sensory perception of red wines. *American Journal of Enology and Viticulture* 57:481–485.

Lawless, H.T. 1998. Commentary on random vs fixed effects for panelists. *Food Quality and Preference* 9: 163–164.

Lawless, H.T. and Heymann, H. 2010. Chapter 10: Descriptive analysis. *Sensory Evaluation of Foods: Principles and Practices*, 2nd edn. New York: Springer.

Lawless, L.J.R., Hottenstein, A., and Ellingsworth, J. 2012. The McCormick Spice Wheel: A systematic and visual approach to sensory lexicon development. *Journal of Sensory Studies* 27: 37–47.

Lin, S., Hsieh, F., Heymann, H., and Huff, H.E. 1998. Effects of lipids and processing conditions on the sensory characteristics of extruded dry pet food. *Journal of Food Quality* 21: 265–284.

Little, R.J. and Rubin, D.B. 1987. *Statistical Analysis with Missing Data.* New York: John Wiley & Sons.

Lotong, V., Chambers, D.H., Dus, C., Chambers, E., and Civille, G.V. 2002. Matching results of two independent highly trained sensory panels using different descriptive analysis methods. *Journal of Sensory Studies* 17: 429–444.

Lund, C.M., Thompson, M.K., Benkwitz, F., Wohler, M.W., Triggs, C.M., Gardner, R., Heymann, H., and Nicolau, L. 2009. New Zealand Sauvignon blanc distinct flavor characteristics: Sensory, chemical, and consumer aspects. *American Journal of Viticulture and Enology* 60: 1–12.

Machado, B. 2009. Revealing the secret preferences for top-rated dry red wines through sensometrics. MS thesis, University of California at Davis, Davis, CA (Advisor: H. Heymann).

Mammasse, N., Cordelle, S., and Schlich, P. 2011. Do we need replications in sensory profiling? *Rosemarie Pangborn Memorial Meeting*, Toronto, Ontario, Canada. September 4–8.

Mammasse, N. and Schlich, P. 2012. Choosing the right panel size for a descriptive or for a hedonic sensory study based on knowledge gained from two sensory databases. *EuroSense Meeting*, Berne, Switzerland. September 9–12.

Martin, N., Molimard, P., Spinnler, H.E., and Schlich, P. 2000. Comparison of odour profiles performed by two independently trained panels following the same descriptive analysis procedure. *Food Quality and Preference* 11: 487–495.

Meullenet, J.-F., Xiong, R., and Findlay, C.J. 2007. *Multivariate and Probabilistic Analyses.* Ames, IA: IFT Press.

Monrozier, R. and Danzart, M. 2001. A quality measurement for sensory profile analysis: The contribution of extended cross-validation and resampling techniques. *Food Quality and Preference* 12: 393–406.

Munoz, A.M. and Civille, G.V. 1998. Universal, product and attribute specific scaling and the development of common lexicons in descriptive analysis. *Journal of Sensory Studies* 13: 57–75.

Murray, J.M. and Delahunty, C.M. 2000. Selection of standards to reference terms in a Cheddar cheese flavor language. *Journal of Sensory Studies* 15: 179–199.

Nachtsheim, R., Ludi, R., and Schlich, E. 2012. Relationship between product specific and unspecific screening tests and future performance of descriptive panelists. *EuroSense Meeting*, Berne, Switzerland. September 9–12.

Naes, T., Brockhoff, P.B., and Tomic, O. 2010. *Design of Experiments for Sensory and Consumer Data*. New York: Wiley.

Naes, T. and Langsrud, Ø. 1998. Fixed or random assessors in sensory profiling? *Food Quality and Preference* 9: 145–152.

Ng, M., Lawlor, J.B., Chandra, C., Hewson, L., and Hort, J. 2012. Using quantitative descriptive analysis and temporal dominance of sensations analysis as complementary methods for profiling commercial blackcurrant squashes. *Food Quality and Preference* 25: 121–134.

Noronha, R.L., Damasio, M.H., Pivatto, M.M., and Negrillo, B.G. 1995. Development of the attributes and panel screening for texture descriptive analysis of milk gels aided by multivariate statistical procedures. *Food Quality and Preference* 6: 49–54.

Owen, J.G. and Chmielewski, M.A. 1985. On canonical variates analysis and the construction of confidence ellipses in systematic studies. *Systematic Zoology* 34: 366–374.

Parker, M., Osidacz, P., Baldock, G.A., Hayasaka, Y., Black, C.A., Pardon, K.H., Jeffrey, D.W., Geue, J.P., Herderich, M.J., and Francis, I.L. 2012. Contribution of several volatile phenols and their glycoconjugates to smoke-related sensory properties of red wine. *Journal of Agricultural and Food Chemistry* 60: 2629–2637.

Paulsen, M.T., Ueland, O., Nilsen, A.N., Ostrom, A., and Hersleth, M. 2012. Sensory perception of salmon and culinary sauces—An interdisciplinary approach. *Food Quality and Preference* 23: 99–109.

Prindiville, E.A., Marshall, R.T., and Heymann, H. 2000. Effect of milk fat, cocoa butter, and whey protein fat replacers on the sensory properties of low-fat and nonfat chocolate ice cream. *Journal of Dairy Science* 83: 2216–2223.

Robinson, A.L., Adams, D.O., Boss, P.K., Heymann, H., Solomon, P.S., and Trengove, R.D. 2011. The relationship between sensory attributes and wine composition for Australian Cabernet Sauvignon wines. *Australian Journal of Grape and Wine Research* 17: 327–340.

Schlich, P. 1998. Comments from Pascal Schlich on Naes and Langsrud. *Food Quality and Preference* 9: 167–168.

Sjostrom, L.B., Cairncross, S.E., and Caul, J.F. 1957. Methodology of the flavor profile. *Food Technology* 11: A20–A25.

Snitkjaer, P., Risbo, J., Skibsted, L.H., Ebeler, S., Heymann, H., Harmon, K., and Frøst, M.B. 2011. Beef stock reduction with red wine—Effects of preparation method and wine characteristics. *Food Chemistry* 126: 183–196.

Sokolowsky, M. and Fischer, U. 2012. Evaluation of bitterness in white wine applying descriptive analysis, time-intensity analysis, and temporal dominance of sensations analysis. *Analytica Chimica Acta* 732: 46–52.

Steinsholt, K. 1998. Are assessors' levels of a split-plot factor in the analysis of variance of sensory profile experiments. *Food Quality and Preference* 9: 153–156.

Stone, H., Sidel, J., Oliver, S., Woolsey, A., and Singleton, R.C. 1974. Sensory evaluation by quantitative descriptive analysis. *Food Technology* 28: 24–28.

Swaney-Stueve, M. and Heymann, H. 2002. A comparison between paper and computerized ballots and a study of simulated substitution between the two ballots used in descriptive analysis. *Journal of Sensory Studies* 7: 527–537.

Verriele, M., Plaisance, H., Vandenblicke, V., Locoge, N., Jaubert, J.N., and Meunier, G. 2012. Odor evaluation and discrimination of car cabin and its components: Application of the "field of odors" approach in a sensory descriptive analysis. *Journal of Sensory Studies* 27: 102–110.

Wakeling, I.N. and MacFie, H.J.H. 1995. Designing consumer trials balanced for first and higher orders of carry-over effect when only a subset of k samples from t may be tested. *Food Quality and Preference* 6: 299–308.

Warmund, M. and Elmore, J. 2008. A sensory lexicon for black walnut (*Juglans nigra* L.) and Persian walnut (*J. regia* L.). *Hortscience* 43: 1102–1103.

Zeppa, G., Bertolino, M., and Rolle, L. 2012. Quantitative descriptive analysis of Italian polenta produced with different corn cultivars. *Journal of the Science of Food and Agriculture* 92: 412–417.

3 Introduction to Multivariate Statistical Techniques for Sensory Characterization

Sébastien Lê

CONTENTS

3.1 INTRODUCTION

Etymologically speaking, *exploratory multivariate analysis* refers to two important concepts: the first one being the concept of exploration, "the action of traveling in or through an unfamiliar area in order to learn about it" (*New Oxford American Dictionary*), and the second one being the concept of multivariate, by definition "involving two or more variable quantities" (*New Oxford American Dictionary*). Therefore, the generic research question for exploratory multivariate analysis could be expressed in the following way: How to explore complex (multivariate) data in order to get meaningful insights?

Hopefully, the answer to that question will be exposed in this chapter, but still we can reveal it summarily: As in all exploration, we need a map; in more statistical terms, we need a fine graphical representation of our data that, if properly used, should lead us to knowledge.

But covering in one chapter an introduction to exploratory multivariate analysis is almost impossible. Choices have to be made, but they certainly need to be motivated. For this overview, we have decided to choose a family of methods that share the same principles and cover a wide range of multivariate data. We have decided to present methods that were also widely used, sometimes incorrectly, but never at full capacity. Old school is still good school, so why would I use fancy methods to understand the data, when I can make it (more) simple? Besides, we really believe that the future is more in the way data are collected, which is all this book is about.

In order to present this family of methods, which will be called *principal axes methods*, we start from a common problem stemming from the intrinsic nature of the data, which is the fact that they are multivariate: how can we possibly graphically represent complexity, simply?

From that visualization problem, stemming originally from a need for exploration, we explain the main principles and features of principal axes methods, in a theoretical framework, mixing different points of view (geometrical, algebraic, and analytical). After all, this is what multivariate is about: using different variables, different perspectives, to understand the same set of objects. No need to say that this chapter is as important as it is complicated to understand.

Then, we present three fundamental methods and their respective specificities and usages: *principal component analysis (PCA)*, *correspondence analysis (CA)*, and *multiple correspondence analysis (MCA)*. Methods are presented through classical sensory issues and data.

Finally, we conclude with an example and its problematic that requires a more sophisticated method, *multiple factor analysis (MFA)*.

3.2   PRINCIPAL AXES METHODS: EXPLORING AND VISUALIZING A MULTIVARIATE DATA SET

In this part, we will see how our generic question "How to explore complex (multivariate) data in order to get meaningful insights?" can be formalized and, to some extent, solved.

3.2.1   NOTATION AND PRESENTATION OF THE PROBLEM

Suppose momentarily, and without loss of generality, that the data are quantitative measures (aka continuous or numerical): for instance, n samples are described or measured by p sensory descriptors (cf. Table 3.2). Let X denote our multivariate data set. $X = (x_{ij})$ can be assimilated to a rectangular matrix of dimension $[n, p]$. A general term of the matrix can be represented as x_{ij}, where i denotes the index of the *rows* (usually rows are assimilated to individuals. In a sensory context, individuals often refer to samples but can also be consumers) and j the index of the *columns* (usually columns are assimilated to variables. Besides, variables often refer to sensory descriptors but can also be usage and attitude questions). Thus, x_{ij} is the value observed on the individual i with respect to the variable j. Each row $i = (x_{i1}, \ldots, x_{ip})$ can be represented by a point in a vector space, denoted R^p, and defined by the directions of the p columns. From a multivariate point of view, $i = (x_{i1}, \ldots, x_{ip})$ is the vector of the p values observed on i. Similarly, each column j can be represented by a point in a vector space, denoted R^n, and defined by the directions of the n rows. From a multivariate point of view, $j = (x_{1j}, \ldots, x_{nj})$ is the vector of the n values observed for variable j. Let N_I denotes the scatterplot of the rows (individuals) in R^p, while N_J denotes the scatterplot of the columns (variables) in R^n.

The problem consists in finding a way to visualize the information, in other words to visualize both scatterplots: N_I and N_J. However, visualizing N_I (resp. N_J) is physically impossible when $p > 3$ (resp. $n > 3$). Therefore, it is necessary to find a subspace in which visualizing N_I (resp. N_J) is not only possible but also relevant and informative (in a sense that will be defined later on).

3.2.2 FROM R^p TO A RELEVANT AND INFORMATIVE SUBSPACE

Ultimately, the information contained in X denotes the distance between rows (resp. columns). This distance, between two rows $i = (x_{i1}, \ldots, x_{ip})$ and $i' = (x_{i'1}, \ldots, x_{i'p})$, is commonly calculated according to the following formula:

$$d^2(i, i') = (x_{i1} - x_{i'1})^2 + \cdots + (x_{ij} - x_{i'j})^2 + \cdots + (x_{ip} - x_{i'p})^2$$

This distance is somehow natural, as it consists of summing up the squared differences between two individuals, with respect to the p variables.

More generally, the weight of each column may be different from one column to the other and the formula can be rewritten as follows:

$$d^2(i, i') = m_1(x_{i1} - x_{i'1})^2 + \cdots + m_j(x_{ij} - x_{i'j})^2 + \cdots + m_p(x_{ip} - x_{i'p})^2$$

This distance is *characteristic* of the information that is contained in X. It defines an indicator that can be understood as a generalization of the notion of variance, when individuals are described by more than one variable. This indicator, denoted $I(N_I)$, is called the *inertia* of the scatterplot N_I (resp. N_J) and can be expressed in the following way:

$$I(N_I) = I_0 = \frac{1}{2n} \sum_{i \in N_I} \sum_{i' \in N_I} d^2(i, i')$$

Considering that inertia is *characteristic* of the information contained in X, $I(N_I)$ and can be interpreted as a direct measure of the information contained in X.

More generally, when rows have different weights (i.e., when they do not have the same importance in the analysis), inertia is equal to

$$I(N_I) = I_0 = \frac{1}{2n} \sum_{i \in N_I} \sum_{i' \in N_I} p_i p_{i'} d^2(i,i')$$

where p_i denotes the weight associated to row i.

Finding an informative subspace leads to finding a subspace in which the information is as preserved as possible. In other words, a subspace in which the distance between i and i' is *as close as possible* to $d^2(i,i')$, for all pairs of rows (i,i'). Typically, the type of good space we are looking for is a 2D space, in other words the map evoked in Section 3.1. In this map, the distance between the rows would be as respectfully displayed as possible.

When the dimension of the subspace equals 1, the problem can be rephrased in the following way: We are looking for an axis in which the distance between i and i', based on this axis only, is *as close as possible* to $d^2(i,i')$, for all pairs of rows (i,i'). Among the different fitting criteria, the simplest one is the least squares criterion.

Let Δu_1 be the axis we are looking for and let u_1 denote a vector unit of Δu_1. Δu_1 is such that the squared distances among the orthogonal projection of N_I onto Δu_1 respect as well as possible the original squared distances in R^p. In other words, it can be shown that if $\langle i, u_1 \rangle$ denotes the scalar product between row i and u_1, then Δu_1 is such that it maximizes the quantity:

$$I_1 = \sum_{i \in N_I} p_i \times \langle i, u_1 \rangle^2$$

This quantity is nothing else than the inertia of the orthogonal projection of the scatterplot N_I onto Δu_1.

Remark: The quantity

$$Ctr_1(i) = \frac{p_i \times \langle i, u_1 \rangle^2}{I_1} \times 100$$

is the contribution of i to the construction of the first axis (this applies to any axes of the analysis).

Suppose $p_i = 1$ for all i, then the inertia on the first axis can be simply written as

$$I_1 = \sum_{i \in N_I} \langle i, u_1 \rangle^2 = (X u_1)' X u_1$$

where $(X u_1)'$ is the transposed form of $X u_1$.

In other words, in order to visualize N_I in a subspace of dimension 1, we have to look for a vector u such that u maximizes $u'X'Xu$ and $u'u = 1$.

Let u_1 be the vector we are looking for; we can easily show that the subspace of dimension 2 in which the distance between i and i' is *as close as possible* to $d^2(i,i')$, in the sense of the least squares criterion, contains u_1. To find that 2D subspace, all we have to do is to find a unit vector u_2 orthogonal to u_1, such that $u_2'X'Xu_2$ is maximum.

More generally, the q-dimensional subspace in which the distance between i and i' is *as close as possible* to $d^2(i,i')$, in the sense of the least squares criterion, is obtained in an analogous way.

From an algebraic point of view, it can be shown that the unit vector u_1, characteristic of the 1D subspace that fits the best N_I in R^p, in the sense of the least squares criterion, is the eigenvector of the matrix $X'X$ associated with the highest eigenvalue denoted λ_1.

More generally, the q-dimensional subspace that fits the best N_I in R^p, in the sense of the least squares criterion, is spanned by the q first eigenvectors of the matrix $X'X$ associated with the highest eigenvalues q.

Finally, the visualization problem leads us to a *singular value decomposition* (SVD) problem of the matrix $X'X$, where $(u_1,...,u_s)$ denotes the sequence of eigenvectors associated with the sequence of eigenvalues $(\lambda_1,...,\lambda_s)$, sorted in descending order, and s denotes the rank of $X'X$.

3.2.3 FROM R^N TO A RELEVANT AND INFORMATIVE SUBSPACE

The problem of representing rows in a relevant and informative space can be easily transposed for columns in R^n. This visualization problem leads us to an SVD problem of the matrix XX', where $(v_1,...,v_s)$ denotes the sequence of eigenvectors associated with the sequence of eigenvalues $(\mu_1,...,\mu_s)$, sorted in descending order, and s denotes the rank of XX'.

3.2.4 LINKING THE TWO SPACES R^P AND R^N

This part is essential as it demonstrates the link between the two spaces R^p and R^n. Actually, this link is of utmost importance to interpret results issued from principal axes methods.

It can be shown that, for all s,

$$\lambda_s = \mu_s$$

It can also be shown that

$$v_s = \frac{1}{\sqrt{\lambda_s}} X u_s$$

and similarly,

$$u_s = \frac{1}{\sqrt{\lambda_s}} X'v_s$$

Let $F_s = Xu_s$ denote the vector of the coordinates of the scatterplot N_I on the axis Δu_s of rank s. Let $G_s = X'v_s$ denote the vector of the coordinates of the scatterplot N_J on the axis Δv_s of rank s. Both vectors can be obtained using the following formulae:

$$Xu_s = F_s = \sqrt{\lambda_s}\, v_s$$

and, similarly,

$$X'v_s = G_s = \sqrt{\lambda_s}\, u_s$$

By multiplying $Xu_s = F_s = \sqrt{\lambda_s}\, v_s$ by u'_s and by summing over s, we obtain the following relation:

$$X = \sum_s \sqrt{\lambda_s}\, v_s u'_s$$

which expresses an important concept. It appears that the original data set X can be reconstituted using all the eigenvectors. In other words, looking at all the F_s and/or all the G_s is equivalent to looking at the whole information contained in the data set X.

3.2.5   INTRODUCING A GEOMETRICAL POINT OF VIEW

Finally, visualizing the scatterplot of the rows N_I consists of, first, finding a sequence of axes $(u_1, \ldots, u_s)$, and then projecting the rows onto these axes, in order to get their coordinates:

$$F_s = Xu_s$$

This geometrical point of view allows to do two things (at least): (1) to assess the quality of representation of a row (or a column) on an axis and (2) to represent the so-called illustrative (aka supplementary) information.

Let $|i|^2$ denote the squared norm of $i = (x_{i1}, \ldots, x_{ip})$. The quality of the representation of i on the axis of rank s can be obtained in the following way:

$$\cos_s^2(i) = \frac{F_s^2(i)}{|i|^2}$$

This indicator lies in [0,1] and equals 1 when i is perfectly well represented. By definition of the quality of representation, it is obvious that for all i,

$$\sum_s \cos_s^2(i) = 1$$

Let $i_+ = (x_{i_+1}, \ldots, x_{i_+p})$ denote an illustrative row. Typically, in a sensory context, that would be a new product that we would like to represent within an *original* product space from a former study. The coordinate of i_+ on the axis of rank s can be obtained in the following way:

$$F_s(i_+) = <i_+, u_s>$$

Similarly, illustrative columns can be represented as well.

Remark: The notion of illustrative information is of utmost importance when interpreting the data, as it will be illustrated in the following sections. This concept is often opposed to the concept of *active* information, which is, by definition, the information of the data set on which the axes are determined. For instance, the active variables are the variables that are used to calculate the distance between the individuals, and *de facto*, the construction of the axes (through the calculation of the distance) depends exclusively on these variables.

3.3 PRINCIPAL COMPONENT ANALYSIS

PCA is applied when individuals are described by quantitative variables. In PCA, the data set to be analyzed is structured in the following way: Rows correspond to individuals or statistical units and columns correspond to variables. By construction, at the intersection of one row i and one column j, x_{ij} is the value measured on i for variable j. In PCA, the role of the rows and the columns is not symmetrical, and the way distances are interpreted in R^p (distance between the individuals) and R^n (distance between the variables) is totally different. In R^p, the closer two individuals are the closer their values, measured over all variables, are. In R^n, the closer two variables are the higher their correlation coefficient (or covariances), calculated over all individuals, is. In PCA, we want to represent the scatterplot of the individuals, N_I, in a 2D subspace, that respect as well as possible the differences among the individuals in R^p. Also, we want to represent the scatterplot of the variables, N_J, in a 2D subspace, which respects as well as possible the correlation coefficients (or the covariances) among the variables in R^n.

Due to the abundance of quantitative data, particularly in sensory evaluation, PCA is certainly one of the most widely used techniques for visualizing and exploring multivariate data. In this section, we demonstrate the

TABLE 3.1

Description of the 12 Perfumes Evaluated by the Expert Panel and the Consumer Panel

Product	Type
Angel	Eau de Parfum
Aromatics Elixir	Eau de Parfum
Chanel N5	Eau de Parfum
Cinéma	Eau de Parfum
Coco Mademoiselle	Eau de Parfum
J'adore	Eau de Parfum
J'adore	Eau de Toilette
L'instant	Eau de Parfum
Lolita Lempicka	Eau de Parfum
Pleasures	Eau de Parfum
Pure Poison	Eau de Parfum
Shalimar	Eau de Toilette

specificities of PCA using data from an expert panel on the one hand and from a consumer panel on the other. While both panels provided a sensory profile on a set of products (cf. Table 3.1), consumers gave also hedonic judgments.

These 12 perfumes were evaluated by the expert panel on a list of 12 attributes (cf. Table 3.2): *Vanilla, Flower Note, Citrus, Woody, Green, Spicy, Heady, Fruity, Sea Freshness, Greedy, Oriental,* and *Wrapping.*

For this data set, we want to represent the proximities among the perfumes described by the 12 attributes. Similarly, we also want to represent the relationships among the sensory descriptors. Finally, we want to combine both representations, in order to get a joint interpretation of the products on the one hand and the sensory descriptors on the other.

Without changes of any kind in terms of interpretation, data can be centered:

$$\forall i, j, x_{ij} \leftarrow x_{ij} - \overline{x}_j$$

which means that x_{ij} is replaced by $x_{ij} - \overline{x}_j$, where $\overline{x}_j$ is the mean of variable j calculated over the n individuals. The usefulness of this transformation is clearly demonstrated in Figure 3.1.

When looking at Figure 3.1, it is easy to see, variable per variable (in other words, within each column), which perfumes take the highest (resp. lowest) values.

Unfortunately, Figure 3.1 does not allow the comparison of individuals across all the variables at the same time. This is due to the fact that

TABLE 3.2

Sensory Profile Data Set of the 12 Perfumes Evaluated by the Expert Panel on 12 Sensory Descriptors

	Spicy	Heady	Fruity	Green	Vanilla	Flower Note	Woody	Citrus	Sea Freshness	Greedy	Oriental	Wrapping
Angel	3.90	7.84	1.92	0.11	7.18	2.49	1.18	0.41	0.14	7.89	4.76	7.55
Aromatics Elixir	6.30	8.31	0.61	0.52	1.82	4.30	2.64	0.60	0.09	0.34	7.45	7.72
Chanel N5	3.73	8.21	1.06	0.44	1.80	6.15	0.95	0.93	0.15	0.63	6.38	7.85
Cinéma	1.08	2.20	5.13	0.21	4.86	5.55	1.02	1.05	0.59	4.38	2.87	5.55
Coco Mademoiselle	0.91	1.14	5.06	0.78	1.95	7.98	0.80	1.24	0.66	2.94	3.09	4.80
J'adore EP	0.26	1.18	6.40	1.56	0.47	8.40	0.91	2.17	1.03	1.30	1.14	3.57
J'adore ET	0.34	1.29	5.63	1.48	0.88	8.18	0.88	1.58	0.28	1.86	0.92	3.33
L'instant	0.74	2.28	3.84	0.30	4.89	7.38	0.98	0.71	0.68	3.38	3.05	5.65
Lolita Lempicka	1.40	4.41	3.35	0.49	8.08	3.03	0.71	0.81	0.15	9.15	3.68	7.64
Pleasures	0.49	0.91	4.46	3.25	0.20	8.25	0.71	1.61	1.17	0.67	1.04	2.63
Pure Poison	1.66	1.90	3.55	0.63	2.06	7.23	1.35	0.60	0.57	1.49	2.40	5.22
Shalimar	6.16	7.89	0.93	0.39	3.25	4.50	2.92	0.88	0.13	1.06	7.62	7.33

	Spicy	Heady	Fruity	Green	Vanilla	Flower Note	Woody	Citrus	Sea Freshness	Greedy	Oriental	Wrapping
Angel	1.65	3.88	-1.57	-0.73	4.06	-3.63	-0.08	-0.64	-0.33	4.96	1.06	1.81
Aromatics Elixir	4.05	4.35	-2.88	-0.33	-1.30	-1.82	1.39	-0.44	-0.38	-2.58	3.75	1.99
Chanel N5	1.48	4.25	-2.44	-0.41	-1.32	0.03	-0.30	-0.12	-0.32	-2.30	2.68	2.11
Cinéma	-1.17	-1.77	1.63	-0.63	1.74	-0.57	-0.24	0.00	0.12	1.45	-0.83	-0.18
Coco Mademoiselle	-1.34	-2.82	1.57	-0.07	-1.17	1.86	-0.45	0.19	0.19	0.01	-0.61	-0.94
J'adore EP	-1.99	-2.78	2.91	0.72	-2.65	2.28	-0.34	1.12	0.56	-1.62	-2.56	-2.17
J'adore ET	-1.91	-2.68	2.13	0.64	-2.24	2.06	-0.38	0.53	-0.19	-1.07	-2.78	-2.41
L'instant	-1.51	-1.68	0.35	-0.55	1.77	1.26	-0.28	-0.34	0.21	0.46	-0.65	-0.08
Lolita Lempicka	-0.85	0.45	-0.14	-0.35	4.96	-3.09	-0.54	-0.24	-0.32	6.23	-0.02	1.90
Pleasures	-1.76	-3.05	0.97	2.40	-2.92	2.13	-0.54	0.56	0.70	-2.25	-2.66	-3.11
Pure poison	-0.59	-2.07	0.05	-0.22	-1.06	1.11	0.09	-0.45	0.10	-1.43	-1.30	-0.52
Shalimar	3.91	3.92	-2.57	-0.46	0.13	-1.62	1.67	-0.17	-0.34	-1.86	3.92	1.60

FIGURE 3.1 Sensory profile data set of the 12 perfumes evaluated by the expert panel on 12 sensory descriptors, when data are centered. The highest values of each attribute are highlighted in dark grey, while the lowest values are highlighted in light grey.

variances are different from one variable to the other. In order to give the same role to each one of the variables, in terms of variance, data have to be scaled to unit variance:

$$\forall i, j, x_{ij} \leftarrow \frac{x_{ij} - \bar{x}_j}{s_j}$$

which means that $x_{ij} - \bar{x}_j$ is divided by s_j, where s_j is the standard deviation of variable j calculated over the n individuals. The usefulness of this transformation is clearly demonstrated in Figure 3.2.

When looking at Figure 3.2, it is easy to see, across all variables, which perfumes are the most peculiar. In the example, *Shalimar* is more *Woody* than the rest of the perfumes, while *Lolita Lempicka* is *Greedy*, as *Shalimar* is located at 2.36 standard deviations from the mean for variable *Woody*, whereas *Lolita Lempicka* is located at 2.25 standard deviations from the mean for variable *Greedy*.

When looking at Figure 3.3, we can see that comparing sensory descriptors, in terms of correlation coefficients, is also facilitated when data are scaled to unit variance.

In PCA, as a preprocessing, data are always centered by the computer and in most cases standardized to unit variance (this is usually the default option in most PCA programs). In other words, when data are only centered, PCA is performed on the covariance matrix; when data are scaled to unit variance, PCA is performed on the correlation matrix. As the weight of a variable is actually equal to its variance, when data are standardized, all variables have the same weight, equal to 1, and therefore play the same role in the analysis.

Remark: When individuals are described by variables with different units of measurement (e.g., this is the case when samples are described by physicochemical variables), scaling variables to unit variance is mandatory, as intrinsic measures depend on the units used, and therefore, results of the PCA will depend on the units as well. When individuals are described by variables with the same unit of measurement, scaling variables to unit variance is open to question. When samples are described by sensory descriptors, we strongly recommend scaling the data: considering that each descriptor plays the same part in the analysis makes sense in such a context. Moreover, it tremendously facilitates the interpretation of the results.

As previously mentioned, the notion of inertia is a generalization of the notion of variance in the case of several variables. By construction, the inertia of N_I is equal to the sum of the variances over the variables:

$$I(N_I) = \sum_j \text{Var}(j)$$

	Spicy	Heady	Fruity	Green	Vanilla	Flower Note	Woody	Citrus	Sea Freshness	Greedy	Oriental	Wrapping
Angel	0.78	1.28	-0.84	-0.87	1.64	-1.79	-0.11	-1.30	-0.92	1.79	0.46	1.00
Aromatics Elixir	1.91	1.44	-1.53	-0.39	-0.52	-0.90	1.96	-0.90	-1.07	-0.93	1.65	1.10
Chanel N5	0.70	1.40	-1.29	-0.48	-0.53	0.02	-0.43	-0.24	-0.90	-0.83	1.18	1.17
Cinéma	-0.55	-0.58	0.87	-0.75	0.70	-0.28	-0.33	0.00	0.33	0.52	-0.36	-0.10
Coco Mademoiselle	-0.63	-0.93	0.83	-0.08	-0.47	0.92	-0.63	0.39	0.54	0.01	-0.27	-0.52
J'adore EP	-0.94	-0.92	1.55	0.85	-1.07	1.13	-0.48	2.27	1.57	-0.59	-1.12	-1.20
J'adore ET	-0.90	-0.88	1.13	0.75	-0.90	1.02	-0.53	1.07	-0.55	-0.38	-1.22	-1.33
L'instant	-0.71	-0.55	0.18	-0.65	0.71	0.62	-0.39	-0.69	0.60	0.16	-0.28	-0.04
Lolita Lempicka	-0.40	0.15	-0.08	-0.42	2.00	-1.53	-0.77	-0.48	-0.91	2.25	-0.01	1.05
Pleasures	-0.83	-1.01	0.51	2.84	-1.18	1.05	-0.77	1.14	1.96	-0.81	-1.17	-1.72
Pure Poison	-0.28	-0.68	0.03	-0.26	-0.43	0.55	0.13	-0.91	0.29	-0.52	-0.57	-0.29
Shalimar	1.85	1.30	-1.36	-0.54	0.05	-0.80	2.36	-0.34	-0.95	-0.67	1.72	0.88

Note: Comparing individuals in terms of number of standard deviations from the mean.

FIGURE 3.2 Sensory profile data set of the 12 perfumes evaluated by the expert panel on 12 sensory descriptors, when data are centered and scaled to unit variance. The highest values of each attribute are highlighted in dark grey, while the lowest values are highlighted in light grey.

	Spicy	Heady	Fruity	Green	Vanilla	Flower Note	Woody	Citrus	Sea Freshness	Greedy	Oriental	Wrapping
Angel	0.78	1.28	−0.84	−0.87	1.64	−1.79	−0.11	−1.30	−0.92	1.79	0.46	1.00
Aromatics Elixir	1.91	1.44	−1.53	−0.39	−0.52	−0.90	1.96	−0.90	−1.07	−0.93	1.65	1.10
Chanel N5	0.70	1.40	−1.29	−0.48	−0.53	0.02	−0.43	−0.24	−0.90	−0.83	1.18	1.17
Cinéma	−0.55	−0.58	0.87	−0.75	0.70	−0.28	−0.33	0.00	0.33	0.52	−0.36	−0.10
Coco Mademoiselle	−0.63	−0.93	0.83	−0.08	−0.47	0.92	−0.63	0.39	0.54	0.01	−0.27	−0.52
J'adore EP	−0.94	−0.92	1.55	0.85	−1.07	1.13	−0.48	2.27	1.57	−0.59	−1.12	−1.20
J'adore ET	−0.90	−0.88	1.13	0.75	−0.90	1.02	−0.53	1.07	−0.55	−0.38	−1.22	−1.33
L'instant	−0.71	−0.55	0.18	−0.65	0.71	0.62	−0.39	−0.69	0.60	0.16	−0.28	−0.04
Lolita Lempicka	−0.40	0.15	−0.08	−0.42	2.00	−1.53	−0.77	−0.48	−0.91	2.25	−0.01	1.05
Pleasures	−0.83	−1.01	0.51	2.84	−1.18	1.05	−0.77	1.14	1.96	−0.81	−1.17	−1.72
Pure Poison	−0.28	−0.68	0.03	−0.26	−0.43	0.55	0.13	−0.91	0.29	−0.52	−0.57	−0.29
Shalimar	1.85	1.30	−1.36	−0.54	0.05	−0.80	2.36	−0.34	−0.95	−0.67	1.72	0.88

Note: Comparison of variables.

FIGURE 3.3 Sensory profile data set of the 12 perfumes evaluated by the expert panel on 12 sensory descriptors, when data are centered and scaled to unit variance. Attributes highlighted with the same color are correlated.

TABLE 3.3
Variance Associated with Each Principal Component

	Eigenvalue	Percentage of Variance	Cumulative Percentage of Variance
Comp 1	7.710	64.250	64.250
Comp 2	2.624	21.871	86.121
Comp 3	0.562	4.686	90.807
Comp 4	0.380	3.168	93.975
Comp 5	0.374	3.120	97.094
Comp 6	0.229	1.912	99.007
Comp 7	0.053	0.442	99.449
Comp 8	0.031	0.257	99.706
Comp 9	0.024	0.197	99.903
Comp 10	0.011	0.088	99.991
Comp 11	0.001	0.009	100.000

Note: The sum of the variances over the 11 components equals 12.

When data are scaled to unit variance, as the variance of each variable equals 1,

$$I(N_I) = \sum_j 1 = p$$

where p is the number of variables.

As shown in Table 3.3, the first (resp. second) dimension carries 64.25% (resp. 21.87%) of the total inertia of the scatterplot N_I (here $I(N_I) = 12$). In other words, the variance of the first component F_1 equals $0.6425 \times 12 = 7.71$.

In PCA, individuals (rows in R^p and, in this particular case, the perfumes that have been assessed) are usually represented by points. Due to the preprocessing, the scatterplot is *centered*: its barycenter is located at the origin of the vector space R^p.

Figure 3.4 shows that the first dimension opposes perfumes such as *Pleasures* and *J'adore EP* to perfumes such as *Angel* and *Shalimar* and the second dimension opposes perfumes such as *Aromatics Elixir* and *Shalimar* to *Lolita Lempicka*.

In PCA, variables (columns in R^n and, in this particular case, the sensory descriptors that have been used) are usually represented by vectors (cf. Figure 3.5). In R^n, the norm of a variable equals its variance, and the cosine of the angle between two variables equals the correlation coefficient between them.

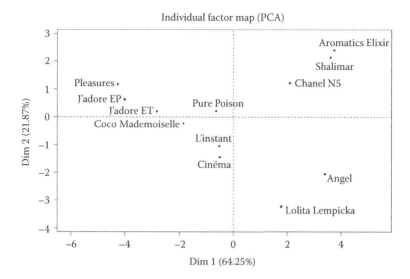

FIGURE 3.4 Representation of the 12 perfumes evaluated by the expert panel on the first and second dimensions of the PCA.

When data are scaled to unit variance, the norm of each variable equals 1; hence, as shown in Figure 3.5, all the variables (sensory descriptors) lie within a circle of radius one. As a vector of R^n, a variable is perfectly represented in a plane if its terminal point (the head of the arrow) touches the circle. In our case, all the variables are almost perfectly represented, which is in accordance with the high percentage of variability carried by the first two dimensions (86.12%, cf. Table 3.3).

When two variables are well represented, the angle between the two variables is also well represented, and therefore, their proximity in terms of correlation coefficient can be assessed by the cosine of that angle. In the example, variables *Spicy* and *Oriental* are well represented, and therefore, we can say that the correlation coefficient between these two variables is rather high (close to 1). In a similar way, *Fruity* and *Greedy* are also well represented and we can guess that the correlation coefficient is rather low (close to −1) between *Oriental* and *Fruity* and close to 0 between *Fruity* and *Greedy*. Figure 3.5 shows an opposition between descriptors such as *Oriental, Heady,* and *Wrapping* and descriptors such as *Sea Freshness* and *Fruity*. It shows also that the second dimension is linked to the descriptors *Greedy* and *Vanilla*.

As previously mentioned, these two representations have to be interpreted jointly. The opposition between *Pleasures* and *J'adore EP*, on one hand, and *Angel* and *Shalimar*, on the other hand, is due to the fact that the first ones were perceived as rather *Fruity* while the second ones were perceived as rather *Heady*.

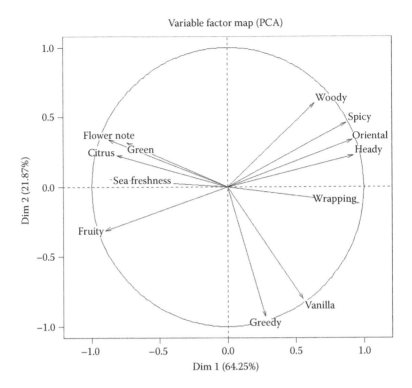

FIGURE 3.5   Representation of the 12 sensory descriptors used by the expert panel on the first and second dimensions of the PCA.

Let us remind that perfumes were also evaluated by a consumer panel on a list of 16 attributes: *Jasmine, Fresh Lemon, Vanilla, Citrus, Anis, Sweet Fruit, Honey, Caramel, Spicy, Woody, Leather, Nutty, Musk, Animal, Earthy,* and *Green*. This information can be easily integrated in the analysis as illustrative (aka supplementary) elements.

As shown in Figure 3.6, integrating illustrative elements (in this case, sensory descriptors from the consumer panel) does not affect the percentage of variability associated with each dimension, due to the active elements only. In the worst case, integrating illustrative elements will not help the end user in interpreting the results.

Figure 3.6 shows that representing both active and illustrative elements can enhance the interpretability of the dimensions. First of all, illustrative elements are rather well represented. Then, homologous descriptors are well correlated, as illustrated by the descriptor *Spicy* for both panels. In other words, the main structure of variability of the expert panel can be interpreted by the consumer panel in a very similar way.

As previously mentioned, hedonic scores were also collected. This information can be integrated as well, as illustrative information. Figure 3.7

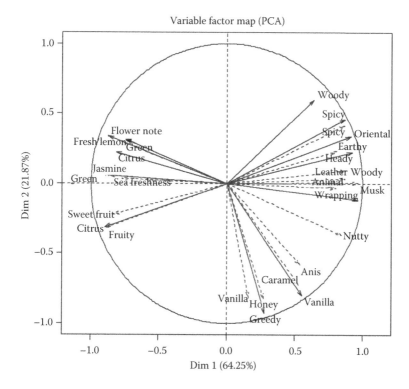

FIGURE 3.6   Representation of the 12 sensory descriptors used by the expert panel (in black); representation of the 16 sensory descriptors used by the consumer panel (in dotted lines), as illustrative information.

shows that there is a consensus among the perfumes, in terms of liking. One side of the first dimension of variability is likely to be more appreciated by almost all the consumers than the other side.

Figure 3.7 shows that this side of the dimension is related to sensory descriptors such as *Citrus, Flower Note, Green, Sea Freshness,* and *Fruity* that could be qualified as drivers of liking by opposition to *Heady, Wrapping,* and *Oriental.*

3.4   CORRESPONDENCE ANALYSIS

CA applies when individuals are described by two qualitative variables (aka categorical variables). In CA, the data set to be analyzed is a *cross tabulation table,* also called *contingency table.* This data set is structured in the following way: Rows correspond to the modalities (aka categories) of one of the qualitative variables; columns correspond to the modalities of the other qualitative variables. By construction, this data set describes the link between the two qualitative variables. At the intersection of one row *i* and one column *j*,

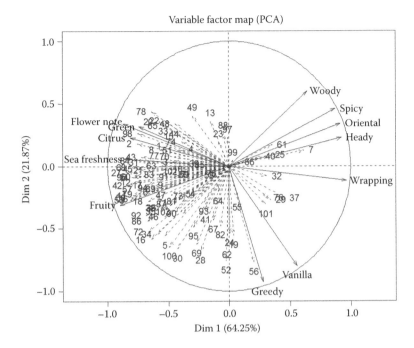

FIGURE 3.7 Representation of the 12 sensory descriptors used by the expert panel (in black); representation of the hedonic scores (in dotted lines), as illustrative information.

x_{ij} is the number of times that the two modalities i and j have been associated with. In CA, the role of the rows and the columns is symmetrical, and the way distances are interpreted in R^p and R^n (distance between the modalities of a given variable) is the same. The closer two modalities of a given variable are the more similar is the way in which they are associated with the modalities of the other variable. In CA, we want to represent the scatterplot of the modalities, N_I, in a 2D space that respect as well as possible the associations in R^p. Also, we want to represent the scatterplot of the modalities, N_J, in a 2D space that respect as well as possible the associations in R^n.

Contingency tables are among the most common structures used for analyzing the link between two qualitative variables, hence the importance of CA. From a theoretical point of view, we will see how CA can be linked to the chi-square test.

To illustrate the method, we consider the following data. Rows correspond to perfumes (12) and columns correspond to words (83) provided by consumers to describe the perfumes. As in PCA, we want to understand the proximities between the perfumes, according to the way in which they were described by the consumers. We also want to understand how the words were used to describe the perfumes.

As shown in Table 3.4, the number of words associated with a perfume is quite different from one sample to the another (from 5 to 15 words). Also, the number of perfumes associated with a word is quite different from one word to the other (from 8 to 49 perfumes). This makes it difficult to directly use such table to compare perfumes on the one hand and words on the other.

Hence, the idea of introducing the notion of row profiles and column profiles. This idea consists in (1) looking at perfumes relatively to the number of words they have been associated with and (2) looking at words relatively to the number of perfumes they have been associated to.

The row profiles in Figure 3.8 are obtained using the following transformation:

$$\forall i, j, x_{ij} \leftarrow \frac{x_{ij}}{\sum_j x_{ij}}$$

In this example, we can see that

1. *Coco Mademoiselle*, *J'adore EP*, *J'adore ET*, and *Pure Poison* have pretty similar row profiles
2. *Aromatics Elixir* and *Chanel N5* have pretty similar row profiles
3. *J'adore EP* and *Chanel N5* have very different row profiles: when a word has been associated with a perfume, it has not been with the other

The column profiles in Figure 3.9 are obtained using the following transformation:

$$\forall i, j, x_{ij} \leftarrow \frac{x_{ij}}{\sum_i x_{ij}}$$

In this example, we can see that

1. *Artificial* and *Grandmother* have pretty similar column profiles
2. *Flowery* and *Fruity* have pretty similar column profiles
3. *Grandmother* and *Flowery* have very different column profiles: when a perfume has been associated with a word, it has not been with the other

The idea that consists in comparing directly the values is not exactly the one that is used in CA. Actually when comparing the profiles, CA takes into account the relative weight of each dimension (rows and columns).

TABLE 3.4

Extract of the Data Set on Which CA Is Performed

	Artificial	Flowery	Grandmother	Food	Fruity	Spicy	Classical	Soap	Total
Angel	0	1	0	1	1	2	0	0	5
Aromatics Elixir	2	1	4	0	1	0	0	0	8
Chanel N5	2	0	2	0	0	0	1	4	9
Cinéma	1	5	0	1	4	1	0	0	12
Coco Mademoiselle	0	7	0	0	3	0	1	1	12
J'adore EP	0	6	0	0	3	0	1	2	12
J'adore ET	1	7	0	0	4	0	1	2	15
L'instant	0	3	2	1	2	0	0	0	8
Lolita Lempicka	0	4	0	1	5	2	0	0	12
Pleasures	0	7	0	0	3	0	0	0	10
Pure Poison	0	6	0	0	2	0	0	1	9
Shalimar	2	2	2	1	0	0	0	0	7
Total	8	49	10	5	28	5	4	10	119

Note: The word grandmother has been used four times to describe Aromatics Elixir.

	Artificial	Flowery	Grandmother	Food	Fruity	Spicy	Classical	Soap	Total
Angel	0.0	20.0	0.0	20.0	20.0	40.0	0.0	0.0	100.0
Aromatics Elixir	25.0	12.5	50.0	0.0	12.5	0.0	0.0	0.0	100.0
Chanel N5	22.2	0.0	22.2	0.0	0.0	0.0	11.1	44.4	100.0
Cinéma	8.3	41.7	0.0	8.3	33.3	8.3	0.0	0.0	100.0
Coco Mademoiselle	0.0	58.3	0.0	0.0	25.0	0.0	8.3	8.3	100.0
J'adore EP	0.0	50.0	0.0	0.0	25.0	0.0	8.3	16.7	100.0
J'adore ET	6.7	46.7	0.0	0.0	26.7	0.0	6.7	13.3	100.0
L'instant	0.0	37.5	25.0	12.5	25.0	0.0	0.0	0.0	100.0
Lolita Lempicka	0.0	33.3	0.0	8.3	41.7	16.7	0.0	0.0	100.0
Pleasures	0.0	70.0	0.0	0.0	30.0	0.0	0.0	0.0	100.0
Pure Poison	0.0	66.7	0.0	0.0	22.2	0.0	0.0	11.1	100.0
Shalimar	28.6	28.6	28.6	14.3	0.0	0.0	0.0	0.0	100.0

FIGURE 3.8 Row profiles corresponding to Table 3.4. Perfumes highlighted with the same color have similar row profiles.

	Artificial	Flowery	Grandmother	Food	Fruity	Spicy	Classical	Soap
Angel	0.0	2.0	0.0	20.0	3.6	40.0	0.0	0.0
Aromatics Elixir	25.0	2.0	40.0	0.0	3.6	0.0	0.0	0.0
Chanel N5	25.0	0.0	20.0	0.0	0.0	0.0	25.0	40.0
Cinéma	12.5	10.2	0.0	20.0	14.3	20.0	0.0	0.0
Coco Mademoiselle	0.0	14.3	0.0	0.0	10.7	0.0	25.0	10.0
J'adore EP	0.0	12.2	0.0	0.0	10.7	0.0	25.0	20.0
J'adore ET	12.5	14.3	0.0	0.0	14.3	0.0	25.0	20.0
L'instant	0.0	6.1	20.0	20.0	7.1	0.0	0.0	0.0
Lolita Lempicka	0.0	8.2	0.0	20.0	17.9	40.0	0.0	0.0
Pleasures	0.0	14.3	0.0	0.0	10.7	0.0	0.0	0.0
Pure Poison	0.0	12.2	0.0	0.0	7.1	0.0	0.0	10.0
Shalimar	25.0	4.1	20.0	20.0	0.0	0.0	0.0	0.0
Total	100.0	100.0	100.0	100.0	100.0	100.0	100.0	100.0

FIGURE 3.9 Column profiles corresponding to Table 3.4. Words highlighted with the same color have similar column profiles.

Let $N = \Sigma_i \Sigma_j x_{ij}$ be the number of associations, perfumes–words. We can introduce the classical notations used in CA and define the notion of row profiles and column profiles in terms of frequencies:

$$f_{ij} = \frac{x_{ij}}{N}$$

For a given column j (resp. row i), let $f_{.j} = \sum_i x_{ij}/N$ (resp. $f_{i.} = \sum_j x_{ij}/N$) denote the sum of the x_{ij} over the rows (resp. over the columns), divided by N.

In fact, the distance used is the so-called chi-square distance that can be expressed in the following way:

$$d^2(i,i') = \sum_j \frac{1}{f_{.j}} \left(\frac{f_{ij}}{f_{i.}} - \frac{f_{i'j}}{f_{i'.}} \right)^2$$

$$d^2(j,j') = \sum_i \frac{1}{f_{i.}} \left(\frac{f_{ij}}{f_{.j}} - \frac{f_{ij'}}{f_{.j'}} \right)^2$$

Finally, our problem consists in visualizing the scatterplot, N_I, of the rows (perfumes) and the scatterplot, N_J, of the columns (words), through their row profiles and their column profiles, respectively.

In CA, it can be shown that the inertia of N_I and N_J are equal, and

$$I(N_I) = I(N_J) = \frac{1}{N}\chi^2$$

where χ^2 denotes the chi-square distance between the two qualitative variables. Hence, graphically analyzing the dispersion (variability) of N_I and N_J, in other words, the shape of N_I and N_J, is equivalent to analyzing the independence between the two qualitative variables.

As shown in Table 3.5, in CA, the eigenvalues are always lower than 1; their sum over the dimensions (components) equals $I(N_I) = I(N_J) = (1/N)\chi^2$. From a theoretical point of view, eigenvalues can be equal to 1. In that case, which never happens practically, it expresses an exclusive association among modalities of one and the other qualitative variable.

As shown in Figure 3.10, in CA, row profiles are represented by points. In R^n, the first dimension opposes perfumes such as *Shalimar* and *Aromatics Elixir* to perfumes such as *Pleasures* and *J'adore EP*. The second dimension opposes perfumes such as *Angel* and *Lolita Lempicka* to the others.

As shown in Figure 3.11, in CA, column profiles are represented by points. In R^p, the first dimension opposes words such as *Grandmother* and *Artificial* to words such as *Flowery* and *Fruity*.

TABLE 3.5

Decomposition of the Inertia, CA on the Words Used by Consumers to Describe the 12 Perfumes

	Eigenvalue	Percentage of Variance	Cumulative Percentage of Variance
Dim 1	0.486	26.935	26.935
Dim 2	0.310	17.196	44.132
Dim 3	0.205	11.358	55.489
Dim 4	0.166	9.205	64.695
Dim 5	0.161	8.914	73.609
Dim 6	0.118	6.532	80.142
Dim 7	0.110	6.077	86.218
Dim 8	0.097	5.383	91.602
Dim 9	0.080	4.427	96.029
Dim 10	0.038	2.117	98.146
Dim 11	0.033	1.854	100.000

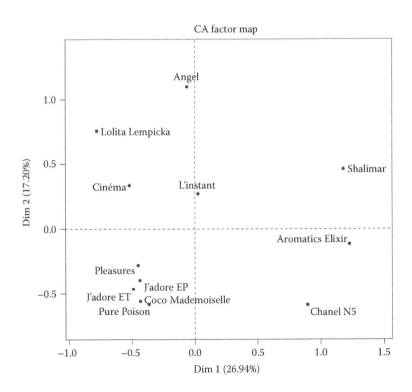

FIGURE 3.10 Representation of the 12 perfumes described by words provided by consumers on the first and second dimensions of the CA.

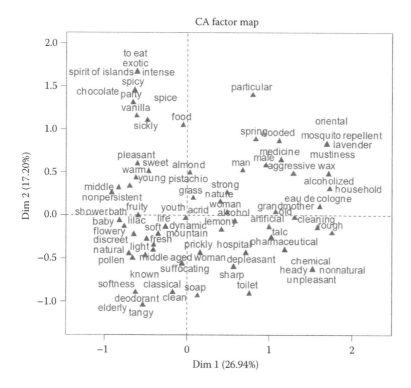

FIGURE 3.11   Representation of the words used by consumers to describe the 12 perfumes on the first and second dimensions of the CA.

Figure 3.12 shows a superimposed representation of the rows and the columns. Although rows and columns belong to different vector spaces (R^n and R^p, respectively), this superimposed representation is often used, due to the symmetrical role of the rows and the columns. Figure 3.12 must be interpreted carefully as, for a given qualitative variable, it should be interpreted modality by modality. For instance, we would say that *Chanel N5* has been rather associated with the words *Hospital*, *Pharmaceutical*, and *Toilet* and has been rarely associated with the words *Spicy*, *Pleasant*, *Sweet*, and *Warm*.

Thanks to the geometrical point of view, illustrative information can be added as well. Figure 3.13 represents French words that have been used as illustrative elements.

3.5   MULTIPLE CORRESPONDENCE ANALYSIS

MCA is applied when individuals are described by qualitative variables through their modalities. In MCA, the data set to be analyzed is structured in the following way: Rows correspond to individuals or statistical units,

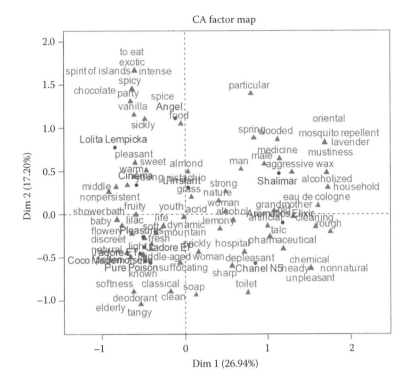

FIGURE 3.12 Superimposed representation of the rows and the columns on the first and second dimensions of the CA; in the example, superimposed representation of the perfumes and the words used by the consumers.

and columns correspond to variables. At the intersection of one row i and one column j, is the modality taken by individual i for variable j.

MCA can be seen as a *mix* between PCA and CA. It is an extension of PCA in the sense that individuals are described by a set of qualitative variables and it is an extension of CA in the sense that it deals with more than two qualitative variables. The main objective of the method is to provide a 2D representation of the individuals that best represents their distances and a 2D representation of the modalities that best represents their mutual associations. As in CA, MCA provides a joint representation of the rows and the columns, more accurately a joint representation of the individuals and the modalities associated with the qualitative variables.

This method is mainly used in a questionnaire context, as questions can be assimilated to qualitative variables. In consumer studies, MCA is used to understand respondents with respect to their profiles in terms of answers. From a consumer point of view, the method provides a graphical representation of the consumers such that the closer two consumers are the

CA factor map

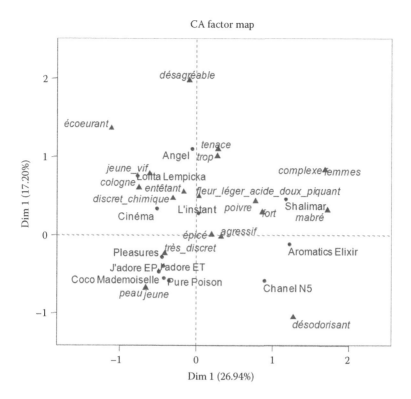

FIGURE 3.13 Superimposed representation of the rows and the illustrative columns on the first and second dimensions of the CA; in the example, superimposed representation of the perfumes and the French words used by the consumers, as illustrative elements. Close to Shalimar, the words complexe (complex) and femmes (women), and close to Coco Mademoiselle, the words peau (skin) and jeune (young).

more similarly they answered to the questionnaire. From a questionnaire point of view, the method provides a graphical representation of the items of the questionnaire such that the closer two items are, the more similar consumers they represent (consumers that have similar profiles in terms of answers to the questionnaire).

To illustrate the method, we consider a consumer study in which 80 respondents are asked to answer to *usage and attitude* questions with respect to makeup. In terms of *usage*, the questions that have been asked are as follows: How do you use makeup on your skin? Your eyes? Your lips? How often do you use makeup? How often do you use the following products: foundation, blusher?

Table 3.6 shows an extract of the data set on which MCA is performed. According to Table 3.6, individuals 75 and 73 (resp. 55 and 53) seem to

TABLE 3.6

Extract of the Data Set on Which MCA Is Performed: How Do You Use Makeup on Your Skin? Your Eyes? Your Lips? How Often Do You Use Makeup? How Often Do You Use the Following Products: Foundation, Blusher?

Individuals	On Skin	On Eyes	On Lips	Freq. Makeup	Freq. Foundation	Freq. Blusher
75	Slightly	Medium	Medium	Every day	Every day	Every day
73	Medium	Medium	Medium	Every day	Every day	Every day
55	Not at all	Medium	Very slightly	≥1 per week	≤1 per year	≤1 per year
53	Very slightly	Medium	Very slightly	≥1 per week	1 and 2 per month	≤1 per year

have a similar profile in terms of behavior; they also seem to behave totally differently than 55 and 53.

In order to calculate distances between individuals on the one hand and distance between modalities on the other, Table 3.6 is transformed into a dummy variables data set. The so-called *complete disjunctive data table* of dimension (n, K) is introduced, where n is the number of individuals and K is the total number of modalities over the p qualitative variables. At the intersection of one row and one column, $x_{ik} = 1$ if individual i has chosen the modality k and 0 otherwise (Table 3.7).

Let I_k be the number of individuals that has chosen modality k. In MCA, the squared distance between two individuals i and i' can be expressed in the following way:

$$d^2(i,i') = C \sum_{k=1}^{K} \frac{\left(x_{ik} - x_{i'k} \right)^2}{I_k}$$

where C denotes a constant that will be determined later.

As $(x_{ik} - x_{i'k})^2$ equals 1 (resp. 0) when i and i' have answered differently (resp. similarly) for item k, this distance is the sum of the differences over all modalities, weighted by $1/I_k$:

1. The distance between two individuals equals 0, if they have the same profile of answers, in other words if they have selected the same items.
2. The distance between two individuals is rather small, if they have a similar profile of answers, in other words if they have in common a rather high number of items.

TABLE 3.7
Extract of the Complete Disjunctive Data Table: How Do You Use Makeup on Your Skin? Your Eyes?

Individuals	On Skin					On Eyes				
	Not At All	Very Slightly	Slightly	Medium	A Lot	Not at All	Very Slightly	Slightly	Medium	A Lot
75	0	0	1	0	0	0	0	0	1	0
73	0	0	0	1	0	0	0	0	1	0
55	1	0	0	0	0	0	0	0	1	0
53	0	1	0	0	0	0	0	0	1	0

3. The distance between two individuals is rather high, if one of the items they do not have in common is specific of one of them, even though they might have in common a lot of items.
4. The distance between two individuals is rather small, if one of the items they have in common is specific of both of them, even though they might not have in common a lot of items.

Let $I_{k \neq k'}$ denote the number of individuals that have chosen either modality k or modality k'. The distance between two modalities k and k' can be expressed in the following way:

$$d^2(k,k') = C' \frac{I_{k \neq k'}}{I_k I_{k'}}$$

where C' denotes a constant that will be explained later.

According to this formula, the distance between two modalities k and k' is all the more small that the number of individuals they have in common is high.

To understand the importance of the weighting, let us consider three modalities k, k', and k'', each composed of 10, 100, and 100 individuals, respectively. If modalities k and k' have no common individuals, then $I_{k \neq k'} = 110$. If modalities k' and k'' have 45 common individuals, then $I_{k' \neq k''} = 110$. However, k and k' have no individuals in common, whereas k' and k'' have in common 45% of their individuals. Modalities k and k' should be more distant than modalities k' and k''. It is therefore important to take into account the sample size for each modality. Finally, in the example,

$$d^2(k,k') = \frac{110}{10 \times 100} > d^2(k',k'') = \frac{110}{100 \times 100}$$

From a theoretical point of view, it can be shown that MCA is equivalent to a CA applied to the *complete disjunctive data table*. Hence, when applying the chi-square distance on this data table, it can be shown that

$$d^2(i,i') = \frac{n}{p} \sum_{k=1}^{K} \frac{(x_{ik} - x_{i'k})^2}{I_k}$$

and

$$d^2(k,k') = n \frac{I_{k \neq k'}}{I_k I_{k'}}$$

Once the distance among individuals is calculated, MCA finds the sequence of axes that best represent the individuals, as previously

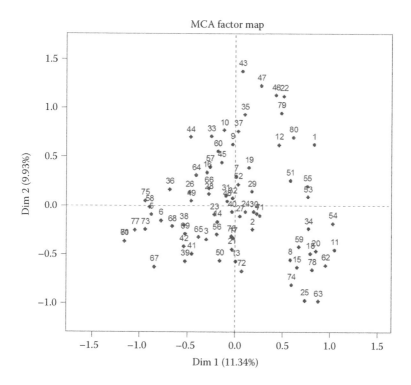

FIGURE 3.14 Representation of the consumers on the first and second dimensions of the MCA: the closer two consumers are, the more similar behavior in terms of makeup usage and attitudes they have.

explained. Figure 3.14 represents the 80 consumers on the two first principal dimensions. With a value of 11.34% (resp. 9.93%), the percentage of variability associated with the first dimension (resp. second dimension) seems rather low. It is often the case with MCA, and these percentages associated with the eigenvalues are rarely interpretable as they are often underestimated. This is due to the coding into dummy variables that induces an artificial orthogonality of the columns of the data set.

The first dimension opposes individuals 75 and 73, on the one hand, to individuals 55 and 53 on the other (cf. Figure 3.14). As we can see in the original data, in Table 3.6, this opposition was expected.

In most cases, the representation of the modalities (cf. Figure 3.15) is not exploitable, due to the high number of modalities to be represented.

Figure 3.16 is an extract of the representation of the modalities (or items) of the questionnaire. To obtain this representation, we first calculate an index of quality of representation based on the sum of the cosine squared of each modality on each dimension over the two first dimensions. Then, we represent those modalities for which the quality of representation is higher than a given arbitrary threshold (equal to 0.2 in this example).

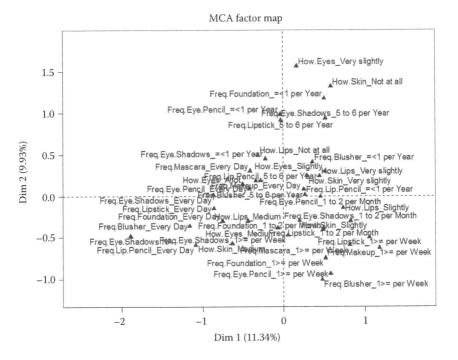

FIGURE 3.15 Representation of the modalities associated with the qualitative variables of the consumers on the first and second dimensions of the MCA; in the example, representation of the items of the questionnaire.

To interpret this graphical output, one of the most common strategies is to list the modalities that are significantly related to each dimension. Table 3.8 lists the 10 most meaningful modalities (items) on the positive side of the first dimension. Those modalities are interrelated, in the sense that on the *left side* of the plane, the same consumers (consumers 75 and 73, for instance) have mainly chosen those modalities. These consumers use makeup on a daily basis.

In the same way, Table 3.9 lists the 10 most meaningful modalities (items) on the negative side of the first dimension. Those modalities are interrelated, in the sense that on the *right side* of the plane, the same consumers (consumers 55 and 53, for instance) have mainly chosen those modalities. These consumers use makeup occasionally.

In the same way as in PCA or CA, illustrative information can be added and represented as shown in Figure 3.17. In the case of illustrative quantitative variables, the information is represented as in PCA, through the correlation circle, whereas in the case of illustrative qualitative variables, the information is represented as in CA, through the common representation of the rows and the columns.

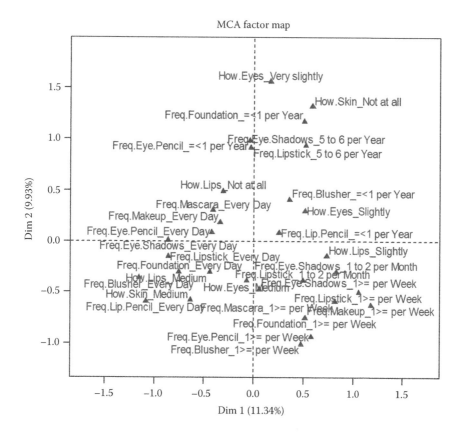

FIGURE 3.16 Representation of the modalities that are well represented on the first plane of the MCA; in the example, representation of the items of the questionnaire.

In the example, the *attitude* questions were considered as illustrative information: For which reasons did you begin to apply makeup? What is the main occasion to apply makeup? For which reasons do you apply makeup? For whom do you make up?

It is interesting to see that from a consumer perspective, the main dimension of variability induced by the active variables only (in the case, the *usage* questions) is related to the notion of seduction: on one side of the plane, consumers who apply makeup every day, for no particular reason, and certainly not to seduce, and on the other side of the plane, consumers who rarely apply makeup, but when they do, we can suppose that it is when they go out, in order to seduce (Table 3.10).

Remark: Despite its seemingly low percentage of variability, the second axis is particularly interesting in terms of illustrative information as it is clearly related to the notion of confidence (Table 3.11).

TABLE 3.8

Ten Most Meaningful Active Modalities (Items of the Questionnaire) on the Positive Side of the First Dimension of the MCA

Frequent lipstick: every day
Frequent makeup: every day
Frequent blusher: every day
Frequent mascara: every day
Frequent foundation: every day
On skin: medium
Frequent lip pencil: every day
On lips: medium
Frequent eye pencil: every day
Frequent eye shadows: every day

TABLE 3.9

Ten Most Meaningful Active Modalities (Items of the Questionnaire) on the Negative Side of the First Dimension of the MCA

Frequent makeup: ≥ 1 per week
Frequent mascara: ≥ 1 per week
Frequent lipstick: ≥ 1 per week
Frequent eye shadows: 1 and 2 per month
Frequent lip pencil: ≤ 1 per year
On lips: slightly
Frequent eye shadows: 5 and 6 per year
Frequent blusher: ≤ 1 per year
Frequent blusher: ≥ 1 per week
On skin: not at all

3.6   MULTIPLE FACTOR ANALYSIS

MFA is applied when individuals are described by variables that can be *a priori* structured into groups of the same type (variables of a given group must all be either quantitative or qualitative). In MFA, the merged data set $X = \begin{bmatrix} X_1 X_2 ... X_J \end{bmatrix}$ is considered, where X_j denotes the data set associated with the variables of group j. Practically, to each group of variables is

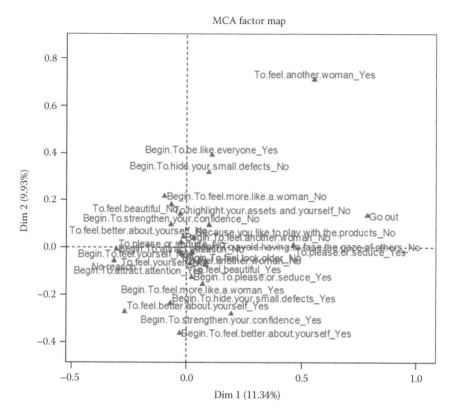

FIGURE 3.17 Representation of illustrative modalities on the first plane of the MCA; in the example, representation of the items of the questionnaire related to the set of *attitude* questions.

TABLE 3.10

Two Most Meaningful Illustrative Modalities on the First Dimension of the MCA

Positive Side	Negative Side
What is the main occasion: No reason	What is the main occasion: Go out
Why? To please or seduce: No	Why? To please or seduce: Yes

associated a multivariate point of view, and this is precisely this multiple points of view that we want to compare when performing an MFA. More precisely, we want to extract the common dimensions to the *J* groups of variables, as well as their specific ones. Once the common dimensions are determined, one of the main objectives of MFA is to provide a representation of the individuals as common as possible to the *J* points of view.

TABLE 3.11

Eight Most Meaningful Illustrative Modalities on the Second Dimension of the MCA

Positive Side	Negative Side
Why? To hide your small defects: No	Why? To hide your small defects: Yes
Begin to hide your small defects: No	Begin to hide your small defects: Yes
Begin to feel more beautiful: No	Begin to feel more beautiful: Yes
To strengthen your confidence: No	To strengthen your confidence: Yes
Everybody: No	Everybody: Yes
Begin to feel better about yourself: No	Begin to feel better about yourself: Yes
On the lips to get attention: No	On the lips to get attention: Yes
To highlight your assets and yourself: No	To highlight your assets and yourself: Yes

To illustrate the method, data that have already been used when presenting PCA is used, where the sensory profiles provided by an expert panel, denoted X_{Expert}, and by a consumer panel, denoted $X_{Consumer}$, are available. Performing an MFA on the merged data set $X = [X_{Expert} X_{Consumer}]$ allows to compare the representation of the perfumes provided by the expert panel (cf. Figure 3.4) to the representation of the perfumes provided by the consumer panel (cf. Figure 3.18). Let us remind that both representations were obtained by performing a PCA on matrices X_{Expert} and $X_{Consumer}$, respectively.

This comparison is done within a single framework, based on a consensus representation issued from both panels. Naturally, in order to get that

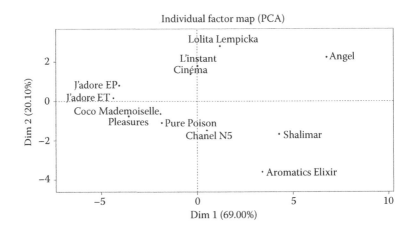

Individual factor map (PCA)

FIGURE 3.18 Representation of the 12 perfumes evaluated by the consumers on the first and second dimensions of the PCA.

consensus, points of view have to be balanced. In MFA, the idea is to give the same weight to each one of the first principal component of each group of variables, without modifying the multivariate structure of each group. In other words, each group of variables is submitted to a homogeneous dilation (aka homothety), such that the variance of the first component of each group (once transformed, i.e., after the dilation) equals 1; *de facto*, the main information of each group has the same weight, and a consensus can be reasonably obtained based on all groups.

Technically, for all j, variables of group j are weighted by $1/\lambda_1^j$, where λ_1^j denotes the first eigenvalue of the multivariate analysis of X_j (a PCA when variables of group j are quantitative, an MCA when they are qualitative, and a CA when X_j is a contingency table). Figure 3.19 and Table 3.12 illustrate how the dilation works.

Even though the respective ranges of Figures 3.18 and 3.19 are different, the two representations are exactly the same in terms of relative distances between the products.

The first eigenvalue of the PCA performed on the weighted variables of $X_{Consumer}$ equals 1 (cf. Table 3.12, second column) and the multivariate structure remains the same (cf. Table 3.12, third and fourth columns).

While the variance of the first component of X_{Expert} equals 7.710 (cf. Table 3.3) and the variance of the first component of $X_{Consumer}$ equals 11.039 (cf. Table 3.12), thanks to its particular weighting, MFA balances the part of each group within a global analysis. In the example, this global analysis is a weighted PCA of the merged data set $X = [X_{Expert} X_{Consumer}]$.

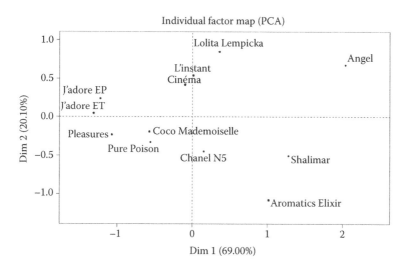

FIGURE 3.19 Representation of the 12 perfumes evaluated by the consumers on the first and second dimensions of the PCA; variables have been weighted by the first eigenvalue.

TABLE 3.12

Variance Associated with Each Component

(Analysis of the Consumers)

| | Eigenvalue | | Percentage | Cumulative |
	PCA	Weighted PCA	of Variance	Percentage of Variance
Comp 1	11.039	1.000	68.995	68.995
Comp 2	3.217	0.291	20.105	89.100
Comp 3	0.714	0.065	4.463	93.563
Comp 4	0.423	0.038	2.643	96.206
Comp 5	0.198	0.018	1.235	97.441
Comp 6	0.129	0.012	0.805	98.246
Comp 7	0.113	0.010	0.707	98.952
Comp 8	0.082	0.007	0.510	99.463
Comp 9	0.041	0.004	0.256	99.719
Comp 10	0.033	0.003	0.209	99.928
Comp 11	0.012	0.001	0.072	100.000

Note: When data are weighted, the variance of the first component equals 1 (cf. second column).

As a principal axes method, MFA provides a representation of the rows, in the example the products, and the columns, in the example the sensory descriptors used by both panels.

Figure 3.20 represents the common information between the expert panel and the consumer panel, the part of each panel being balanced. As expected, the main axis of variability opposes *Shalimar* and *Aromatics Elixir*, on one side of the plane, to *J'adore EP*, *J'adore ET*, and *Pleasures*, on the other side of the plane: this opposition corresponds to the one shown in Figures 3.4 and 3.18.

Figure 3.21 is of particular interest when comparing the expert and the consumer panels, as we can see how homologous descriptors are correlated and how different sensory descriptors can be interpreted in the same way by the two panels. The descriptor *Spicy* seems to be understood in the same way by both panels, which is not quite the case for the descriptor *Citrus*. The descriptor *Greedy* used by the expert panel is very close to the descriptor *Honey* used by the consumer panel.

As in all principal axes methods we have seen so far, MFA can also handle illustrative information. After all, the core of MFA is a weighted PCA.

Figure 3.22 is one of the most important outputs provided by MFA as it represents how the separate multivariate analysis of each group is linked to the consensus issued from MFA. To obtain this figure, for all *j*, the

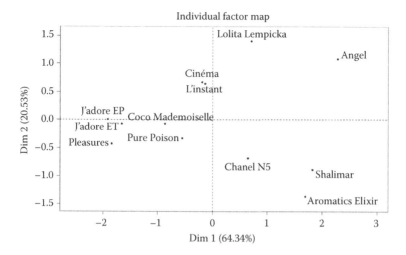

FIGURE 3.20 Representation of the 12 perfumes evaluated by the experts and the consumers on the first and second dimensions of the MFA, the part of each panel being balanced.

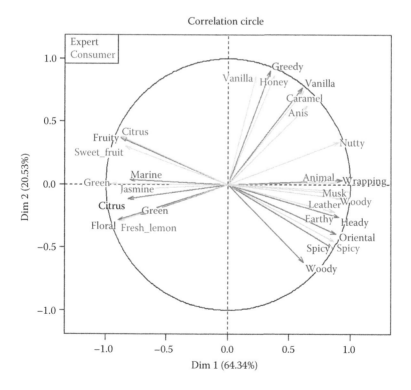

FIGURE 3.21 Correlation circle, expert panel, and consumer panel on the first and second dimensions of the MFA.

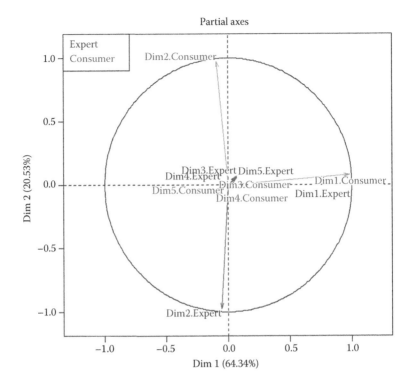

FIGURE 3.22   Representation of the separate analyses on the first and second dimensions of the MFA.

components of the separate multivariate analysis of the variables of group *j* are projected as illustrative information on the axes issued from MFA performed on all *J* groups.

As it can be seen in Figure 3.22, the first component of the PCA performed on X_{Expert} and the first component of the PCA performed on $X_{Consumer}$ are closely correlated to the first component of the MFA performed on $X = [X_{Expert} X_{Consumer}]$. The same comment applies for the second component.

Finally, MFA provides two very specific graphical outputs that are the representation of the groups (cf. Figure 3.23) and the partial representations of the individuals (cf. Figure 3.24).

In the representation of the groups, the coordinate of a group of variables *j* on an axis of rank *s* is obtained by calculating the *Lg* measure between the variables of the group *j* and F_s and the coordinates of the individuals on the axis of rank *s*. Due to the weighting used in MFA to balance the part of each group, the coordinates of the groups of variables lie in [0, 1]. The coordinate of a given group is all the more close to 1 than the variables of this group are highly correlated with the dimension issued from

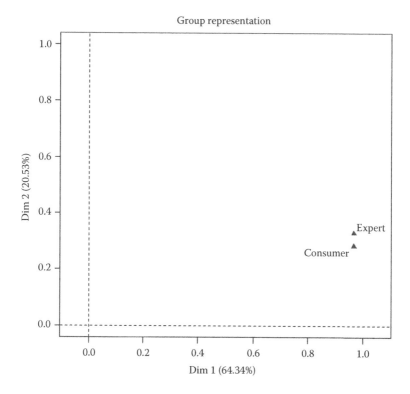

FIGURE 3.23 Representation of the groups of variables, expert panel, and consumer panel on the first and second dimensions of the MFA.

the MFA (either positively or negatively). Hence, the closer two groups are, the closer the structures they induce on the individuals are.

As shown in Figure 3.23, the structure on the perfumes is induced by experts and the consumers are really close.

This result is confirmed by the representation of the perfumes described by experts and consumers.

It can be seen in Figure 3.24 how close the two points of view for each perfume are. This is particularly true for perfumes such as *J'adore EP* and *J'adore ET*. In Figure 3.24, it also can be seen that the differences between experts and consumers for *Angel* (resp. *Shalimar*) are due essentially to the first (resp. second) axis and must be interpreted with respect to that axis.

3.7 PRACTICAL RECOMMENDATIONS AND CONCLUSION

From a practical point of view, as all these principal axes methods are based on the notion of distance. The main issue, which may be the only issue, the analyst should be concerned about is the status of the columns

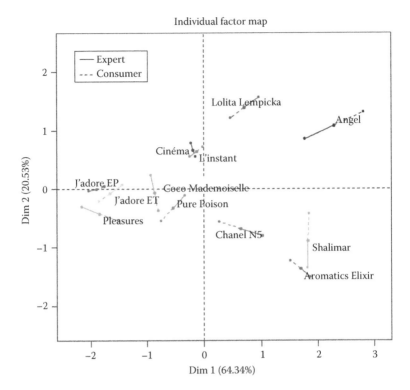

FIGURE 3.24 Partial representations of the 12 perfumes described by the experts and the consumers on the first and second dimensions of the MFA.

(or groups of columns in the case of MFA), whether they should be considered as active or illustrative. The analyst should bear in mind that active columns are by essence the ones that actively induce the distance among the rows and that to a set of active columns corresponds a multivariate point of view. Is this set of active columns, or in other words, this point of view, that the analyst wants to use to answer the research problem? That is the question.

What has been just discussed for the columns applies as well for the rows to a lesser extent, at least in practice.

Once the analysis is completed, the analyst should pay attention to the notion of contribution and the notion of quality of representation, both concepts applying for the rows and the columns, in order to properly interpret his or her results. If possible, the analyst should come back to the raw data, in order not to overinterpret his or her results.

Finally, he or she should keep in mind that clustering methods are often complementary to principal axes methods.

The core methods that have just been presented offer almost infinite possibilities for those analysts who deeply understand them and who have good skills for revisiting methods. Among the many different ways of using them in a sensory context, let us recall the visualization of the variability of the products with confidence ellipses; the analysis of so-called holistic methods such as Napping, (hierarchical) sorting, and sorted Napping; taking into account the existence of complex structure on the variables, such as a hierarchy structure on the variables (with hierarchical MFA); and the analysis of free comments and CATA data.

4 Ideal Profiling

Thierry Worch and Pieter H. Punter

CONTENTS

4.1   IDEAL PROFILE METHOD

4.1.1   INTRODUCTION

The methodologies applied for product optimization are diverse, but they share the common underlying assumption that an *ideal product* exists. A core assumption underlying this thinking is that consumer disliking of the product is a weighted linear combination of the absolute attribute level deviations from the ideal product. More formally, this can be expressed as

$$h_{jp} = H_{jp} - \sum_{a=1}^{Att} b_a (z_{jpa} - y_{jpa}) \qquad (4.1)$$

where

    h_{jp} is the consumer liking judgment for product p
    b_a is the relative importance of deviations on attribute a for consumer's overall liking judgment
    y_{jpa} is the consumer perception of product p on attribute a
    z_{jpa} is the ideal level of attribute a that would generate maximum liking H_{jp}

When the deviations from ideal are zero for the attributes impacting liking, liking is maximized.

The ideal profile method (IPM) belongs to the family of descriptive methods like quantitative descriptive analysis (QDA) (Stone and Sidel 2004), free-choice profiling (Williams and Langrons 1984), and is also similar to consumer-based methodologies such as just-about-right (JAR) scaling (Rothman and Parker 2009). It is similar to QDA® except that the products are not rated by trained assessors but by naive consumers; it resembles JAR scaling but instead of an implicit ideal, the consumers are asked to rate their ideal intensities explicitly. The different methodologies differ in the way the information is obtained, as shown in Table 4.1 (Van Trijp et al. 2007).

4.1.2   HISTORICAL BACKGROUND

In sensory analysis, one of the most common tasks consists in describing the way a set of products is perceived. To do so, assessors are asked to rate

TABLE 4.1

Comparisons between IPM, JAR, and QDA

	IPM	JAR	Conventional QDA
h_{jp}: Overall liking	Measured	Measured	Measured by consumers in parellel
b_a: Attribute importance	Calculated	Calculated	Calculated
y_{jpa}: Attribute perception	Measured	Not available	Measured
z_{jpa}: Attribute ideal point	Measured	Not available	Calculated
$z_{jpa} - y_{jpa}$: Attribute deviation	Calculated	Measured	Calculated

their perception of the products tested on a list of predefined attributes. This practice is also known as descriptive analysis (such as QDA®, Stone et al. 1974) and results in sensory profiles of the products. One of the goals of such practice is to obtain a product space, a map in which the products that are perceived as similar are close to each other and apart from those that are perceived as different. In the literature, it is recommended to use experts or trained panelists for such task.

Although this methodology is extensively used, some alternative methods have been developed. These methods differ according to the points of view adopted by the user. For instance, if the user is interested in the assessors' personal opinion, each assessor should be free in the choice of attributes used to describe the products. This is the point of view adopted when *free-choice profile* (Williams and Langrons 1984) or *flash profile* (Dairou and Sieffermann 2002; Sieffermann 2002) methods are used.

If the user is interested in the assessors' overall perception of the products, holistic methodologies are preferred. Among holistic approaches, we can mention *Napping®* (Pagès 2005), *ultra flash profile* (Perrin et al. 2008), *(hierarchical) free-sorting task* (Lawless 1989; Cadoret et al. 2009; Cadoret et al. 2011), or *sorted Napping* (Pagès et al. 2010). These methodologies require a presentation of all the products simultaneously.

A common point to all these methodologies is that no (or only short) training is required. They are hence qualified as rapid methodology (Dehlholm et al. 2012). For these methodologies, consumers are often used instead of trained or expert assessors.

Since consumers are the final deciders of the market place success, the interest in their opinion became increasingly important through the decades. Therefore, they have been more and more involved in the product development process. Moskowitz (1996), Husson et al. (2001), and, more recently, Worch et al. (2010b) have shown in different studies that consumers can profile products while meeting the requirements of discrimination, consensus, and reproducibility of a sensory panel. This is particularly true when the attributes that are evaluated are not complex and understandable by naïve consumers.

In parallel, other studies have shown that consumers can use an internal imagined product as reference to compare products (Booth et al. 1987).

From these findings, new types of questions have appeared. For instance, consumers can be asked to compare a set of products to their internal representation of the ideal, that is, the product they would like to have. Such comparison is done when using *JAR* scales. In JAR scaling, consumers are asked to rate the intensity of the products on each attribute by indicating whether the intensity of that attribute is *just about right*, *too strong*, or *too weak* compared to their internal ideal.

If consumers can rate the perceived intensities of the products in function of an internal or imagined ideal that works as a reference, one can also expect them to be able to rate their ideal explicitly.

Moskowitz (1972) was the first one considering this idea. He proposed to extend the classical sensory evaluation by giving the consumers the opportunity to suggest on a scale the degree to which they would alter products for the given attribute set so that it would be closer to their representation of their ideal. Depending on the study, the subject was asked either to rate the ideal directly (explicit) or to rate the perceived intensity relatively to this ideal (implicit). Some years later, Szczesniak et al. (1975) proposed a derivative of the texture profile technique (Brandt et al. 1963) using consumers. In their study, apart from providing descriptions of the texture of the products, consumers were also requested to rate the ideal intensity on the specific texture attributes. Hoggan (1975) applied a similar technique for optimizing beers, by including taste attributes as well. Ancestors of the IPM were born.

The procedure to estimate an ideal product with consumers was taken up by the Massy University in New Zealand in the late 1970s (Cooper et al. 1989). In the late 1980s, the method was rediscovered in the Netherlands by OP&P Product Research and in the United Kingdom by SRL. Although the methodology has been used intensively in the past 25 years by several Dutch, Australian, and New Zealand market research agencies (OP&P Product Research, SRL, Sensory Solutions, Colmar Brunton, etc.), it had not been really accepted by the sensory community. The main argument was that consumers cannot give valid sensory profiles. Since several studies proved the contrary, the global opinion on consumers profiling changed. More recently, Worch (2012) explored the ideal data provided from consumers and proposed a new methodology to analyze them. This methodology is referred to as the ideal profile analysis (IPA) (Worch et al. 2014).

It should be noted that the alternative methodology (JAR scaling) has been intensively used (Coombs 1964; Meullenet et al. 2007; Rothman and Parker 2009) by market researchers since the 1960s.

4.1.3 Methodology

In the IPM, consumers are asked to rate products on both the perceived and ideal intensities for a list of sensory attributes. In this case, P products tested yield P sensory profiles and P ideal profiles per consumer. Additionally, consumers rate the products on overall liking.

In practice, during the test according to the IPM, the following questions would be asked: if the first question is "please rate the sweetness of this product," the second question would be "please indicate your ideal sweetness for this product" (Figure 4.1). Additionally, they would rate each product on several hedonic attributes.

The sensory and ideal intensities can be rated on line scales or category scales, as long as the scales for perceived and ideal intensities are the same.

For such methodology, consumers are required. By using consumers, we are closer to the market since they are the final deciders of product success. However, with this advantage comes the requirement that the attributes used to describe the products should be simple and understandable by consumers. They cannot be technical or chemical. As for any descriptive test, the list of attributes should also cover the relevant appearance, odor, taste, mouthfeel, and aftertaste aspects. Depending on the purpose of the test, the number of attributes asked to the consumers can vary between

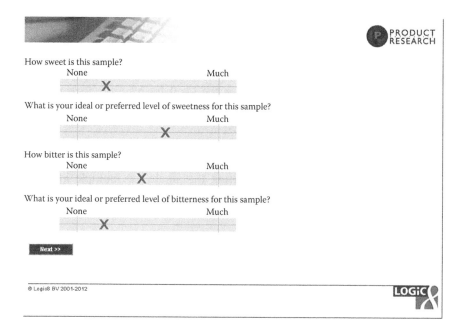

FIGURE 4.1 Example of the ideal profile question on a computer screen using EyeQuestion http:// www.logic8.nl.

10 and 30. In practice, we recommend to generate and discuss the attribute list with the product developers and marketers to guarantee sufficient coverage of the product perceptual space. Classical methodology such as focus group could also be used for the generation of the list.

Since naïve and untrained consumers are used, the panel size should be at least 60–100 (Moskowitz 1997).

4.1.4   ALTERNATIVE PROCEDURES

Historically, asking for the ideal intensity was done only once, either at the beginning or at the end of the test. Such methodology seems suitable and provides adequate results. It can be mentioned that by asking the ideal only once for each attribute at the end of the test, comparison with the ideal while evaluating the products on overall liking is somewhat avoided. In other words, such procedure reduces the eventual halo effect of the ideal questions on liking, if any.

However, when consumers rate products on liking, they compare the actual perception of the product with their ideal in practice. Whether this comparison is implicit or explicit does not make a large difference in terms of halo effect. Moreover, the same situation occurs when JAR scales are used.

For those reasons, we would still recommend asking the ideal every time a product is tested and this is because of multiple reasons:

- By asking the ideals for each product, we mimic the JAR scale procedure where for each attribute, the deviation from the ideal is asked for each product.
- It is known that the ideal rating is influenced by the tested product. By repeating the measurement of the ideal for each product, this influence can be corrected by considering the average score for each consumer over the products.
- The extra information collected (variability of the ideal ratings between products, within consumer) is a very important information that is necessary for checking the single vs. multiple ideal procedure (see Section 4.3.2) and for the construction of the ideal map (see Section 4.4.1).

As a counterpart, the test seems more demanding since consumers are asked much more questions. However, this seems not to be an issue in practice.

4.1.5   NOTATION AND PRELIMINARY REMARK

Let P denote the number of products tested, A the number of attributes used to describe the products, and J the number of consumers who

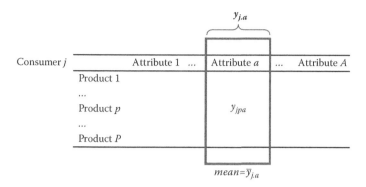

FIGURE 4.2 Organization and notation of the sensory data provided by each consumer.

participated in the test. The following notation will be used to describe the data obtained from *IPM* (vectors are in bold):

y_{jpa}: intensity perceived by the consumer j for the product p and the attribute a.

$\boldsymbol{y}_{j \cdot a} = \{y_{jpa}; p = 1 : P\}$ vector of intensities perceived by the consumer j for the P products and the attribute a.

$\overline{y}_{j \cdot a}$: average over the index p; average intensity perceived by the consumer j on attribute a over the P products (Figure 4.2).

z_{jpa}: ideal intensity of the attribute a provided by the consumer j after testing the product p.

$\boldsymbol{z}_{j \cdot a} = \{z_{jpa}; p = 1 : P\}$ vector of ideal intensities of the attribute a provided by the consumer j for the P products.

$\overline{z}_{j \cdot a}$: average over the index p; average ideal intensity of the attribute a provided by the consumer j over the P products (Figure 4.3).

h_{jp}: hedonic judgment provided by the consumer j for the product p.

$\boldsymbol{h}_{j \cdot} = \{h_{jp}; p = 1 : P\}$ vector of hedonic judgments provided by the consumer j for the P products.

As it has been already mentioned, consumers rate as many times their ideal profile as they test products. If the set of products belongs to the same category and type (see Section 4.3.2), people would rate their ideal attributes of the tested products in a consistent way (the ideal ratings being very similar, if not the same). Consequently, each consumer would be assigned a unique ideal profile that corresponds to the averaged ideal rating he or she gave for each attribute. This averaged ideal profile is denoted $\overline{z}_{j \cdot \cdot} = \{\overline{z}_{j \cdot a}; a = 1 : A\}$ and is defined by

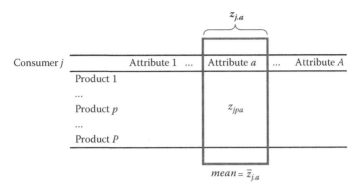

FIGURE 4.3 Organization and notation of the ideal data provided by each consumer.

$$\overline{z}_{j\cdot a} = \frac{1}{P}\sum_{p=1}^{P} z_{jpa} \quad \text{and} \quad \overline{z}_{j\cdot\cdot} = \{\overline{z}_{j\cdot a};\ a = 1:A\} \tag{4.2}$$

In sensory science, it is known that consumers differ, both physiologically (i.e., in their perception of the products) and psychologically (i.e., they differ in their use of the scale of notations, some scoring only low *vs.* high, or using all *vs.* a part of the scale). By studying the ideal product of the consumers, we are interested in the structural differences between the ideals of consumers (i.e., what consumers want) but not in the differences due to different scale usage.

Since consumers do not use the scale in the same way and since this difference in the use of the scale is a major source of variability in consumer studies, it is important to correct their averaged ideal profiles before comparing them. The correction is done by translating the consumers' averaged ideal profile according to the averaged perceived intensities they also provided. It is done by subtracting the averaged perceived intensities over the P products for each consumer and each attribute from his or her averaged ideal profile (Equation 4.3). This corrected ideal profile is noted $\tilde{z}_{j\cdot\cdot}$:

$$\tilde{z}_{j\cdot a} = \overline{z}_{j\cdot a} - \overline{y}_{j\cdot a} \quad \text{and} \quad \tilde{z}_{j\cdot\cdot} = \{\tilde{z}_{j\cdot a};\ a = 1:A\} \tag{4.3}$$

As before, $\tilde{z}_{j\cdot\cdot} = \{\tilde{z}_{j\cdot a};\ a = 1:A\}$ represents the vector of ideal intensities averaged over the index p and corrected according to the averaged intensities perceived by the consumer j considered for the A attributes. This vector is called *corrected averaged ideal profile of the consumer j*.

An additional step (in standardizing this difference) can be performed. In that case, the quantity $\tilde{z}_{j\cdot a}$ is divided by the standard deviation $\sigma_{j\cdot a}^{(y)}$ of the perceived intensities provided by the consumer j for the attribute a:

$$\text{Standardized } \tilde{z}_{j \cdot a} = \frac{\tilde{z}_{j \cdot a}}{\sigma_{j \cdot a}^{(y)}} \quad (4.4)$$

For the rest of the document, the translated (but not standardized) ideal ratings will be used. These methodologies will be illustrated using a real case study.

4.1.6 MATERIAL

The dataset used for illustration is from a study on pasta sauce, which is available for download from the CRC Web site: http://www.crcpress.com/product/isbn/9781466566293. Thirteen sauces were tested by 112 Dutch consumers using the IPM. Each consumer tested each product in a monadic sequential way balanced for order and carryover effects (MacFie et al. 1989) and rated them on both perceived and ideal intensities on 23 attributes using a 100 mm line scale with anchor points at 10 and 90. The list of the attributes is given in Table 4.2. Consumers also rated the products on overall liking using a 9-point category scale.

For confidentiality reasons, more information concerning the products cannot be given here.

The methodology proposed in the next sections describes step by step a solution on how the ideal data obtained from consumers can be analyzed. This complete methodology, also known as the *IPA*, includes steps for checking the consistency of the data as well as their use (from the definition of an ideal product of reference to guidance on improvement reported to product developers) in an optimization procedure.

4.1.7 SOFTWARE

Data were collected using the EyeQuestion software, an all-in-one software solution for sensory and consumer research from Logic8, and analyzed

---

TABLE 4.2
List of Attributes Used to Describe the Pasta Sauces

Color	Fruity taste	Smooth mouthfeel
Gloss	Sweetness	Thick mouthfeel
Thickness appearance	Saltiness	Burn mouthfeel
Smooth appearance	Sourness	Intensity aftertaste
Tomato odor	Bitterness	Sweet aftertaste
Herbal odor	Herbal taste	Sour aftertaste
Sour odor	Spiciness	Spicy aftertaste
Tomato taste	Specific taste	

---

with R2.15.1 (R Development Core Team 2012) and the FactoMineR v1.19 (Lê et al. 2008) and SensoMineR v1.15 packages (Lê and Husson 2008).

4.2 CONSISTENCY OF THE IDEAL DATA

With the IPM, consumers are providing ideal ratings explicitly. Although it is now accepted that consumers can also be used to profile products, the impact these sort of data and in particular the ideal data, can have would be questioned (since it comes from consumers who are asked to describe a fictive product). And this information being of great importance for industries (its main use is to guide products and portfolio optimization), its *quality* needs to be checked.

The procedure of measuring the consistency of the ideal data provided by consumers is done through the evaluation of the consistency of the data obtained from consumers through their different descriptions (i.e., sensory, hedonic, and ideal) of the products.

But before proposing a methodology for checking the consistency of the ideal data, we need to go back to the definition of an ideal.

4.2.1 WHAT IS AN *IDEAL*?

The Oxford online dictionary (http://oxforddictionaries.com) defines an *ideal* as

- Satisfying one's conception of what is perfect; most suitable
- Existing only in the imagination; desirable or perfect but not likely to become a reality
- Representing an abstract or hypothetical optimum
- A standard or principle to be aimed at

To summarize, an ideal can then be seen as a standard or goal to achieve that is highly satisfactory and that would be the best of its kind. Hence, the ideal seems to have two components: one related to the physical description of the product (i.e., the ideal used as reference) and one related to liking (i.e., the ideal as an optimum).

By using these two components, such definition can be extended to sensory science. In that case, an ideal product is a product with particular sensory characteristics, which would maximize liking. Such product should be used as reference to match in the optimization procedure.

In this definition, we can see that the ideal product is still defined by using two components, one related to sensory (i.e., the ideal product seen as a reference to match) and one related to the hedonic (i.e., the ideal product seen as an optimum, as a highly satisfactory product). For that reason, the methodology for checking the consistency of the ideal information provided by consumers proposed in the next sections is done in two steps: first from a sensory point of view (Worch et al. 2012a) and second from a hedonic point of view (Worch et al. 2012b).

4.2.2 Sensory Consistency

In 2012a, Worch et al. defined the sensory consistency of an ideal product as follows: "the profile of an ideal product is consistent if it shares similar sensory characteristics as the profile of the most appreciated product."

In other words, the ideal product provided by a consumer should have similar sensory properties as the product he or she likes the most. At the attribute level, this means that if a consumer says he or she likes better a product perceived as sweeter and less bitter (i.e., the sweeter the better, and the less bitter the better), one would expect that this consumer would rate his or her ideal as rather intense in sweetness and as rather not intense in bitterness.

This concept is highlighted in Figure 4.4. The representation of liking scores as a function of the perception of an attribute provided by a consumer helps defining the area where the ideal should be positioned. For a

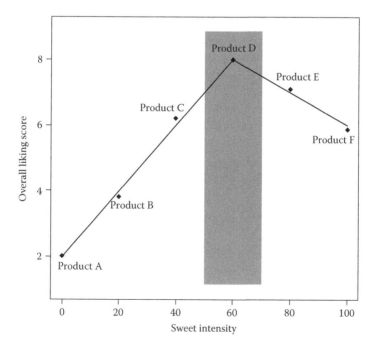

FIGURE 4.4 Example of the relationship between the perception of an attribute (here sweetness) and liking ratings provided by a consumer and the expectation of the ideal rating provided.

consistent consumer, it is expected to be at the intensity rating associated with maximum liking. This should be the case for all the attributes that influence liking.

Since checking this for each consumer and each attribute becomes quickly exhaustive, we propose a more general methodology that allows checking for this consistency both at the panel and consumer level.

4.2.2.1 At the Panel Level

4.2.2.1.1 Principle and Statistical Analysis

At the panel level, the sensory consistency is evaluated by looking at a general trend in the individual ideal descriptions. If consumers described their ideal with similar sensory characteristics as their most appreciated product, then consumers are considered consistent at the panel levels.

To check for that, the relation between *sensory*, *hedonic*, and *ideal* descriptions is evaluated. This evaluation is done indirectly, by first focusing on the *ideal* description.

The consumers are represented in the same space according to their ideal profiles, two consumers being close if they share similar ideal profiles and apart from consumers with different ideal profiles. In this ideal space, groups of consumers sharing similar ideals are identified. These groups can be characterized by using simultaneously the *sensory* and the *hedonic* descriptions of the products tested (i.e., by associating the *sensory* and *hedonic* information related to the products with the *ideal* information). Such solution can be obtained by associating each homogeneous group of consumers (*ideal*) with the closest product tested (*sensory*) and with the most liked product (*hedonic*).

In this case, the *ideal* descriptions are consistent if each group of consumers sharing similar ideals is associated with the same product based on its *sensory* and *hedonic* descriptions.

In practice, such solution is obtained by aggregating to the ideal information two blocks of data: the *hedonic* on one hand and the *sensory* on the other hand (Figure 4.5). The three blocks of data are then analyzed 2×2.

First, the *ideal* product space is created. Next, the link between *ideal* profiles and *hedonic* judgments is studied. In this case, each group of consumers sharing a similar ideal is linked to the products they like most. Next, the link between the *ideal* and the *sensory* profiles of the products is studied. Consumers are linked to the products for which the sensory profiles are similar to the averaged ideal profile they provided. Finally, the link between *hedonic* judgments and *sensory* profiles, within the *ideal* space, is measured. This link is strong if consumers, whose ideal profiles

	Attribute 1	...	Attribute a	...	Attribute A	Product 1	...	Product p^h	...	Product P
Consumer 1										
...										
Consumer j			$\tilde{z}_{j,a}$					$h_{jj^h} - \bar{h}_{j.}$		
...										
Consumer J			(a)					(b)		
Product 1										
...										
Product p^d			$\bar{y}_{.p^d a} - \bar{y}_{..a}$							
...										
Product P			(c)							

FIGURE 4.5 Sensory consistency measured through the relationship between the ideal, sensory, and hedonic data. (a) Represents the corrected average ideal profiles provided by the consumers active. (b) Represents the hedonic judgments of the actual products supplementary variables. (c) Represents the sensory profile of the actual products supplementary entities.

are close to the sensory profile of a product p are also the consumers who like p more than the other products.

Finally, the sensory consistency of the ideal data provided by consumers is evaluated at the panel level through the relative position of the corresponding products obtained from the different blocks (*sensory* and *hedonic*) within the ideal space.

In practice, the procedure used to evaluate the sensory consistency of the ideal data consists in performing a principal component analysis (PCA) on the table of corrected averaged ideal profiles (Figure 4.5a). In this analysis, one statistical entity represents the corrected average ideal profile from one consumer.

In this ideal product space, the hedonic judgments of the products are projected as supplementary variables. Each vector of hedonic judgments (noted p^h) is previously centered by consumer. Hence, each supplementary variable corresponds to the preference for one product (Figure 4.5b). In this analysis, the link between *ideal* and *hedonic* data is measured through the correlation coefficient between the corrected averaged ideal ratings of each attribute and the preference for each product. If the supplementary variable p^h (corresponding to the "preference" of product p) is highly correlated with the ideal attribute a, the more the ideal profile has a high score for the attribute a, the more the consumer prefers p^h compared to the other products.

Simultaneously, the sensory profiles of the products (noted p^d) are projected as supplementary entities in the ideal product space (Figure 4.5c). In this case, each product is considered as a particular consumer who would have the product under consideration as ideal. The link between the ideal profiles and the products is measured by the distance in the space: a consumer, whose ideal is close to a product p^d in the ideal product space, described an ideal profile that is close to the sensory profile of the product p^d.

From the panel point of view, the consistency of the *ideal* data is measured through the direct correspondence between identical entities (i.e., the products) described according to their *hedonic* (p^h) and *sensory* (p^d) aspects within the *ideal* product space.

Since the ideal product has two components, one related to sensory and the other related to hedonic, the relationships between *ideal* data and *hedonic* judgments (Figure 4.5a and b) or the *ideal* data and the *sensory* descriptions (Figure 4.5a and c) are straightforward. On the contrary, there is no formal link between *hedonic* judgments and *sensory* descriptions of the products (Figure 4.5b and c). However, empirically, one can expect that consumers who appreciate a product p^h more should also provide an ideal profile closer to the sensory profile of p^d then to any other product. Similarly, from the sensory point of view, if the preference of a product p^h is strongly correlated with the description of the ideal attribute a, one can expect that the product p^h should be described as more intense on that attribute a than the other products. In both cases, if the different descriptions are consistent with each other, the projection in the ideal product space of the product p^d will be closer to the ideal of consumers who preferred p^h to the other products.

4.2.2.1.2   Illustration

The ideal space obtained by PCA on the corrected average ideal profiles shows that consumers do not share the same ideal (Figure 4.6). The first dimension opposes the consumers (e.g., 14 and 42) who described their ideal sauce as more *sour* and *burning* than consumers who clearly reject these sensory characteristics (e.g., 52 and 32). The second dimension opposes consumers (e.g., 66 and 82) who described their ideal sauce as *smoother* and with more intense *tomato taste* and *odor* as other consumers (e.g., 53 and 111).

On the ideal space, the projection of liking ratings for each consumer as supplementary variables (Figure 4.7b) shows that consumers who have an ideal which is more *sour* and *burning* (e.g., consumers 14 and 42) prefer products 13 and 6 more than the other consumers. The consumers who rejected these characteristics (e.g., consumers 52 and 32) prefer products 1, 4, and 12 more than the other consumers. On the second dimension, no clear difference in terms of preference can be observed. Consumers who described their ideal with stronger *tomato taste* and *tomato odor* and as *smoother* do not have a clear preference pattern for one product over the other consumers.

Let's consider each product as a particular consumer who would have that product as an ideal. By doing so, we can characterize each homogeneous group of consumers with the products tested that are projected close to them. In this case, Figure 4.7a shows that the consumers who described their ideal as more *sour* and *burning* are relatively closer to products 6, 7, 13, and 2. On the other hand, the consumers who rejected these attributes have an ideal with sensory characteristics that is relatively closer to products 1, 4, and 12. On the second dimension, no clear distinction can be observed.

As a conclusion, by relating the sensory and hedonic description of the tested products within the ideal space, it appears that the consumers who preferred products 2, 6, and 13 more than the other consumers also provided an ideal profile with sensory characteristics that are relatively closer to these products. Inversely, the consumers who preferred products 1, 4, and 12 more than the other consumers also rated their ideal product with sensory profiles relatively closer to these products than to any other.

The correlation coefficient measured between the configuration of the products from the hedonic perspective (i.e., loadings of the supplementary

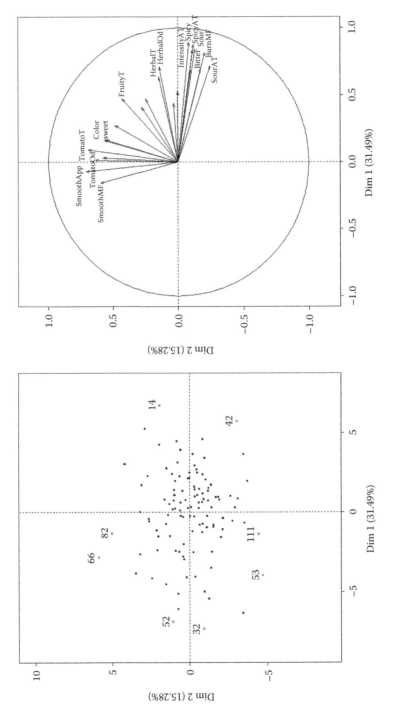

FIGURE 4.6 Ideal space obtained for the pasta sauce data.

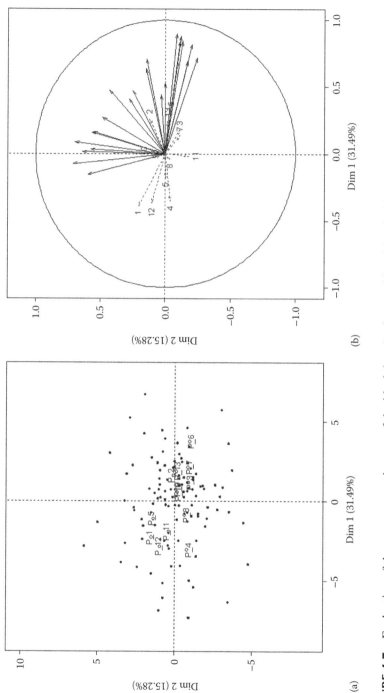

FIGURE 4.7 Evaluation of the sensory consistency of the ideal data (at the panel level) by double projection as illustrative of the sensory and hedonic information of the product tested, within the ideal space, for the pasta sauce data: (a) Projection of the sensory profiles of the products within the ideal space and (b) correlation between the hedonic judgments of the products and the ideal attributes.

variables) and the configuration of the same products from the sensory perspective (i.e., scores of the supplementary entities) within the ideal space for each dimension confirms this result. Indeed, on the two first dimensions, the correlation coefficients are high and positive (0.93 and 0.55, respectively). Since the threshold value for the one-tailed test of the correlation coefficient with 11 degrees of freedom is 0.48, these two coefficients are significant at 5%.

We can conclude here that globally (at the panel level), consumers rated their ideal with the same characteristics than the most appreciated products. The consumer panel is consistent from a sensory point of view.

4.2.2.2 At the Consumer Level

4.2.2.2.1 Principle

By definition, to be consistent, a consumer should provide a high ideal rating for an attribute a if that consumer said he or she appreciated the product more when he or she perceived it as stronger for a. Inversely, if the less he or she perceived the intensity of an attribute a' and the more he or she liked the product, we would expect that consumer to rate a low ideal for a'.

In order to check the consistency at the consumer level, we will check for each consumer that the linear drivers of liking (*vs.* disliking) correspond to the attributes for which his expectations in terms of ideal intensities are high (*vs.* low).

To do so, a vector of drivers of liking/disliking is calculated for each consumer. In practice, this vector of drivers of liking/disliking is obtained by measuring for each consumer and each attribute the correlation coefficient $r(y_{j \cdot a}; h_{j \cdot})$ between his perceived intensities and his appreciation of the product. This vector is then compared to the corrected ideal profiles obtained from the same consumer. The correlation coefficient $R^j_{y,h,z}$ between the averaged corrected ideal profile and the vector of drivers of liking/disliking is used (Equation 4.5):

$$R^j_{y,h,z} = r(r(y_{j \cdot a}; h_{j \cdot}); \tilde{z}_{j \cdot \cdot})$$

(4.5)

with $r(y_{j \cdot a}; h_j)$ the vector of drivers of liking/disliking associated to the consumer j and $\tilde{z}_{j \cdot}$ his average corrected ideal profile.

This coefficient is high and positive if the consumer considered has a higher appreciation for the products with a high intensity for the attribute considered as a driver of liking. Inversely, it is high and negative if the consumer in question has a lower appreciation for the products with a strong intensity for that attribute, considered then as a driver of disliking. Likewise, it is close to 0 if that attribute does not influence his appreciation of the products.

As the *ideal* description is making the link between *sensory* and *hedonic*, one can expect that consumers would associate a strong ideal intensity (*resp.* weak) to the attributes that drive linearly liking (*resp.* drivers of disliking). Hence, consumers are consistent if they are associated to a coefficient $R^j_{y,h,z}$ close to 1. The test checking for the significance of this correlation coefficient is a one-tailed test.

For simplicity, only linear relationships are considered in this methodology. This is in agreement with the methodology for evaluating the sensory consistency at the panel level since it uses linear relationships between ideal and sensory on one hand and between ideal and hedonic on the other hand.

In the empirical situation where all the products tested are defined on one side of the saturation curves (some products being not intense enough while some others are too intense, a maximum being reached) representing most cases in our experience, the proposed methodology is suitable. However, if for an attribute a some products are considered as not intense enough while some others are too intense, a saturation effect is observed. In this particular case, our methodology will show limitations. It is then advised to consider quadratic relationships between the perception of the attributes and liking ratings. The use of dummy variables is then advised (Xiong and Meullenet 2006).

4.2.2.2.2 Illustration

In order to evaluate the sensory consistency of each consumer separately, a vector of linear drivers of liking is calculated for each consumer. This vector of drivers of liking is obtained by calculating the correlation coefficient between the vector of perceived intensity of each attribute and the vector of ideal ratings. For each consumer, the vector of drivers of liking (containing A elements, one correlation coefficient per attribute) is then correlated to the vector of corrected average ideal intensity provided by the same consumer. For consistent consumers, this correlation coefficient is expected to be large and positive. In this example, the correlation coefficient $R^j_{y,h,z}$ is significant at 5% if it is larger than the critical value 0.31 (one-tailed test of the correlation coefficient with 21 degrees of freedom).

In this example, the distribution of the individual correlation coefficients $R^j_{y,h,z}$ (Figure 4.8) shows that the large majority of consumers is consistent from a sensory point of view. Indeed, most of the consumers have an individual correlation coefficient $R^j_{y,h,z}$ larger than 0.31 (median of 0.62).

4.2.2.3 Conclusion

From our experience, it can be concluded in most cases that the ideal information provided by the consumers is consistent from a sensory point of view.

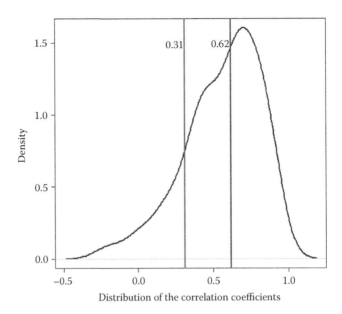

FIGURE 4.8   Distribution of the correlation coefficient measuring the sensory consistency of the ideal data at the consumer level for the pasta sauce data.

It should be noted that it often appears at the panel level that the projections of the products as supplementary entities and variables (corresponding, respectively, to the sensory and hedonic perspective) are not always well represented. From the sensory point of view, this is due to the fact that the ideal products from the consumers are more variable than the products tested. From the hedonic point of view, this is due to the fact that we are looking at a correlation between preference ratings and ideal attributes involving a very large number of values (i.e., as many as there are consumers), which means that it is difficult to have high correlation coefficients.

Since we are interested in general trends, such result is not worrying because the interpretation is more visual than numerical.

At the consumer level, the correlation coefficient is generally high for the large majority of consumers. However, for a small (but nonnegligible) proportion of consumers, inconsistency would be concluded. This is a classical result in consumer testing: indeed, it is well known that in consumer tests, around 20% of the consumers are considered as random noise and are not providing relevant information. Through this methodology, these consumers could be identified and eventually discarded from the analysis. However, we recommend keeping the entire panel as it is.

4.2.3 HEDONIC CONSISTENCY

In the previous section, we have just seen how to check the consistency of the ideal data from a sensory point of view. Since "ideal" information also has a component related to liking, it is also important to check that ideal data are consistent from a hedonic point of view. In this case, we would expect that ideal products correspond to an optimum, that is, a product that is more liked than the products tested, if it happens to exist.

4.2.3.1 Methodology

To check for the hedonic consistency, it is necessary to compare for each consumer the products tested with the ideal product from a hedonic point of view. Since liking of the ideal profile (also called *liking potential of the ideal product*) is unknown, it has to be estimated.

To do so, a model explaining liking scores as a function of the perception of the product is defined for each consumer. Each individual model so defined is then applied to the averaged ideal ratings from the same consumer, and the liking potential of that averaged ideal product is estimated. This estimated liking potential is compared to the liking scores given to the products by that consumer.

In this case, an ideal profile is considered *consistent* (in terms of the liking potential of the ideal profile) if the estimated liking potential is superior to the liking scores given to the products.

A summary of this procedure is given in Figure 4.9.

In sensory analysis, it is well known that attributes are often highly correlated. Hence, the individual models defined should take into account the multicollinearity between attributes. Models based on PLS regression or principal component regression are hence usually considered. Here, the quadratic model used in external preference mapping is considered (Danzart 1998, 2009).

For each consumer, a PCA is performed on the sensory profiles of the products he or she provided. The two first sensory dimensions of each individual PCA are extracted and liking scores are regressed on these dimensions using the model presented in Equation 4.6. A procedure to select the

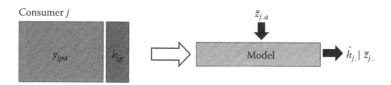

FIGURE 4.9 Schematization of the procedure used to check for hedonic consistency of ideal data.

best model is performed, so that only significant (at 5%) dimensions are kept in the final model:

$$h_{jp} = Dim_1^{(j)} + Dim_2^{(j)} + Dim_1^{2(j)} + Dim_2^{2(j)} + Dim_1^{(j)} : Dim_2^{(j)} \quad (4.6)$$

Once defined, the individual model thus obtained is applied to the averaged ideal profile generated by this consumer. In practice, the ideal profile of the consumer is projected in the sensory space associated with that consumer and the coordinates of the averaged ideal product are extracted. The individual model is applied on these coordinates and the liking potential of the averaged ideal product for that consumer is estimated.

Finally, the liking potential of the ideal product is compared to the actual liking scores assigned for the tested products for each consumer. In order to simplify the interpretation, liking potential of the ideal product of each consumer is standardized according to the liking scores given to the products tested. To do so, we subtract the averaged liking score given to the tested products from the liking potential. This difference is divided by the standard deviation of the liking score:

$$\text{Standardized liking potential}_j = \frac{\hat{h}_j \mid \bar{z}_{j..} - \bar{h}_{j.}}{\sigma_{h_{j.}}} \quad (4.7)$$

For consistent ideal information (from a hedonic point of view), the standardized liking potential should be large. However, a low standardized liking potential is not only caused by nonconsistent ideal information. Since the methodology is based on modeling, it can also be caused by a lack of fit of the individual model. For that reason, it is important to consider the quality (fit) of the individual models as well. The quality can either be evaluated using cross validations or by coefficients such as the adjusted R^2. Due to the small amount of data (we create one model per consumer), the use of the adjusted R^2 is preferred.

If the adjusted R^2 coefficients are high (in practice, we consider $R^2 = 0.5$ as threshold), the model fits the liking scores well. In such situation, a low standardized ideal score would mean that the ideal product provided by the consumer is not consistent from the hedonic point of view. Otherwise, if both the adjusted R^2 coefficient and the standardized liking potential are low, it cannot be decided whether it is related to nonconsistent ideal information or to the poor fitting property of the individual model.

4.2.3.2 Illustration

The ideal profiles obtained from consumers would be consistent from a hedonic point of view if they are more appreciated than the tested products

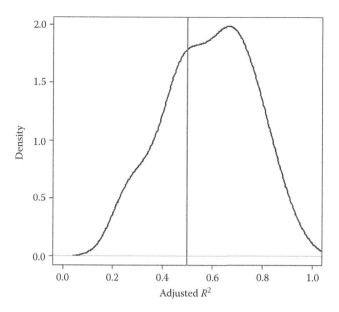

FIGURE 4.10 Distribution of the adjusted R^2 coefficient associated with the individual models for the pasta sauce data.

(if these ideals happen to exist). Since the liking potential of the ideal profiles cannot be measured, it is estimated. Individual models are hence created for each consumer. But before checking for the consistency through the estimation of the liking potential of the ideal profiles, it is important to verify that the individual models fit the liking scores.

To check for that, the distribution of the adjusted R^2 coefficients associated with the individual models is inspected (Figure 4.10). In this example, it seems clear that the individual models fit the liking scores for the majority of the consumers well. Indeed, at the panel level, the median is high (>0.60). More precisely, a very large proportion of consumers is associated to an adjusted R^2 coefficient larger than 0.5.

Since the individual models fit the data, one could be confident in the estimation of the liking potential of the ideal profiles for the majority of the consumers. The comparison of the distributions of the liking potential of the ideal profiles on one hand and the liking ratings given to the products tested on the other hand shows that at the panel level, the ideal seems consistent, the potential liking being larger than the actual liking scores (Figure 4.11). Indeed, the median of the liking scores is around 6 across products and consumers, while the median of the liking potential of the ideal products is close to 7.

It can be noted that the lower estimates of the liking potential are lower than the smallest liking scores given to the products tested. This suggests

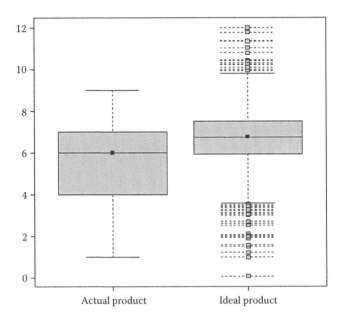

FIGURE 4.11   Box plots of the liking scores provided to the products tested and of the liking potential of the ideal products showing the hedonic consistency of the ideal data at the panel level for the pasta sauce data.

that some consumers are not consistent from a hedonic point of view. Since these outliers are a minority, results are still satisfactory.

The representation of the standardized liking potential of the ideal profile as a function of the adjusted R^2 associated with the individual model is given in Figure 4.12. For the majority of the consumers, the standardized liking potential and the adjusted R^2 are both larger than 0.5. Hence, the ideal profiles are consistent from the hedonic point of view.

It still has to be noted that for a threshold of 1.64, which corresponds to the usual critical value at 95% of the normal distribution $\mathcal{N}(0,1)$, only three consumers would be consistent. In this example, this threshold seems to be a bit too high. It can also be noted that quite a large proportion of consumers is associated to an individual model with an adjusted R^2 lower than 0.5. However, for most of these consumers, the standardized liking potential of their average ideal product is still large.

In this example, the large standardized liking potentials suggest that consumers are consistent from both the sensory and hedonic point of view. Since we are confident about the quality of the data, we can go further with the optimization procedure. But before that, we propose another procedure that allows checking for the consistency of the data. Although this procedure is *quick and dirty*, it is convenient.

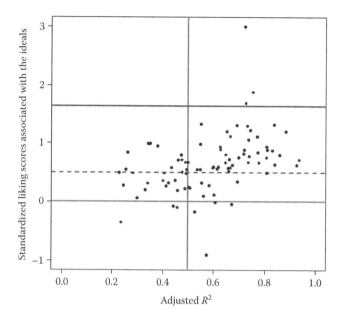

FIGURE 4.12 Representation of the standardized liking potential in function of the adjusted R^2 coefficients to check for the hedonic consistency of the ideal data at the consumer level for the pasta sauce data.

4.2.4 ALTERNATIVE: A QUICK AND DIRTY WAY TO CHECK FOR THE CONSISTENCY

Since checking for the consistency of the ideal data is complex and since it needs a good understanding of the methodologies to avoid misinterpretations, we propose a simpler alternative.

4.2.4.1 Methodology

In order to check for the consistency of the ideal information provided by the consumers, we could look visually if the ideal products would point in the direction of the most appreciated products. To do so, we will adopt a methodology very similar to the one checking for the sensory consistency of the ideal data at the panel level except that the focus is on the correspondence between consumers through the ideal and liking perspectives. For that, we position ourselves in the sensory space.

The sensory space of testing the products is created by PCA on the table crossing the products and the sensory attributes (Figure 4.13a). In this space, both ideal (Figure 4.13c) and liking (Figure 4.13b) information provided by consumers are then projected as illustrative.

	Attribute 1	...	Attribute a	...	Attribute A	Cons. 1	...	Cons. j	...	Cons. J
Product 1										
...										
Product p			$\bar{y}_{pa} - \bar{y}_{.a}$							
...										
Product P										

(a)

h_{jp}

(b)

Cons. 1 Product 1		
...		
Cons. j Product p^d	$\tilde{z}_{jpa}$	
...		
Cons. J Product P		

(c)

FIGURE 4.13 Organization of the data used for the definition of the sensory space (a) with projection of the liking scores as supplementary variables (b), and of the corrected ideal profiles as supplementary entities (c).

In this case, each element projected as supplementary corresponds to one consumer *via* his or her ideal profile (supplementary entity) or his or her liking ratings (supplementary variable).

In this analysis, the main idea is to check whether ideal and liking information follow similar trends. For consistent ideal data, the two projections should be going in the same direction: indeed, the ideal products from consumers should be close to the products they like the most. However, to simplify the methodology, no one-to-one comparison is performed here. For that reason, the interpretation is purely visual and gives an overview without providing a clear measure of the consistency. It is for this reason that this methodology is qualified as *quick and dirty*.

4.2.4.2 Illustration

The majority of the projections of the individual ideal profiles provided by consumers are projected in the negative part of the first dimension of the sensory space. The projections of the liking scores provided by consumers also show that the large majority of the consumers like the products situated on the negative part of the first dimension (Figure 4.14).

Since the double projection within the sensory space of the ideal profiles (as supplementary entities) and of the liking scores of the tested products (as supplementary variables) provided by consumers are going in the same direction, it can be concluded that consumers describe their ideal with similar characteristics as the products they like the most.

The high correspondence between the two configurations of consumers suggests that the panel is consistent. However, since this correspondence is not checked on a one-to-one basis, there is no measure of fit or quality of the data.

4.2.5 Conclusion

Since ideal information is obtained from consumers who are describing a fictive product, it is critical to check the consistency of this information. This is particularly true since this information plays an important role in the product optimization procedure. Improving products based on ideal information that is not consistent might lead to critical consequences. For that reason, we would recommend to check always for the consistency of the data before starting the optimization procedure.

From our experience, the sensory consistency of the ideal data at the panel level is sometimes hard to reach. This is an artifact of the methodology, which is quite demanding. Indeed, the proposed methodology looks at the relationship between sensory and hedonic within the ideal space by using an indirect approach. Such methodology is not optimal since the

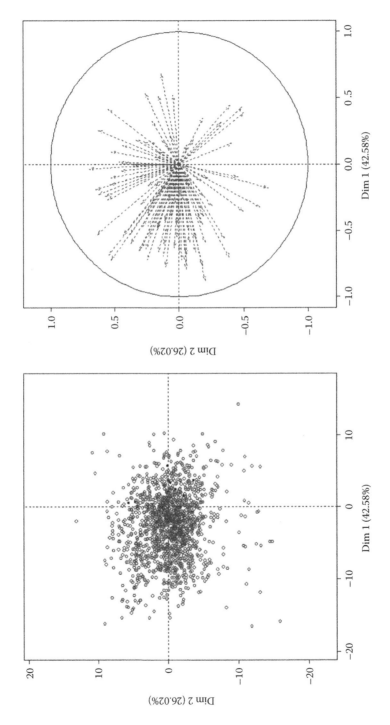

FIGURE 4.14 *Quick and dirty* procedure to check for the consistency of the ideal data obtained for the pasta sauce data.

projection is not optimizing the relationship between tables, although it highlights relationships when these are dominant.

For hedonic consistency, the level of the threshold to define when a product can be considered as acceptable is questionable and depends on the user's point of view. For *easy* users, the threshold can be set as 0, meaning that liking potential should be larger than the averaged liking score for each consumer (minimum required for an ideal product). When the threshold is set at 1.64, which corresponds to the critical value at 95% of the normal distribution $\mathcal{N}(0,1)$ only a small percentage of the consumers is usually considered as consistent. This might indicate the fact that a threshold of 1.64 is too large for this approach. In practice, considering a standardized liking potential between 0.2 and 0.5 seems to be a good alternative. In such situation, the ideal product is still among the most liked products.

When ideal information is considered consistent both from a sensory and hedonic point of view, the user can be confident in the guidance of improvement provided to the product developers. On the other hand, when the consistency of the ideal data is questionable, the user might be cautious when providing guidance.

It also has to be said that data could be considered as not consistent from sensory or hedonic point of view and still be. This would be the case when consumers provide ideal products that are not optimum according to liking but to some other criteria such as, for example, health issues. In this case, information provided can be consistent although it is in disagreement with the definition we provided (which is liking-oriented). As an example, this would be consumers who like more the fatter/sweeter products but who would provide as ideal low-sugar or low-fat products.

4.3 DEFINING HOMOGENEOUS GROUPS

4.3.1 OF CONSUMERS

When it comes to product optimization, defining homogeneous groups of consumers is an important step to include in the procedure, whatever the methodology used. If clusters of consumers are not considered, it can lead to nonsense and to the creation of products, which would be rejected (at least not fully liked) by all the consumers. In that case, the product optimized would not reach its primary goal since it will not be successful.

To define homogeneous groups of consumers, the usual clustering methods (hierarchical or nonhierarchical cluster analysis using similarity or dissimilarity matrices) can be used on the liking information provided by consumers.

Since ideal information has a liking component, it is also possible to cluster consumers as a function of their ideals, rather than as a function of their liking. However, since consumers with similar liking patterns can have quite different ideals (by still being both consistent), we would recommend defining homogeneous groups of consumers according to liking.

For its good visual representation, the authors would recommend the use of hierarchical clustering on principal components (Husson et al. 2010, 2011). Since the clustering analysis is well known and is not specific to this technique, no further details will be given here.

Note: The *quick and dirty* procedure proposed to check the consistency of the ideal data already provides an idea whether clusters can be found. Indeed, if the large majority of the liking variables are highly positively correlated, meaning that all the consumers are pointing in the same direction of the space, the panel is probably made of one unique cluster (except for a small proportion of consumers).

4.3.2 OF PRODUCTS

In many cases, the methodologies used for product optimization consider that each cluster should be associated to one single ideal for the entire product set. It is, for instance, the case of methods such as the landscape segmentation analysis (Ennis 2005), the Euclidean distance ideal point model (Meullenet et al. 2008), or the Bayesian statistical model integrated with characteristics (Nestrud 2012) where each consumer is associated to one unique ideal product for the entire product set. This assertion is true if all the products point to one unique ideal. This would be the case when products are close in their sensory profiles. However, when the product set tested involves products with large differences in their sensory profiles, this assertion could not be verified. In this case, consumers might associate the product set to different ideals, and the product set might be partitioned in different subcategories, each product from the same subcategory pointing to the same ideal. This would be the case, for example, if milk and dark chocolates are tested together: consumers might like them both and have separate ideals both for dark and milk chocolates. In this case, considering one unique ideal per consumer (mixing milk and dark sensory properties) might lead to a reference that would not be appreciated.

Unfortunately, defining *a priori* subcategories of products is difficult and subjective. Since the line between subcategories of products can be very thin, we propose a methodology based on the ideal: two products are from the same subcategory if, at the panel level, consumers associate them with the same ideal. This procedure is performed at the panel level since it searches for a systematic shift across consumers of the ideal ratings (Worch and Ennis 2013).

With the IPM, consumers are asked to rate their ideal for each tested product. If the majority of consumers uses a single ideal to evaluate all the products, then averaged ideal ratings related to each product will not be different. On the other hand, if a product belongs to a different subcategory and hence has a different ideal, a systematic shift is observed in the averaged ideal for that product, compared to the averaged ideals for the other products.

By considering all attributes simultaneously, this systematic shift can be highlighted multidimensionally within the sensory space. To do so, the sensory space of the products tested is created by performing a standardized PCA on the table crossing the products in rows and the sensory attributes in columns (Figure 4.15a). To avoid giving importance to non-discriminating attributes (Borgognone et al. 2001), a selection *a priori* of the attributes of interest can be done. This selection involves a two-way ANOVA (i.e., including the *product* and *consumer* effects) performed on each sensory attribute (i.e., perceived intensity): all attributes with a product effect not significant at 20% are excluded in the construction of the product space.

Since we are interested in finding a systematic shift of the ideal ratings across consumers, the averaged ideal profiles are calculated for each product (i.e., averaged over the consumers). P tested products yield P averaged ideal products. These P averaged ideal products are projected as supplementary entities on the sensory space (Figure 4.15b). An important

	Attribute 1	...	Attribute a	...	Attribute A
Product 1					
...					
Product p			$\bar{y}_{.pa}$		
...					
Product P					(a)
Ideal product 1					
...					
Ideal product p			$\bar{z}_{.pa}$		
...					
Ideal product P					(b)
Cons. 1 product 1					
...					
Cons. j product p			z_{jpa}		
...					
Cons. J product P					(c)

FIGURE 4.15 Organization of the data used for the single *vs.* multiple ideals procedure. (a) Represents the sensory profiles of the products that are used for the creation of the sensory space active. (b) Represents averaged ideal product used to evaluate the shift in the ideal ratings illustrative. (c) Represents individual ideal data used for the construction of the confidence ellipses around the ideal products illustrative.

point here is that the P averaged ideal products are only used to check for the possibility of multiple subcategories within the tested products—they should not be used for product or portfolio optimization purposes since they are not necessarily meaningful.

If there is a systematic shift in the consumer ideals as we move from one product to another, this shift will be observed in the relative position of the projection of the averaged ideal products. When consumers associate all the products to one unique ideal, no systematic shift is observed and all the averaged ideal products are projected in a small area (ideally, all the projections would be overlapping). On the contrary, if consumers associate the products with multiple ideals, the projections of the averaged ideal products will be spread on the sensory space.

Since the distance between products is somewhat subjective in multivariate analysis, a test evaluating the significance of those distances is needed. The solution proposed here consists in using consumer's variability to create confidence ellipses around the averaged ideal products (Husson et al. 2005; Figure 4.15c). If all the confidence ellipses are overlapping, no systematic shift is observed across products. In such case, it can be concluded that consumers associate the product set to one unique ideal. On the contrary, if the confidence ellipses are separated, a systematic shift is observed across products. It can then be concluded that consumers associated the product set to more than one ideal.

These confidence ellipses can either be constructed according to partial bootstrap or total bootstrap (Cadoret and Husson 2012). In our case, since we are dealing with low dimensionality, the two techniques return similar results and only partial bootstrap is used. It consists in creating fictive panels of consumers by selecting randomly with replacement consumers from the original panel and by estimating the average point of each fictive panel by using the barycentric property of the PCA. By reproducing these steps many times, confidence ellipses containing 95% of the projections obtained from the fictive panels are created.

The confidence ellipses are usually represented on the first two dimensions of the PCA, as these dimensions explained the maximum of variance. However, differences between products might be highlighted on further dimensions, especially if the first two dimensions together only explain a low proportion of the total variance. In this case, limiting the interpretation to the first two dimensions might bring wrong conclusions (e.g., single ideal while products are differentiated on the third or fourth dimension).

In order to avoid such misinterpretation, a Hotelling T^2 test is performed. This test is a good addition to the visual inspection of the confidence ellipses since it checks for significant differences between products in a multivariate way. This test includes all the dimensions associated with an eigenvalue larger than 1.

Based on visual inspection and on results of the Hotelling T^2 test, conclusions are drawn on whether the panel associated the product set with one or with multiple ideals. In the case where the consumer panel associates the product set to multiple ideals, each homogeneous subgroup of products should be optimized separately. The methods proposed by Worch et al. (2010a) such as the PLS on dummy variables and the fishbone method can be used. In this case, each subgroup of products would be optimized according to its corresponding ideal product. When the number of products belonging to a same subcategory is large enough (in practice, a minimum of six products is advised), techniques that assume a single ideal product per consumer, such as the family of ideal point modeling methods (e.g., LSA), can be applied to this product subset.

A cautionary note is needed here: when we conclude that the product set contains products from different subcategories, we must consider the balance of the experimental design when we form subsets of the original dataset. Thus, we recommend analyzing the design matrix for artifacts in the sequence effects that could be caused by subsetting. In the event that the design is no longer well balanced after some products are removed from the main set, we recommend optimizing all products individually using PLS on dummy variables.

4.3.3 ILLUSTRATION

4.3.3.1 Clustering

The *quick and dirty* procedure to check for the consistency of the ideal data already gives an insight on the segmentation of the panel. Since the majority of consumers pointed in the same direction, it indicates that most of the consumers agree that the less *sour* and *less* burning are the best products.

To confirm that, a PCA is performed on the table of liking scores crossing in rows the consumers and in columns the products (Figure 4.16). A hierarchical clustering on principal components is performed on the consumers' configuration thus obtained. This methodology considers the position of the consumers on the principal components and performs hierarchical clustering analysis on it. The dendrogram associated to this cluster analysis suggests the existence of two groups (Figure 4.17). These two groups are separated along the first dimension of the PCA (Figure 4.18): the first cluster, containing 60 consumers, likes products 2, 6, and 13 more, while the second cluster containing 52 consumers, likes products 1, 4, 8, and 12 more.

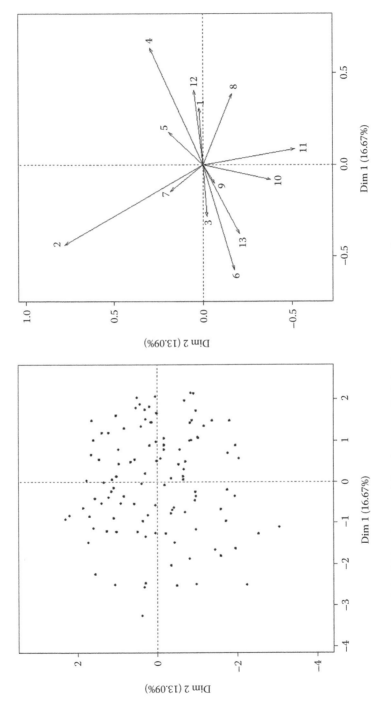

FIGURE 4.16   PCA of the liking data used for cluster analysis in the pasta sauce example.

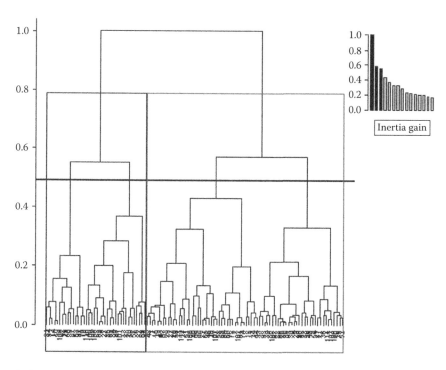

FIGURE 4.17 Dendrogram used for the cluster analysis in the pasta sauce example.

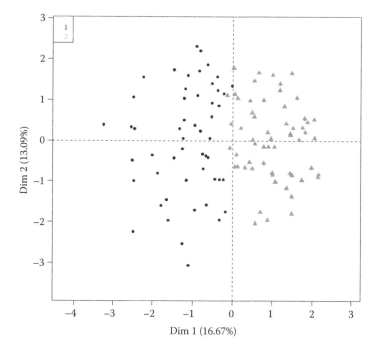

FIGURE 4.18 Results of the cluster analysis in the pasta sauce example.

4.3.3.2 Single vs. Multiple Ideals

Once the clusters of consumers are defined, it is important to check whether the products belong to one or many subcategories. The procedure checking for multiple ideals is performed. Since the ellipses are all represented in the small area of the space (Figure 4.19), it can be concluded that consumers associated the product set to one unique ideal. However, on the second dimension, it can be seen that product 7 and products 9 and 10 are separated. Since the separation with the rest of the groups is not clear, one unique ideal is considered here. The Hotelling T^2 test confirms this result: some significant differences between some pairs of products (Table 4.3) are observed. However, we would suggest associating the product set with one unique ideal product.

Once homogeneous groups of consumers and products are defined, the user can provide guidance on improvement. To do so, the sensory profile of the ideal product, that is used as reference to match, needs to be defined.

For the rest of the manuscript, to simplify the interpretation, we will consider one unique cluster of consumers and we will consider that the products are associated with the same subcategory of products (i.e., consumers associated the tested products with one unique ideal).

4.4   DEFINING THE PRODUCT OF REFERENCE

It is well known that consumers differ in their ideals. However, it is not possible for companies to satisfy each consumer individually by creating their own ideal. For that reason, it is important to define an ideal product that will be used as reference to match in the optimization procedure.

A first solution consists of considering the averaged ideal product for each homogeneous cluster of consumers. In this case, we assume that consumers within each cluster point to a similar ideal. This solution is the one considered in the early literature (Hoggan 1975; Szczesniak et al. 1975; Cooper et al. 1989).

In this case, the ideal product will be liked by all consumers although it does not correspond to their exact ideal. As an alternative, we would like to propose another solution: instead of satisfying (up to a certain level) all consumers as much as possible, we will consider the product that corresponds to the ideal of a maximum of consumers as product of reference. This solution is the one obtained with the Ideal Mapping (*IdMap*) technique. Depending on the point of view adopted by the user, any of these two methods can be used.

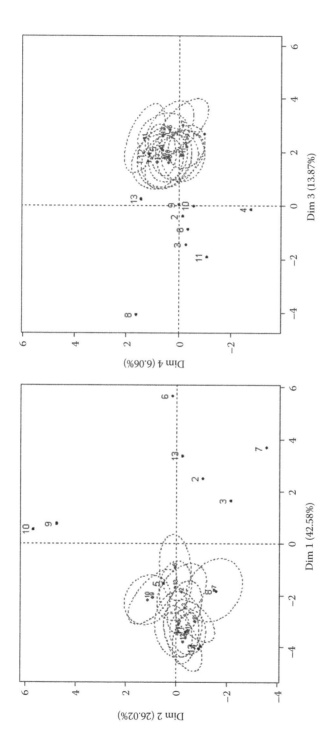

FIGURE 4.19 Results of the single *vs.* multiple ideals procedure obtained for the first four dimensions with the pasta sauce data. The confidence ellipses are obtained at the 95% confidence level.

TABLE 4.3
P-Value Associated with the Hotelling T^2 Test Checking for the Significance between Pairs of Products

	1	2	3	4	5	6	7	8	9	10	11	12	13
1	1.000												
2	0.000	1.000											
3	0.001	0.325	1.000										
4	0.093	0.004	0.065	1.000									
5	0.243	0.009	0.027	0.479	1.000								
6	0.000	0.567	0.016	0.000	0.003	1.000							
7	0.000	0.008	0.003	0.002	0.016	0.007	1.000						
8	0.000	0.000	0.003	0.227	0.043	0.000	0.001	1.000					
9	0.065	0.034	0.007	0.009	0.126	0.140	0.000	0.000	1.000				
10	0.000	0.003	0.001	0.010	0.019	0.022	0.000	0.000	0.871	1.000			
11	0.092	0.028	0.429	0.918	0.224	0.001	0.003	0.172	0.003	0.002	1.000		
12	0.909	0.001	0.012	0.432	0.530	0.000	0.001	0.085	0.001	0.000	0.397	1.000	
13	0.004	0.410	0.340	0.069	0.117	0.467	0.027	0.017	0.221	0.054	0.140	0.011	1.000

4.4.1   Methodology of the *IdMap*

The IdMap technique (Worch et al. 2012c) is a methodology derived from external preference mapping. As a starting point, it takes the sensory space of the products derived from a PCA performed on the averaged sensory profiles. Instead of regressing the hedonic ratings provided by each consumer on the sensory space, the ideal ratings are used. In this case, the averaged ideal profile from each consumer is projected as supplementary on the product space (Figure 4.20). In the sensory space (obtained by PCA in Figure 4.20a), the variability of the ideal ratings within consumers is then taken into consideration and a confidence ellipse is constructed around the average ideal profile of each consumer (Figure 4.20c). For the same reason as in the multiple ideal procedure (see Section 4.3.2), confidence ellipses are obtained by partial bootstrap, except that in this case, the permutation test is performed on the tested products and not on the subjects. Such procedure corresponds to answering the question: "Where would the corrected averaged ideal product of a consumer be projected if instead of rating it according to the P products, he or she would rate it according to P' products?"

Once the ellipses are created, the product space is partitioned. For each point of the resulting sensory space, the amount (in percentage) of ellipses

	Attribute 1	...	Attribute a	...	Attribute A	
Product 1						
...						
Product p			$\bar{y}_{pa}$			
...						
Product P						(a)
Consumer 1						
...						
Consumer j			$\bar{z}_{j.a}$			
...						
Consumer J						(b)
Cons. 1 product 1						
...						
Cons. j product p			z_{jpa}			
...						
Cons. J product P						(c)

FIGURE 4.20   Organization of the data used for the single *vs.* multiple ideals procedure. (a) Represents the sensory profiles of the products that are used for the creation of the sensory space active. (b) Represents averaged ideal product of each consumer illustrative. (c) Represents individual ideal data used for the construction of the confidence ellipses around each consumer illustrative.

covering that area is computed. In other words, for each point of the space, the proportion of consumers having an ideal in that particular area of the space is computed. To facilitate the interpretation of the results, a color code can be associated with each zone of the space: the larger the proportion of consumers in an area of the space, the darker the color. A surface plot similar to the one proposed by the PrefMap is thus obtained.

Finally, the ideal map is constructed including contour lines associated with the proportion of consumers.

In external preference mapping, the coordinates corresponding to the maximum proportion of consumers overlapping can be extracted. By using the inverse formula of the PCA (or reversed regression), a potential profile of the ideal product can be estimated based on the coordinates on the map. This is the solution proposed by Moskowitz et al. (1977) in their eclipse method.

Such procedure can also be applied to the *IdMap* solution. But since the individual ideal profiles are already provided by consumers, such ideal profile can be calculated directly. The averaged ideal shared by the maximum of consumers is used.

4.4.2 Illustration

The projection of the individual ideal profiles on the sensory space shows that the majority of the consumers have their ideal product in the left part of the first dimension. Moreover, the size of the ellipses suggests that most consumers have a similar representation of their ideal since the ellipses have relatively similar sizes (Figure 4.21).

By gridding the space and counting the number of consumers sharing a similar ideal on that point for each point of the space (i.e., which corresponds to counting for each point of the space the number of ellipses overlapping on that area), we can see that the majority of consumers have their ideal in the left part of the first dimension (Figure 4.22). However, the ideal product shared consensually by a maximum of consumers corresponds to a proportion of 27% of the total panel size.

The sensory profile of the ideal product can then be computed by considering the average ideal profile for the consumers sharing a similar ideal. The sensory profile thus obtained (Table 4.4) is considered as reference to match and is further used to guide on improvement.

Note that with the IdMap, the maximum proportion obtained is low compared to the results obtained with PrefMap. This difference is due to the divergence of point of view adopted by these two methods: with PrefMap, the standard acceptance level corresponds to a product that is more liked than average by a consumer, while in the IdMap, it corresponds

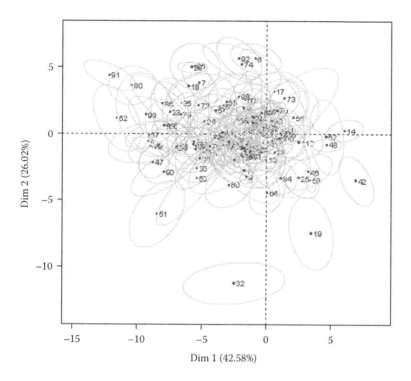

FIGURE 4.21 Representation of the consumers' ideal zone within the sensory space for the pasta sauce data. The confidence ellipses are obtained at the 95% confidence level.

to the ideal product. For more information concerning the level, please look at Worch et al. (2012c) and Delarue et al. (2010).

4.5 GUIDANCE ON IMPROVEMENT

4.5.1 QUICK SOLUTIONS

The optimization procedure of the tested products is done by comparing the sensory profiles of the products to the sensory profiles of a referent ideal product. To do so, Szczesniak et al. (1975) proposed two graphical comparisons of the tested products' profiles with the ideal profile. In both cases, the attributes are ordered in the descending order of average intensity ratings. A connecting line is drawn between the individual characteristics resulting in a profile. In the first graph, the profiles for several products including the ideal (i.e., the reference to match) are represented together on the same chart allowing direct comparison. In the second graph, the ideal intensities define a straight line set at the "0" rating, and

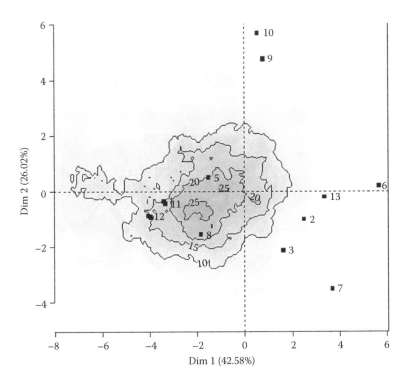

FIGURE 4.22   *IdMap* solution obtained for the pasta sauce data.

TABLE 4.4
Sensory Profile of the Ideal of Reference Obtained with the IdMap Technique

Reference	Ideal	Reference	Ideal	Reference	Ideal
color	61.04	fruityT	41.07	smoothMF	57.42
gloss	61.20	sweet	46.85	thickMF	56.24
thickApp	60.76	salt	44.33	burnMF	35.33
smoothApp	59.78	sour	40.41	intensityAT	46.06
tomatoOd	54.76	bitter	33.79	sweetAT	41.10
herbalOd	43.59	herbalT	42.32	sourAT	28.49
sourOd	37.33	spicy	39.48	spicyAT	37.57
tomatoT	56.99	specificT	26.15		

the tested products are represented accordingly using their deviation from the ideal. The latter case is similar to a JAR situation. A similar procedure has been used by Hoggan (1975). The perceived intensity of the beer to optimize was compared to the perceived intensity of the competitor (and brand leader on the market) on a list of attributes. These comparisons of the intensities were done graphically using bar charts. For each attribute, a tick mark corresponding to the ideal level provided by the consumer panel was added on the graph, this ideal level being considered as the reference to match to improve the beer. In these two cases, the optimization is done by direct comparison.

With the IPM, since the perceived and ideal intensity is measured on each attribute for each consumer, the deviation from ideal can be computed (Szczesniak et al. 1975; Cooper et al. 1989). The information considered is similar to JAR. Hence, the methodology developed to analyze JAR data can be applied to these deviations. A large overview of the methodologies analyzing JAR data can be found in Meullenet et al. (2007). This also includes the adaptation of the use of PLS regression on dummy variables proposed by Xiong and Meullenet (2006) and adapted for the IPM by Worch et al. (2010a).

Instead of considering the raw ideal intensities or the deviation between perceived and ideal intensity, Cooper et al. (1989) also proposed to use the ratio between the perceived and ideal intensity. In this case, the deviation from the ideal is expressed in percentage of change to adopt. However, with this procedure, the authors point out three main issues:

1. Depending on the ratio considered, the percentage of change can vary. Let's consider an absolute sample score of 4 and an ideal score of 5. The ratio will either be 0.8 (corresponding to a percentage of change or 20%) or 1.25 (corresponding to a percentage of change of 25%) whether the ideal intensity is considered as the numerator or the denominator in the ratio.
2. For attributes with negative hedonic connotation (i.e., attributes for which the absence is the ideal level), it can be problematic since a score of 0 for the ideal intensity will return an infinite ratio of the product to the ideal score.
3. This procedure is working well when the products are close to the ideal and show some weaknesses when the ideals are situated far from the ideals.

To avoid the first issues, they propose a log transformation of the ratio. In that case, log(Sensory/Ideal) becomes log(Sensory)—log(Ideal) or more precisely log(Sensory)—a constant. And this formula is true whether the ratio is considering the ideal intensity as numerator or denominator. For the second issue, the ideal scores of 0 are replaced by scores close to 0 (0.1 can be used in practice).

In addition to their graphical representation, Szczesniak et al. (1975) proposed to compare the products with the ideal by submitting the data to factor analysis. In this case, the comparison is based on all the evaluated attributes simultaneously. This methodology has also been considered by Cooper et al. (1989).

In most of these methodologies, the potential link between the perception of an attribute and the appreciation of the products is not taken into consideration. However, it seems important to consider it since optimizing products on attributes that are not driving liking is of minor interest. The procedures presented in the next section take into consideration whether or not attributes are drivers of liking and guide on improvement by suggesting which attributes should be changed in priority to have a larger impact on liking.

4.5.2 MORE ADVANCED SOLUTIONS

4.5.2.1 Eclipse Method

The first methodology called the eclipse method was proposed by Moskowitz et al. (1977). It requires the experimenter to have a series of products for which the different formulations (or recipes) are known. Consumers are then asked to evaluate each product and to rate the perceived intensity on a list of attributes using the method of magnitude estimation (Moskowitz and Sidel 1971). For each attribute, a separate regression equation predicting the magnitude of that attribute based on the formulation (i.e., physical ingredients) of the products is estimated. Since the evaluation of the products is performed by consumers, the experimenter can also ask them to rate the intensity they would like to have, for each attribute on the same scale, which would correspond to the profile of their ideal product. Finally, the experimenter can reverse the regression equations defined previously and predict the formulation of a product that would come close to the ideal product (also known as reversed regression).

Note: This notion of reversed regression is sometimes used in external preference mapping studies to estimate the profile of the optimum product.

4.5.2.2 Fishbone Method

As stated in the introduction (Section 4.1.1), the decrease in overall liking depends on both the deviation from ideal and the relevance of the individual attributes on overall liking (Equation 4.1).

Although the deviation can be calculated easily for each consumer and each product (by subtracting the ideal intensity of reference z_a from the perceived intensity y_{jpa}), we still need to determine the impact of each attribute on liking.

The fishbone method takes the relevance of each attribute for overall liking into account.

To estimate the impact of each attribute on liking, one could consider regressing the attributes on liking. However, because of the high collinearity between attributes, this solution is not recommended. For that reason, regression on principal dimensions from PCA is preferred.

First, a PCA is performed on the sensory descriptions of the product. In this PCA, all dimensions associated with an eigenvalue larger than 1 are extracted, the liking scores are regressed on those dimensions, and the regression weight of each dimension on overall liking is computed.

From these regression weights, the impact of each individual attribute on liking is computed. To do so, for each attribute, the regression weight for each dimension is multiplied by the attribute loading on that dimension. The final attribute weight is obtained by summing up the weighted impact of that attribute over all dimensions. This final weight represents the relative importance of that attribute for liking (β_a).

Next, for each product, the absolute differences d_{pa} between perceived and ideal intensity of reference are computed for each attribute. These differences are divided by the perceived intensities and multiplied by the corresponding attribute-regression weights (Equation 4.8):

$$d_{pa} = \beta_a \times \left| \frac{(z_{ref,a} - \bar{y}._{pa})}{\bar{y}._{pa}} \right| \qquad (4.8)$$

This results in the corrected deviations from ideal. In order to express the difference related to one attribute relatively to the others, each individual difference d_{pa} is divided by the sum (noted D_p) of the differences of the A attributes:

$$D_p = \sum_{a=1}^{A} d_{pa} \qquad (4.9)$$

This sum represents the total amount of possible change for that product (corrected for the relative importance of that attribute for liking).

Next, the estimated gain in liking is computed. This is the difference between the maximum liking a product could get (we would consider 9 on a 9-point scale) and the actual liking score. In order to get proportions, this difference is divided by the actual liking score (Equation 4.10). This results (noted egl_p) in the estimated gain in liking in percentages for that product. As an example, for a liking score of 6 on a 9-point scale, the estimated gain in liking is $(9-6)/6 = 50\%$:

$$egl_p = \frac{(H_{max} - \bar{h}._p)}{\bar{h}._p} = \frac{(9 - \bar{h}._p)}{\bar{h}._p} \qquad (4.10)$$

Finally, for each attribute, the potential increase in liking (noted pl_{pa}) is computed by dividing the corrected deviation from ideal by the total amount of possible change for that product multiplied by the estimated gain in liking (Equation 4.11). For each product, this results in a percentage of improvement for each attribute when that attribute would be ideal (i.e., when the perceived intensity is set at its ideal level):

$$pl_{pa} = \frac{d_{pa}}{D_p} \times egl_p \qquad (4.11)$$

Results for each product are presented as fishbone plots (see Section 4.5.3 for an example). The fishbone method has been used extensively in product optimization projects for many different product categories.

4.5.3 APPLICATION

The first step of the analysis consists of extracting the underlying principal components from the PCA on the *product × subject* dataset. In our example, seven relevant dimensions were extracted. These dimensions and their major attributes related to them are provided in Table 4.5.

Overall liking is regressed on these dimensions, and the significant ones, as well as their corresponding regression weights, are shown in Figure 4.23. Finally, after computing the potential increase in liking for each product and each attribute according to the methodology presented in Section 4.5.2.2, the so-called fishbone plots are created (Figure 4.24).

In this figure, diamonds represent deviations from ideal (right axis) and bars represent the relative increase in liking when the attribute would be ideal (left axis). However, only attributes that contribute 2% or more are considered as being meaningful, and hence shown. The attributes are ordered from right to left in decreasing importance for liking.

In our example, product 1 needs less improvement than product 3 (Figure 4.24). For product 1, all the relevant attributes are too low compared to the ideal (diamonds are pointing upward meaning that the actual intensities are lower than ideal). Moreover, the order of the attributes suggests that improving *spicyAT* has the largest impact on liking, while improving *herbalT* has the least impact on liking.

For product 3, most attributes are perceived as too intense (most of the diamonds pointing downward meaning that the actual intensities are higher than ideal). In this case, the most impactful attributes to change to increase liking are *tomato* (need more), *tomatoOd* (need more), and *sourAT* (need less).

It has to be noted that the fishbone plot is not a recipe for change but it shows the impact of deviations from ideal on consumer overall liking. However, interactions between attributes are not taken into account here.

TABLE 4.5
Underlying Factor Structure of the Sensory Data

Dimension 1: Spiciness	Dimension 2: Bitter/Sour	Dimension 3: Thickness	Dimension 4: Herbal	Dimension 5: Sweet/Fruity	Dimension 6: Smoothness	Dimension 7: Tomato
Spicy AT	Bitter	Thick App	Herbal O	Sweet	Smooth App	Tomato O
Burn MF	Sour AT	Thick MF	Herbal T	Sweet AT	Smooth MF	Tomato T
Intensity AT	Sour	Color	Sour O	Fruity T	Gloss	
Spicy	Specific T					
	Salt					

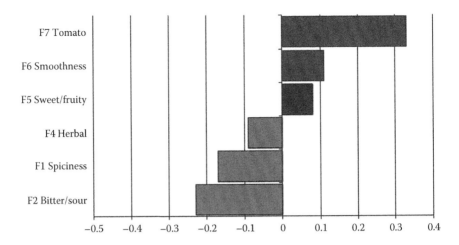

FIGURE 4.23   Regression weights of the factors contributing significantly to overall liking.

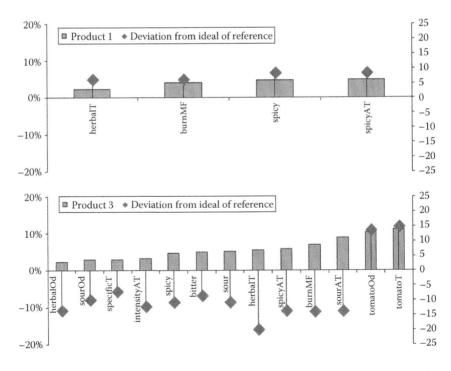

FIGURE 4.24   Guidance for improvements for the products 1 and 3 using the *fishbone* plots.

In practice, the most efficient way to interpret the results is by inspecting the fishbone plot and the spider plots in combination with an actual tasting of the products tested.

4.6 CONCLUSION

The IPM is a methodology that has been used for many years to help improving a wide range of products, from food to beverages and cosmetics. From our experience, it is a very effective tool for both product optimization and for defining new niche of products. Indeed, the IPM is an excellent method that helps understanding consumers and their needs (in terms of product sensory characteristics).

Since this valuable information (drivers of liking, guidance on improvement, etc.) can also be provided by other tools, one can wonder why using the IPM over other methodologies (such as the JAR scale).

To our point of view, the major quality of the IPM lies in the richness of the data collected and the analyses related to them. Compared to JAR scale, more information concerning the products and consumers is collected since perceived and ideal intensities for each product and each attribute are measured. In this case, the exact difference between perceived and ideal can be computed for each product and each attribute. In particular, in terms of analyses and information extracted, it provides the following:

- A methodology to check for the consistency of the data. To our point of view, this analysis is very important since it *measures the quality* of the data collected. Such procedure is rather unique for consumer data (to our knowledge, no similar analysis exists for JAR data).
- A procedure for determining whether the consumers associated the product set with one or multiple ideals. In many cases, methodologies are making the strong assumption that consumers have one unique ideal. Although this assumption is often verified, situations where it is not the case also exist. In those cases, considering one unique ideal can be misleading. It is hence important to check whether this assumption is verified before optimization. Again, this procedure is unique to the IPM.
- A complete optimization solution, from the determination of the ideal of reference (solution from the *IdMap*, which is not necessarily defined within the sensory space) to the guidance on improvement obtained through the *fishbone method*, which provides useful information to product developers.

REFERENCES

Booth, D.A., Conner, M.T., and Marie, S. 1987. Sweetness and food selection: Measurement of sweeteners' effects on acceptance. In *Sweetness*, ed. J. Dobbing, pp. 143–160. London, U.K.: Springer-Verlag.

Borgognone, M.G., Bussi, J., and Hough, G. 2001. Principal component analysis in sensory analysis: Covariance or correlation matrix? *Food Quality and Preference* 12: 323–326.

Brandt, M.S., Skinner, E.Z., and Coleman, J.A. 1963. Texture profile method. *Journal of Food Science* 28: 404.

Cadoret, M. and Husson, F. 2012. Construction and evaluation of confidence ellipses applied at sensory data. *Food Quality and Preference* 28: 106–115.

Cadoret, M., Lê, S., and Pagès, J. 2009. A Factorial Approach for Sorting Task data (FAST). *Food Quality and Preference* 20: 410–417.

Cadoret, M., Lê, S., and Pagès, J. 2011. Statistical analysis of hierarchical sorting data. *Journal of Sensory Studies* 26: 96–105.

Coombs, C.H. 1964. *A Theory of Data*. New York: Wiley.

Cooper, H.R., Earle, M.D., and Triggs, C.M. 1989. Ratios of ideals—A new twist to an old idea. In *Product Testing with Consumers for Research Guidance. ASTM STP 1035*, ed. L.S. Wu, pp. 54–63. Philadelphia, PA: American Society for Testing and Materials.

Dairou, V. and Sieffermann, J.M. 2002. A comparison of 14 jams characterized by conventional profile and a quick original method, the Flash Profile. *Journal of Food Science* 67: 826–834.

Danzart, M. 1998. Quadratic model in preference mapping. *Fourth Sensometric Meeting*, Copenhagen, Denmark, August 1998.

Danzart, M. 2009. Cartographie des préférences. In *Evaluation sensorielle, manuel méthodologique*, 3rd edn., ed. SSHA, pp. 443–450. Paris, France: Lavoisier.

Dehlholm, C., Brockhoff, P.B., Meinert, L., Aaslyng, M.D., and Bredie, W.L.P. 2012. Rapid descriptive sensory methods: Comparison of free multiple sorting, partial napping, napping, flash profiling and conventional profiling. *Food Quality and Preference* 26: 267–277.

Delarue, J., Danzart, M., and Sieffermann, J.M. 2010. Revisiting the definition of preference in preference mapping studies. In *KEER 2010 Proceedings*, Paris, France.

Ennis, D.M. 2005. Analytic approaches to accounting for individual ideal point. *IFPress* 82: 2–3.

Hoggan, J. 1975. New product development. *MBAA Technical Quarterly* 12: 81–86.

Husson, F., Josse, J., and Pagès, J. 2010. Principal component methods—Hierarchical clustering—partitional clustering: Why would we need to choose for visualizing data? Technical Report—Agro campus, retrieved from http://www.agrocampus-ouest.fr/math (retrieved October 30, 2012).

Husson, F., Lê, S., and Pagès, J. 2011. Clustering. In *Exploratory Multivariate Analysis by Example Using R*, eds. F. Husson, S. Lê, and J. Pagès, pp. 169–204. London, U.K.: CRC Press.

Husson, F., Le Dien, S., and Pagès, J. 2001. Which value can be granted to sensory profiles given by consumers? Methodology and results. *Food Quality and Preference* 12: 291–296.

Husson, F., Le Dien, S., and Pagès, J. 2005. Confidence ellipse for the sensory profiles obtained with principal component analysis. *Food Quality and Preference* 16: 245–250.

Lawless, H.T. 1989. Exploration of fragrance categories and ambiguous odors using multidimensional scaling and cluster analysis. *Chemical Senses* 14: 349–360.

Lê, S. and Husson, F. 2008. SensoMineR: A package for sensory data analysis. *Journal of Sensory Studies* 23: 14–25.

Lê, S., Josse, J., and Husson, F. 2008. FactoMineR: An R Package for multivariate analysis. *Journal of Statistical Software* 25: 1–18.

MacFie, H.J., Bratchell, N., Greenhoff, K., and Vallis, L.V. 1989. Designs to balance the effect of order of presentation and first-order carry-over effects in hall tests. *Journal of Sensory Studies* 4: 129–148.

Meullenet, J.F., Lovely, C., Threlfall, R., Morris, J.R., and Striegler, R.K. 2008. An ideal point density plot method for determining an optimal sensory profile for Muscadine grape juice. *Food Quality and Preference* 19: 210–219.

Meullenet, J.F., Xiong, R., and Findlay, C.J. 2007. *Multivariate and Probabilistic Analyses of Sensory Science Problems*, 1st edn. Ames, IA: IFT Press, Blackwell Publishing.

Moskowitz, H.R. 1972. Subjective ideals and sensory optimization in evaluating perceptual dimensions in food. *Journal of Applied Psychology* 56: 60–66.

Moskowitz, H.R. 1996. Experts versus consumers: A comparison. *Journal of Sensory Studies* 11: 19–37.

Moskowitz, H.R. 1997. Base size in product testing: A psychophysical viewpoint and analysis. *Food Quality and Preference* 8: 247–255.

Moskowitz, H.R. and Sidel, J.L. 1971. Magnitude and hedonic scales of food acceptability. *Journal of Food Science* 36: 677.

Moskowitz, H.R., Stanley, D.W., and Chandler, J.W. 1977. The eclipse method: Optimizing product formulation through a consumer generated ideal sensory profile. *Canadian Institute of Food Science Technology Journal* 10: 161–168.

Nestrud, M. 2012. Product landscaping the Bayesian Way: Uncovering the evaluative dimensions of consumers. In *Oral Presentation at the Third Society of Sensory Professionals Conference*, Jersey City, NJ, October 10–12.

Pagès, J. 2005. Collection and analysis of perceived product inter-distances using multiple factor analysis: Application to the study of 10 white wines from the Loire Valley. *Food Quality and Preference* 16: 642–649.

Pagès, J., Cadoret, M., and Lê, S. 2010. The Sorted Napping: A new holistic approach in sensory evaluation. *Journal of Sensory Studies* 25: 637–658.

Perrin, L., Symoneaux, R., Maître, I., Asselin, C., Jourjon, F., and Pagès, J. 2008. Comparison of three sensory methods for use with the Napping® procedure: Case of ten wines from Loire Valley. *Food Quality and Preference* 19: 1–11.

R Development Core Team 2012. R: A language and environment for statistical computing. *R Foundation for Statistical Computing*, Vienna, Austria. http://www.R-project.org/ (retrieved September 2012).

Rothman, L. and Parker, M. 2009. *Just-About-Right JAR Scales: Design, Usage, Benefits and Risks*. West Conshohocken, PA: ASTM International, Manual MNL-63-EB.

Sieffermann, J.M. 2002. Flash profiling. A new method of sensory descriptive analysis. In *AIFST 35th Convention*, Sydney, New South Wales, Australia, July 21–24.

Stone, H. and Sidel, J.L. 2004. *Sensory Evaluation Practices*. San Diego, CA: Academic Press.

Stone, H., Sidel, J., Oliver, S., Woosley, A., and Singleton, R.C. 1974. Sensory evaluation by quantitative descriptive analysis. *Food Technology* 28: 24–34.

Szczesniak, A., Loew, B.J., and Skinner, E.Z. 1975. Consumer texture profile technique. *Journal of Food Science* 40: 1253–1256.

Van Trijp, H.C., Punter, P.H., Mickartz, F., and Kruithof, L. 2007. The quest for the ideal product: Comparing different methods and approaches. *Food Quality and Preference* 18: 729–740.

Williams, A.A. and Langron, S.P. 1984. The use of free-choice profiling for the evaluation of commercial ports. *Journal of Science of Food Agriculture* 35: 558–568.

Worch, T. 2012. The *Ideal Profile Analysis:* From the validation to the statistical analysis of Ideal Profile data. PhD document, retrieved from www.opp.nl/uk/.

Worch, T., Crine, A., Gruel, A., and Lê, S. 2014. Analysis and validation of the Ideal Profile Method: Application to a skin cream study. *Food Quality and Preference*, 32: 132–144.

Worch, T., Dooley, L., Meullenet, J.F., and Punter, P.H. 2010a. Comparison of PLS dummy variables and Fishbone method to determine optimal product characteristics from ideal profiles. *Food Quality and Preference* 218: 1077–1087.

Worch, T. and Ennis, J.M. 2013. Investigating the single ideal assumption using Ideal Profile Method. *Food Quality and Preference* 29: 40–47.

Worch, T., Lê, S., and Punter, P. 2010b. How reliable are the consumers? Comparison of sensory profiles from consumers and experts. *Food Quality and Preference* 21: 309–318.

Worch, T., Lê, S., Punter, P., and Pagès, J. 2012a. Assessment of the consistency of ideal profiles according to non-ideal data for IPM. *Food Quality and Preference* 24: 99–110.

Worch, T., Lê, S., Punter, P., and Pagès, J. 2012b. Extension of the consistency of the data obtained with the Ideal Profile Method: Would the ideal products be more liked than the tested products? *Food Quality and Preference* 26: 74–80.

Worch, T., Lê, S., Punter, P., and Pagès, J. 2012c. Construction of an Ideal Map *IdMap* based on the ideal profiles obtained directly from consumers. *Food Quality and Preference* 26: 93–104.

Xiong, R. and Meullenet, J.F. 2006. A PLS dummy variable approach to assess the impact of jar attributes on liking. *Food Quality and Preference* 17: 188–198.

5 Use of Just-About-Right Scales in Consumer Research

Richard Popper

CONTENTS

5.1 INTRODUCTION

Just-about-right (JAR) scales are used in consumer research to identify whether an attribute is present in a product at a level that is too high or too low or whether it is "just about right." For example, in order to determine consumers' preferred level of sweetness in a soft drink, consumers might be asked to taste a prototype formulation and to rate its sweetness on a scale ranging from "much too high" to "much too low" (Figure 5.1). In addition to sweetness, other attributes of interest might include strength of flavor and carbonation, and JAR scales could be included in the questionnaire to assess these attributes. Consumers' answers will provide an indication of whether there is opportunity to improve the prototype and suggest the direction for any potential formulation change.

JAR scales are particularly common in research designed to optimize *sensory* attributes. Sensory attributes frequently exhibit *satiety*—that is,

Much too high	☐
Somewhat too high	☐
Just about right	☐
Somewhat too low	☐
Much too low	☐

FIGURE 5.1   Example of a five-point JAR scale.

more is not necessarily better. For any one attribute, there are intensity levels that consumers will find "too low" and other intensities they will find "too high." A JAR scale asks respondents to judge the intensity they experience in the product relative to the level they desire.

There are several research methods, besides JAR scaling, for determining the ideal level of product attributes. Perhaps the most rigorous would be an approach that presents consumers with alternative formulations that vary in their characteristics according to an experimental design and asks consumers to rate their liking for each product. Using statistical models, the researcher is able to infer the formulation that maximizes liking, often with great precision (Gacula et al. 2009). When products are not varied according to an experimental design, optimization is still possible using one of several nonexperimental approaches. Ideal point models, dating back to Coombs (1964), have been used to understand consumer preferences in the absence of experimental designs (Meullenet et al. 2007) and rely on applying multivariate statistical models to liking ratings (or rankings) of multiple products. Other nonexperimental optimization approaches include correlational approaches that relate overall liking ratings to sensory intensity ratings collected from trained sensory panelists or consumers (Meilgaard et al. 1999).

The emergence of JAR scales in sensory and market research is not well documented. Moskowitz (1972) discussed JAR scales as an alternative to sensory intensity scales for identifying optimal formulations of food products. Shepherd and others (Shepherd et al. 1984; McBride and Booth 1986) used JAR scales in studies in which a single food ingredient (e.g., the concentration of salt in soup) was varied systematically. The authors found that the JAR ratings (e.g., of saltiness) were approximately logarithmically related to ingredient concentration (e.g., % salt) and used regression to estimate the concentration corresponding to an average response of "just right."

Today, JAR scales are a tool commonly used for guiding product development. JAR scale data are used to suggest modifications to prototypes, and follow-up studies are often conducted to confirm that these changes were effective in improving the product. JAR scales find their most frequent use in nonexperimental product optimization studies. In these studies, products are not varied systematically but instead comprise one or more prototypes or in-market products. Unlike other optimization approaches, JAR scale data can aid product development even when only a single product is assessed, making JAR scales a cost-effective tool.

5.2 CONSTRUCTION OF JAR QUESTIONS

While JAR scales are seemingly simple tools, care is needed in selecting the scale type, the attributes, and the appropriate verbal anchors in order to ensure meaningful data are collected with this technique.

5.2.1 SCALE TYPE

The most commonly used JAR scale is the five-point scale (Figure 5.1), though seven-point, nine-point, and continuous line scales have also been proposed (Rothman and Parker 2009). A scale with more than five points, while offering more response options, does not necessarily increase sensitivity—in fact, a longer scale may be harder for respondents to use and may increase the level of random variation in the data. Analysis and reporting of JAR scales often relies on a reduction of the data to just three categories—just right, not enough, and too much—and from that vantage point, scales with more than five points are hardly warranted. In research with special populations, a three-point scale might be most appropriate, for example, in working with children, the elderly, or cultures in which ratings scales are unfamiliar. Regardless of the type of scale, there is value in standardization. The interpretation of JAR scale results often relies on company internal benchmarks or rules of thumb (see below). Since different scale lengths can yield different results, the comparison across studies requires consistency in the scale employed. Global research poses additional challenges, because even with scales of the same length, differences in the interpretation of the verbal anchors can lead to differences in scale use.

5.2.2 ATTRIBUTE SELECTION AND VERBAL ANCHORS

As in all consumer research, the attribute used for a JAR question should be unambiguous and easily understood by consumers. Consumers have been shown to provide reliable and detailed sensory attribute evaluations (Worch et al. 2010). However, compared to product developers and sensory analysts, consumers may have difficulty in understanding the meaning of some attributes (e.g., *mouthfeel* of a beverage) or may be unable to isolate and differentiate among related flavor notes in a flavor complex, such as *onion* vs. *garlic* flavor. While these attributes and distinctions may be important to optimizing a product, their impact on liking may need to be assessed by other means, for example, using panels trained in descriptive analysis who are capable of making reliable judgments on these attributes.

Given the bipolar nature of the JAR scale, thought is required regarding how to label the scale end points. For example, a scale that ranges from "much too sweet" to "much too sour" implies that a product could not be "too sour" and "too sweet"—yet some products could be both (e.g., certain confections). In addition, such a scale implies that a product that is "too sweet" is "not sour enough," which may not be a valid inference. Except for cases where attributes are clearly semantic opposites ("too thin" vs. "too thick"), it is best to restrict each JAR question to one sensory attribute and to ask separate JAR questions for attributes that are not semantic opposites.

A JAR scale assumes the consumer has an internal *ideal* against which to compare the actual product experience. This assumption may not always be warranted. Attributes for which the consumer does not have an ideal may be rated "just right" by default or may lead to random responding. Another problem are attributes that the consumer perceives as intrinsically negative, such as bitterness. Consumers may avoid rating a coffee as "not bitter enough," in the belief that any level of bitterness is a negative. This may or may not be the case—a certain level of bitterness may indeed contribute to liking. Attributes that are perceived as negatives do not lend themselves to JAR scales as readily as more "neutral" attributes and are better asked as intensity scales. This is one example of how the consumer perceptions of the intrinsic desirability (or lack thereof) of certain attributes may influence a JAR response. An opposite bias may exist for attributes that have a positive halo (e.g., the tendency to rate the amount of chocolate chips in a chocolate chip cookie as "not enough").

5.2.3 QUESTIONNAIRE DESIGN

JAR scales are meant to elicit information that will aid in explaining a product's overall appeal, which is often measured using the standard nine-point hedonic scale or the five-point purchase intent scale. When JAR scales are used, it is customary not to use sensory intensity scales for the same attributes, in part to manage the overall length of the questionnaire. When ratings of sensory intensity are of interest, it is possible to ask respondents to rate sensory intensity and, separately, to rate their ideal product on intensity scales (van Trijp et al. 2007). A *difference from ideal* measure can then be calculated by computing the difference between actual and ideal intensities for each attribute. JAR scales, in contrast, ask consumers for a direct judgment regarding the direction and degree of difference from ideal; the actual and ideal intensities are not separately measured.

No firm guidelines exist regarding the number of JAR scales to include in a questionnaire. It is not uncommon to encounter questionnaires with 10 or more such scales. Respondent fatigue is always a consideration in questionnaire design and should be a consideration in determining the number of JAR questions included.

5.3 ANALYZING JAR DATA

JAR scale data are usually summarized by the percentage of respondents choosing each of the response categories of the JAR scale. Often, combining some of the response categories simplifies the presentation of the results. For example, it is common practice to summarize results of a five-point JAR scale with three percentages: the percent "just right," the percent "too high" (combining "somewhat too high" and "much too high"), and the percent "too low" (combining "somewhat too low" and "much too low"). A product's profile across a number of JAR attributes can then be graphically displayed in the form of stacked bar chart (Figure 5.2). Several product profiles can be compared by side-by-side presentations of bar charts. An alternative graphical display format, though seldom used, is the triangular or ternary plot (see Figure 5.3). In this plot, the attribute is located with reference to three axes, representing the percent "just right," "too high," and "too low," respectively.

The statistical analysis of JAR data can take many forms, depending on the analysis objective, and the ASTM manual on JAR scales provides examples of a multitude of approaches (Rothman and Parker 2009). The basic analyses focus on simple hypothesis testing, for example, whether products differ in the distribution of JAR scale responses. This can be accomplished using a variety of nonparametric tests (Rothman and Parker 2009, Appendix G).

Averages computed on the basis of assigning numbers to the categories of the JAR scale are often not appropriate summaries, since they do not take the bipolar nature of the scale into account. For example, two products could have a similar average score on a JAR scale but could differ in the underlying distribution of "too much" and "not enough" judgments. One-number summaries, such as averages, may be appropriate after recoding of the JAR scale. After recoding a JAR response as "1" and all other responses as "0," it is possible to apply tests such as analysis of variance (ANOVA) and post hoc mean comparison procedures to determine the statistical significance of differences in the percentage JAR among products. Such an approach, while perhaps violating some of the statistical distributional assumptions underlying these parametric tests, in practice works well enough to provide effective summaries.

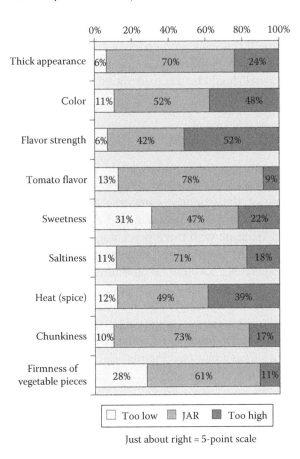

Just about right = 5-point scale

FIGURE 5.2 JAR profile of a vegetable salsa (bar chart).

Another recoding of the JAR scale is the one based on *folding* the JAR scale about the midpoint. A folded JAR scale is one in which the direction of the deviation from the center of the scale is ignored. In the case of a five-point JAR scale, this would mean coding the "just right" category as "3," coding the responses on either side of "just right" as "2," and coding the two extreme categories as "1" (regardless of direction). This recoding provides a three-scale step measure of "just rightness" and can be used as the basis for calculating averages and for performing other analyses, such as a consumer segmentation (Vigneau et al. 2004). Note that on the folded scale two products can have the same mean but can differ in the distribution on either side of "just right," since the folded scale ignores the direction of non-JAR responses. Folded results correlate very strongly with simple recoding of just right responses to "1" and all other responses to "0."

Despite the availability of an arsenal of statistical tests, in many practical applications, statistical testing of JAR data is considered of

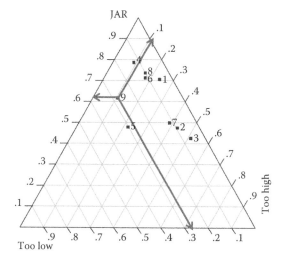

FIGURE 5.3 JAR profile of a vegetable salsa (ternary plot). *Note:* The coordinates for attribute 9 (firmness of vegetable pieces) are 28% too low, 61% just right, and 11% too high.

secondary interest or is even ignored all together. Instead, the interpretation of the results relies on rules of thumb, based on common sense or a history of testing in a particular product category, both of which can be used to define benchmarks for good performance. For example, many companies consider "75% just right" an indication that a product performs well on an attribute (assuming a five-point JAR scale was used). Differences in performance across products above that 75% benchmark would be considered of little practical importance, even though they may be statistically significant. This 75% guideline can be fine-tuned to accommodate differences among product categories. For example, past testing may indicate that the appropriate benchmark for confections and ice cream is 80% and for nonfat dairy products 60%. Similarly, other rules of thumb are used to identify problem areas. For example, 20% (or higher) of responses on one side of just right (e.g., "too strong") may be an indication that the product needs improvement in that aspect, especially if additional analyses, such as penalty analysis (see below), corroborate this conclusion. A high proportion of responses on both sides of "just right" may signal the presence of segments in the consumer population being tested (see below).

Rules of thumb are especially useful when only a single product is tested and no control or reference product is included for comparison. In these circumstances, such rules provide the only guide to the interpretation of JAR scale results. It is important to keep in mind, however, that such historically based interpretations of JAR scale results may not apply when the product tested is from a category very different from the one on which the past testing history is based or when there is a difference in research methodology that may impact the results. For example, the percent "just right" in a monadic (single product) test can be higher than the percent obtained in tests where respondents evaluate several products sequentially. Consequently, even a good product may not reach 75% JAR on an attribute in a multiproduct test, whereas it would in a monadic test. Benchmarks are most applicable when test conditions are standardized, and to the extent that they are not, judgment will have to be exercised as to their applicability.

5.4   PENALTY ANALYSIS

The summary of response proportions on a JAR scale provides a first indication of improvement opportunities for a product. However, before recommending formulation changes to address a seeming deficiency, it is important to establish a link between performance on the attribute and overall product performance, as measured by overall liking or purchase intent. In the discussion that follows, overall liking is used as an example of a criterion measure against which to evaluate the impact of attribute performance. While overall liking is one of the most common measures used for this purpose, other measures can be used, such as purchase intent or performance relative to expectations. Some attributes may not be important drivers of overall product performance. For example, consumers might agree that a product is too light in color, but improving color may not lead to greater liking, because color may be a minor consideration for consumers, compared to the product's flavor or texture. Therefore, in identifying a product improvement opportunity and in prioritizing among several such opportunities, it is critical to link attribute performance to acceptability. In the following sections, it is assumed that the number of consumers included in the study is sufficiently large to draw meaningful conclusions regarding such product improvements. In many industrial applications, this means a sample size of 100 consumers or more.

Penalty analysis has emerged as the most widely applied technique for linking performance on JAR attributes with overall liking (or another criterion measure). Penalty analysis is a method for determining if those respondents who find the product not "just right" on an attribute rate it lower in overall liking. With the help of penalty analysis, it is possible to identify product modifications that may bring a product closer to the (implied) ideal, thereby increasing overall liking, and to prioritize among such potential product improvements. In addition to its wide applicability, penalty analysis has the advantage of being computationally simple.

Penalty analysis is a *within product* analysis and is applicable regardless of the number of products tested (even if only one). When multiple products are tested, a penalty analysis is conducted separately for each product. Penalty analysis is most often conducted on the total respondent sample, though it is possible to conduct the analysis at the level of any subgroup of respondents, provided the sample size of the subgroup is large enough. For example, in a two-product test in which a preference judgment is obtained after the second product, it may be of interest to conduct a penalty analysis separately for each preference group in order to understand which attributes contribute to the differences in liking between the two groups.

Penalty analysis is conducted on all JAR attributes, one attribute at a time, and, for each attribute, quantifies the impact of being "too high" and "too low" on overall liking. In the example calculation given below, it is assumed that overall liking is the criterion measure of interest and that JAR scale response categories have been combined to yield three response categories: too weak, just right, and too strong. The labels "too high" and "too low" are meant generally to represent the two sides of the just right scale—depending on the verbal anchors used to define each JAR scale, the wording for the end points will differ.

The penalty calculations proceed as follows (see Table 5.1 for an example):

Step 1. For a particular JAR attribute, determine the average overall liking ratings for three groups of respondents: those who rated the product "just right" on that attribute, those who rated the product

TABLE 5.1

Example of Penalty Calculation

	Penalty Analysis for Flavor Strength JAR			
	Respondents (%)	Overall Liking	Mean Drop	Weighted Penalty
Too low	6.4	6.8	1.1	0.07
JAR	42.0	7.9		
Too high	51.6	5.1	2.8	1.44
Total	100	6.4		

Penalty (mean drop) calculation:
  Mean drop for "too low" = 7.9 − 6.8 = 1.1.
  Mean drop for "too high" = 7.9 − 5.1 = 2.8.
Weighted penalty calculation:
  Weighted penalty for "too low" = 0.064 × 1.1 = 0.07.
  Weighted penalty for "too high" = 0.516 × 2.8 = 1.44.

"too high," and those who rated the product as "too low." For ease of expression, designate the overall liking averages for these three groups of respondents as OL (JAR), OL (too high), and OL (too low).

Step 2. Calculate the *mean drop* in overall liking for the respondents rating the product either too strong or too weak, as follows:

Mean drop for "too strong," OL (JAR)-OL (too high)

Mean drop for "too weak," OL (JAR)-OL (too low)

The mean drop or *penalty* is a measure of how much overall liking decreases when the product is viewed as too high or too low on that attribute. The greater the mean drop, the greater the impact of that product deficiency on overall liking.

Each JAR attribute yields two penalties (one for each side of just right). Note that there is no reason that the two mean drops associated with an attribute should be similar—for example, it may be much worse for a product too be viewed as "too high" than "too low." Also note that the mean drop calculation does not take into account the number of consumers who rate the product as either too weak or too strong. The number of respondents feeding into the overall liking averages can be quite different for the two "tails" of the attribute.

Step 3. Calculate the penalties for each JAR attribute included in the questionnaire, in the manner described in Steps 1 and 2.

Step 4. Summarize the penalty calculations information in tabular or graphical form.

Step 5. Prioritize areas of greatest potential for improvement. Those attributes with the greatest potential for product improvement will have both a larger percentage allocated to a problem (e.g., "too much salt") and a large mean drop.

One type of graphical summary is shown in Figure 5.4. The figure plots the mean drops (vertical axis) for each attribute against the percent of consumers rating the product as too strong or too weak on that attribute (horizontal axis). Note that the horizontal axis starts at 20%, not 0%. This convention reflects the belief that attribute skews with fewer than 20% of responses are likely to be inconsequential for the purpose of the penalty assessment and, depending on the respondent base size, may be unreliable. The cutoff may be lowered depending on the respondent base size—with large base sizes, percentages less than 20% may be plotted.

While it is possible to subject a mean drop to significance testing (via t-test, against the null hypothesis that the mean drop = 0), many researchers forgo such testing. Instead, some organizations utilize guidelines for what degree of mean drop merits attention. For example, when the mean drop is

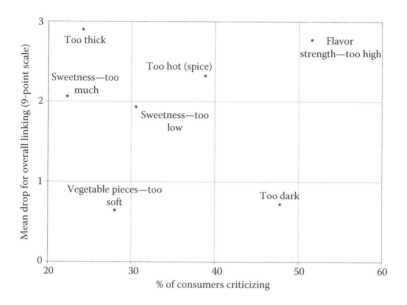

FIGURE 5.4   Penalty chart.

calculated on the basis of the nine-point overall liking scale, a mean drop of 1.0 is often considered a threshold for what constitutes a meaningful decline in liking for the subgroup of respondents critical of an attribute; a mean drop less than 1.0, by such a guideline, would be considered negligible. Of course, such a cutoff depends on the scale used to calculate a mean drop; a smaller numerical cutoff would be used if, for example, mean drops were calculated on the basis of a seven-point liking scale.

The interpretation of a penalty analysis not only relies on the mean drop but also considers the percentage of respondents who voiced a particular concern (e.g., "flavor too strong"). Figure 5.4 leads to an interpretation based on where in the plot the attribute falls, taking these two aspects into account. Attributes that fall in the upper right-hand corner are those that are most concerning, since they represent attributes that have the highest skews and are associated with the greatest mean drop. Attributes in the lower left corner are those for which there is minimal concern. Such a *quadrant* interpretation is strengthened by a standardization of the axis ranges used to plot the results: by setting the ranges for the vertical and horizontal axes to be the same for each product and each study, the user begins to associate regions of the plot with the degree to which an attribute has (negative) impact on liking.

The goal of penalty analysis is to prioritize attributes in terms of their impact on overall liking (or another criterion measure), thereby suggesting priorities for product development or reformulation. Figure 5.4 leads to such a prioritization by focusing the greatest attention on those attributes in the upper right-hand corner of the chart (flavor too strong). Nonetheless, there are still ambiguities in interpretation. How does one assess the importance of an attribute in the upper left-hand corner of the chart (too thick), one with a large mean drop based on a relatively small number of consumers? And how does one weigh the importance an attribute with a small mean drop, but a large number of consumers voicing that particular concern (too dark)?

One answer to such interpretive difficulties is to further summarize a penalty analysis by multiplying the percentage of respondents voicing a concern by the associated mean drop (Table 5.1). Such a *weighted* penalty collapses the two dimensions of Figure 5.4 into a single dimension. The weighted penalty has a convenient interpretation: it represents the increase in overall liking (or whatever measure is used to calculate the mean drop) that would occur if all those critical of a certain attribute (e.g., "flavor too strong") could be converted to rating the product "just right" on that dimension.

Using the weighted penalty, attributes can be ranked for their likely impact on product improvement (Table 5.2). A cutoff can also be defined, below which a weighted penalty can be considered negligible. In line with earlier guidelines to consider attributes for modification only if the mean drop is at least 1.0 (on a nine-point liking scale) and at least 20% of consumers express the concern, a weighted penalty of at least 0.20 would be needed for the attribute to be a candidate for reformulation.

While such a one-dimensional representation of the results of a penalty analysis facilitates ranking of the attributes, critical information may be lost by this transformation. The weighted penalty measure will assign similar importance to an attribute with a small number of critical consumers and a large mean drop as to one with a large number of critical

---

TABLE 5.2

Ranking of Product Improvements Based on Weighted Penalties

Attribute	% Consumers Criticizing	Mean Drop	Weighted Penalty
Flavor strength—too high	51.6	2.8	1.44
Too hot (spice)	38.9	2.3	0.90
Too thick	24.2	2.9	0.70
Sweetness—too low	30.6	1.9	0.59
Sweetness—too high	22.3	2.1	0.46
Too dark	47.8	0.7	0.34
Vegetable pieces—too soft	28.0	0.6	0.17

consumers and small mean drop. However, the researcher may not view these two instances as the same. The weighted penalty gives equal numeric importance to the penalty and the percentage of respondents indicating the product problem and this equality of importance may not be justified. A characteristic that many consumers fault but do not heavily penalize may be one that represents a degree of *wishful thinking*—many consumers will want *more* of a characteristic such as chocolate chips in a chocolate chip cookie but, assuming that the number of chips is in line with their general expectations, may not penalize the product heavily for *not enough choco-late chips*. On the other hand, a finding that a small percentage of consumers are very critical of an attribute (e.g., "too thick" in Figure 5.4) may be an indication of a lack of homogeneity in the population of consumers tested, which may have implications for the product reformulation strategy. Should a product be reformulated to satisfy this segment of consumers, who may want a fundamentally different product? These interpretive nuances suggest that a two-dimensional interpretation of the outcome of a penalty analysis provides a deeper understanding of the product improvement opportunities and should be retained, and that the weighted penalty be used as an additional guide to interpretation.

Another interpretive difficulty pertains to polarized attributes. For example, a substantial (>20%) proportion of consumers may rate a product "too strong" and "too weak" on the same attribute (e.g., see similar penalties for "too low" and "too high" sweetness in Figure 5.4). This can be due to a number of reasons. Consumers may disagree on their ideal level for an attribute: one segment preferring a weaker level and another segment preferring a stronger level. Alternatively, polarization may also be the result of a problem with the *quality* of an attribute, rather than its *quantity*. For example, in rating the cherry flavor in a beverage, consumers may dislike the particular character of cherry flavor (perhaps it is an artificial type of flavor), not necessarily the level of flavor. Those rating the product "too strong" may be expressing wish for less of an artificial type of flavor, and those rating the product "too weak" may be expressing a wish for more of a *true* cherry flavor. Finally, polarization can also result if the consumers do not understand the attribute they are asked to rate, leading to random responses to that particular question.

Polarized attributes make it difficult to provide clear product direction, and any recommendation must be viewed cautiously. Considerations in such cases include the size of the difference in skew between the two sides of just right and the size of the associated penalties. It seems reasonable to recommend modification according to the larger skew on the JAR scale only if that skew exceeds by 10%–15% the skew toward the other side of just right (and provided that penalty analysis confirms the greater impact of this side of the JAR attribute). However, when the skews and penalties are similar, the researcher will need to conclude that there is no clear direction for product reformulation.

The analyses and interpretive issues described previously lead to a general strategy for the use of JAR scales to provide formulation direction:

1. Attributes with a high proportion of "just right" responses (e.g., 75% +) and less than 20% of responses on one or the other side of "just right" can be considered as not substantially benefiting from modification.
2. Consider an attribute a candidate for improvement if it has at least 20% responses to one side of "just right" and if that skew exceeds by at least 10% the skew on the other side of just right.
3. Consult the penalty analysis to confirm an attribute's impact to a criterion measure such as overall liking, and use penalty analysis to prioritize among multiple attributes for reformulation.

5.5   ESTIMATING THE EFFECTS OF MODIFYING SEVERAL ATTRIBUTES

Penalty analysis is useful in prioritizing product modifications and provides an estimate of the potential improvement in overall liking that could be achieved if an attribute were to be viewed as "just right." This *one attribute at a time* approach does not take into account the potential improvement in response to the simultaneous modification of more than one attribute. For example, one may wish to identify the effect of modifying *two* attributes that a penalty analysis has identified as having a substantial impact on overall liking (e.g., *flavor strength* and *spice level* in Figure 5.4). A cross-tabulation of the responses to the JAR scales for flavor strength and spice level is used to identify those respondents who rated the product "just right" on both attributes. The average overall liking rating of those respondents, when compared to the total sample average, provides an indication of the potential increase in liking that could be achieved if both attributes were to be perceived as "just right."

Simulation offers another approach to estimating the effect of one or more formulation changes. The simulation process (Market Facts 2000) involves generating a series of random samples to simulate the outcome of having conducted the study multiple times. Each sample is built by randomly drawing respondents (with replacement) from the pool of respondents in the original study, with the number of random draws equaling the number of respondents in the original study. This bootstrapping process is repeated 1000 times or more, creating a collection of samples differing from each other by virtue of random sampling. The bootstrapped samples can then be used to study the effect of changes in one or more attributes that the penalty analysis has identified as important. For example, to estimate the effect on overall liking of decreasing the percent "too strong" in two attributes by 10%, one would find within the collection of boot-strapped samples those that show a 10% reduction in skew in those attributes. The average overall liking score of the bootstrapped samples that fit that description then serves as an indication of the potential improvement in liking that would result if those reductions in "too strong" were in fact achieved.

In general, in estimating potential product improvements, the researcher must consider the source: all such analyses represent *what-if* scenarios, based on responses to a single product. This is true for a penalty analysis conducted on individual attributes. Predicting the outcome of multiple attribute modifications requires additional caution.

5.6 LIMITATIONS OF JAR SCALE–BASED FORMULATION RECOMMENDATIONS

The analysis of JAR data does not provide the developer with a measure of the amount of formulation change needed to reduce the skew in a particular attribute (Moskowitz 2004). The size of the skew is not a reliable indication of the magnitude in the ingredient change needed; the sensitivity of the JAR scale to changes in formulation is not known. Even if the change in sensory *intensity* as a function of an ingredient change were known, JAR ratings are more complex than sensory intensity ratings and may be influenced by the degree to which respondents are forgiving of deviations from just right. For this reason, a JAR-based analysis provides at most a *direction* for change (more or less of an attribute) and an indication via penalty analysis of the possible impact of the change on product acceptability.

As mentioned previously, the analysis of JAR scales treats all attributes as independent of one another. In reality, a product's sensory impact is the result of the interplay of several attributes. Consequently, changes in one attribute are likely to impact other attributes, and improvements in one attribute may have positive (or negative) consequences on another attribute. In a cheese cracker, for example, increases in saltiness alone may increase perception of cheese flavor. Consequently, a cracker that consumers consider "too weak in cheese flavor" and "not salty enough" may benefit from increases in saltiness, but may not require the addition of cheese flavor. Improvements in one attribute may have unintended—and possibly undesirable—consequences for other attributes. Attribute interactions are often known to the product formulator and sensory analyst, and their expert knowledge is often key in devising a successful product reformulation strategy based on JAR scale data.

The assumption underlying the use of JAR scales for product optimization is that the product formulation changes based on an analysis of JAR scales will lead to an improved product. However, a product with the best "just right" score may not always be the one that is most liked. Epler et al. (1998) varied the amount of sugar in lemonade and determined the optimal sugar level in two ways: based on a JAR scale for sweetness and based on liking. The optimal sugar level was somewhat higher based on liking than based on the JAR: consumers liked the sweeter formulations even though they rated them as somewhat too sweet. Similar results were reported by Popper et al. (1995) for aspartame-sweetened beverages. While the difference found by Epler et al. (1998) in the optimal sugar concentrations according to the two methods was small, it was large enough to make a difference in a follow-up preference test. A separate group of consumers, when presented with the two formulations, preferred the product optimized on the basis of liking over the one optimized on the basis of the JAR scale. This suggests that the optimal JAR response profile may not always be one that is maximally "just right."

One possible explanation for instances where the maximally liked product is not the one with the best JAR profile is that JAR scales may be subject to response bias. In the cases of sweetness, for example, respondents may perceive a sweeter formulation as potentially unhealthy. When rating such a product, they may indicate that the product is "too sweet" because they are aware of the potentially negative consequences of consuming such a product on a regular basis. At the same time, consumers may actually like the taste of the sweeter product, a fact reflected in their liking ratings. Similar response biases may be operating in the case of other attributes where social desirability biases affect attribute ratings. Bower and Baxter (2003) attempted to confirm such a hypothesis by comparing JAR and liking ratings of sweetness among two groups of respondents that differed in their concern with healthy eating but came to no clear conclusion regarding such attitudinal effects.

The question whether JAR scale–based optimization leads to similar results as optimization by other means has been examined in cases involving more than just one attribute. Moskowitz et al. (2003) found that applying regression techniques to predicting which formulation was optimal yielded different results depending on what criterion was optimized: optimizing a product to be "just right" on all attributes resulted in a different predicted optimum than when optimization based on overall liking. On the other hand, Marketo and Moskowitz (2004) found the two predictions to yield similar results. Lovely and Meullenet (2009) compared several approaches to optimizing strawberry yogurt, including preference mapping using trained panel ratings, and proceeded to test out the predictions of each approach with products formulated according to the recommendations of each method. Among their optimization approaches was one based on JAR scales (i.e., identifying the formulation that was predicted to be closest to just right across multiple attributes). In their validation study, the formula based on JAR scale optimization was liked as much, if not more, than the formulas based on more complicated preference mapping approaches. Similarly, van Trijp et al. (2007) found broad agreement in formulation direction when comparing preference mapping, ideal point profiling, and penalty analysis based on JAR scales.

It is worth noting a key difference between a penalty analysis approach to identifying drivers of liking and one based on an analysis across products, as exemplified in preference mapping or other correlational analyses. The attributes that drive the appreciation of any one product (as revealed by penalty analysis) may not be the same as those that drive differences in liking among several products. For example, texture may emerge as a key driver based on a penalty analysis of JAR scale data for *hardness*, but JAR responses (e.g., % "too hard") may not correlate with overall liking across several products if all products are similarly "too hard." In practice, attributes may drive liking in both ways: those attributes that show the greatest potential for increasing liking by moving a product closer to ideal are the same attributes that explain most of the differences in liking between products.

A separate concern with the use of JAR scales is how their inclusion on the questionnaire might impact other questions, in particular ratings of overall liking. Earthy et al. (1997) found that including JAR scales changed overall liking ratings compared to when JAR scales were not asked. Popper et al. (2004) extended this finding and showed that JAR scales, but not sensory intensity scales, had biasing effect on overall liking, even though the same attributes were being rated. However, subsequent studies (Popper et al. 2005) showed no effect of attribute questions on overall liking or similar effects for JAR and intensity scales. The effect of attribute questions—intensity, JAR, or otherwise—on overall liking or preference is an area that deserves further study.

5.7 CONCLUSION

JAR scales offer the opportunity to gain insight into the strengths and weaknesses of a product and, when coupled with statistical analysis, provide a guide toward product improvement. Consumers find JAR scales easy, even natural, to use. Nonetheless, it is clear that JAR scales also suffer several limitations as noted previously.

There is no substitute in optimization work for providing respondents a full range of product alternatives to best explore the effect of manipulations in formulation. This is not so much a limitation of JAR scales, as it is a caveat on their use. There is risk in assuming that JAR scales obtained from a limited number of prototypes or in-market products are a substitute for a more systematic exploration of product differences, based on experimental or nonexperimental designs that span a wide range in sensory characteristics. With an awareness of their limitations, however, JAR scales and associated analytics offer an efficient approach to product optimization.

REFERENCES

Bower, J.A. and I.A. Baxter. 2003. Effects of health concern and consumption patterns on measures of sweetness by hedonic and just-about-right scales. *Journal of Sensory Studies* 18: 235–248.

Coombs, C.H. 1964. *A Theory of Data*. New York: John Wiley & Sons.

Earthy, P.J., J.H. McFie, and H. Duncan. 1997. Effect of question order on sensory perception and preference in central location trials. *Journal of Sensory Studies* 12: 215–237.

Epler, S., E. Chambers, and K. Kemp. 1998. Hedonic scales are a better predictor than just-about-right scales of optimal sweetness in lemonade. *Journal of Sensory Studies* 13: 191–197.

Gacula Jr., M.C., J. Singh, J. Bi, and S. Altan. 2009. *Statistical Methods in Food and Consumer Research*, 2nd edn. San Diego, CA: Academic Press.

Lovely, C. and J.F. Meullenet. 2009. Comparison of preference mapping techniques for the optimization of strawberry yogurt. *Journal of Sensory Studies* 24: 457–478.

Market Facts. 2000. Modelling simulated data. *Decision Systems Newsletter* 2: 1–4.

Marketo, C. and H. Moskowitz. 2004. Sensory optimization and reverse engineering using JAR scales. In *Data Analysis Workshop: Getting the Most Out of Just-About-Right Data. Food Quality and Preference* 15: 891–899.

McBride, R.L. and D.A. Booth. 1986. Using classical psychophysics to determine ideal flavor intensity. *Journal of Food Technology* 21: 775–780.

Meilgaard, M., G.V. Civille, and B.T. Carr. 1999. *Sensory Evaluation Techniques*. Boca Raton, FL: CRC Press.

Meullenet, J.F., R. Xiong, and C.J. Findlay. 2007. *Multivariate and Probabilistic Analyses of Sensory Science Problems*. Ames, IA: Blackwell Publishing.

Moskowitz, H.R. 1972. Subjective ideals and sensory optimization in evaluating perceptual dimensions of food. *Journal of Applied Psychology* 56: 60–66.

Moskowitz, H.R. 2004. Just about right (JAR) directionality and the wandering sensory unit. In *Data Analysis Workshop: Getting the Most Out of Just-About-Right Data. Food Quality and Preference* 15: 894–896.

Moskowitz, H.R., A.M. Munoz, and M.C. Gacula. 2003. *Viewpoints and Controversies in Sensory Science and Consumer Product Testing*. Trumbull, CT: Food & Nutrition Press.

Popper, R., P. Chaiton, and D. Ennis. 1995. Taste test vs. ad-lib consumption based measures of product acceptability. In *Second Pangborn Sensory Science Symposium*, University of California, Davis, CA, July 30–August 3.

Popper, R., W. Rosenstock, M. Schraidt, and B.J. Kroll. 2004. The effect of attribute questions on overall liking ratings. *Food Quality and Preference* 15: 853–858.

Popper, R., M. Schraidt, and B.J. Kroll. 2005. When do attribute ratings affect overall liking ratings? In *Second Pangborn Sensory Science Symposium*, University of California, Davis, DA, 7–11 August.

Rothman, L. and M.J. Parker. 2009. *Just-About-Right (JAR) Scales*. West Conshohocken, PA: ASTM International.

Shepherd, R., C.A. Farleigh, D.G. Lang, and J.G. Franklin. 1984. Validity of a relative-to-ideal rating procedure compared with hedonic rating. In *Progress in Flavor Research*, ed. J. Adda, pp. 103–110. Amsterdam, the Netherlands: Elsevier.

van Trijp, H., P. Hunter, F. Mickartz, and L. Kruithof. 2007. The quest for the ideal product. *Food Quality and Preference* 18: 729–740.

Vigneau, E., E.M. Qannari, and P. Courcoux. 2004. Analysis of just-about-right data: Segmentation of the panel of consumers. In *Data Analysis Workshop: Getting the Most Out of Just-About-Right Data. Food Quality and Preference* 15: 897–899.

Worch, T., S. Le, and P. Punter. 2010. How reliable are the consumers? Comparison of sensory profiles from consumers and experts. *Food Quality and Preference* 21: 309–318.

6 Free-Choice Profile Combined with Repertory Grid Method

Amparo Tárrega and Paula Tarancón

CONTENTS

6.1 INTRODUCTION

Free-choice profiling (FCP) is a sensory technique that, as other descriptive techniques (generic descriptive analysis, Flavor Profile®, Quantitative Descriptive Analysis®, and Spectrum method®), can be used to describe a product in terms of their sensory characteristics, that is, appearance, flavor, aroma, or texture (Oreskovich et al. 1991).

The basic principle of FCP is that each assessor uses his/her own list of sensory characteristics to evaluate the products.

FCP was developed in 1981 by Williams and co-workers as a new profile analysis approach suitable for assessors with low degree of training, thus reducing or avoiding the time and the costs of training and performance maintaining of conventional panels. FCP assumes that individuals do not differ in how they perceive the sensory characteristics of the products but do differ in the way they label or express them, and also assumes that these individuals, using its personal vocabulary, are able to consistently score a set of products. FCP was first applied in a study on commercial wines (Williams and Langron 1984) that illustrated the use of the technique and demonstrated that there was no need of using precisely defined sensory descriptors to determine relationships and differences among products. Later, Guy et al. (1989) proposed the use of FCP in consumer research as a technique to obtain direct and spontaneous information about what sensations consumers perceived. Although trained assessors use precisely defined terms of sensory descriptors and the results generally show a high degree of reliability and precision, they do not always reflect what consumers perceive. When consumers are asked to define their own terms, they use more common or easily understandable vocabulary but possibly difficult to be interpreted by other consumers (Piggott et al. 1990). Since, as mentioned previously, FCP allows each assessor the use of her or his individual vocabulary, this technique is fully suitable for obtaining and analyzing data generated by consumers. Guy et al. (1989) applied FCP for consumer profiling of Scotch whisky, showing that FCP was used by consumers in a consistent and meaningful way and allowed to investigate the dimensions used by consumers to discriminate among products and to identify consumer perceptions of certain specific attributes in the product.

FCP has been used to describe sensory differences among commercial or experimental products, including a wide variety of food types: alcoholic beverages (Piggott et al. 1990); meat patties (Beilken et al. 1991); cheeses (Parolari et al. 1994; González Viñas et al. 2001; Bárcenas et al. 2003); orange gels (Costell et al. 1995); cooked hams (Delahunty et al. 1997); almonds (Guerrero et al. 1997); grape jellies (Tang and Heymann 2002); orangebased lemonades (Lachnit et al. 2003); honeys (González Viñas et al. 2003); passion fruit juices (Deliza et al. 2005); dairy desserts (González-Tomás and Costell 2006a,b), orange juices (Pérez-Aparicio et al. 2007); dairy beverages (Arancibia et al. 2011); and biscuits (Tarancón et al. 2013).

Most studies working with FCP, especially those with consumers, include the repertory grid method (RGM) for generating the individual list of terms and descriptors. As stated by McEwan and Thomson (1989), assessors are sometimes unable to describe what they perceive when samples are presented in the isolation of the sensory testing booths. Subjects with lack of sensory experience cannot generate sufficient and suitable terms to fully describe sensory differences (Piggott et al. 1990; Guerrero et al. 1997). The RGM is a set of techniques related to Kelly's personal construct theory that are used to investigate individual constructs (Gains 1994), and it is particularly suited to develop consumer-related vocabulary. It involves triadic or dyadic comparisons of products in one-on-one interviews, in which assessors describe differences and similarities between products using their own words. McEwan et al. (1989) compared the conventional FCP procedure with the combined RGM–FCP procedure when investigating the sensory characteristics of chocolate and showed that perceptual maps of samples obtained from both approaches were very similar, as were the interpretations of the main perceptual dimensions. However, even if RGM is more time-consuming than natural eliciting vocabulary, most researchers in recent studies preferred using RGM probably because it is a more structured procedure that not only facilitates consumers' task, increasing the number of terms elicited, but also assures the researcher that most of the differences among samples are explored and covered.

In the interpretation of FCP data, the statistical technique named generalized Procrustes analysis (GPA) has played a relevant role. This technique, such as analysis of variance or principal component analysis (PCA), can be applied to analyze conventional descriptive data in which all assessors use the same descriptors and results can be averaged. But it can also be applied to data from FCP, in which the number of attributes and their meaning can vary from assessor to assessor, making it impossible to obtain average values. Several studies using FCP for sensory evaluation also include the analysis of the relationships between sensory perception and instrumental data. Different approaches have been used for establishing these relationships. For instance, Williams and Langron (1983) used multiple regressions to relate color measurements and appearance evaluation of wines obtained by FCP. The relationship between sensory and instrumental data sets has also been studied by means of GPA. With this approach, González-Tomás and Costell (2006a) determined the relationship between FCP sensory data and instrumental color (CIELAB system) and rheological measurements of vanilla dairy desserts. Results indicated that, for both color and texture, the perceptual configurations showed good agreement with the corresponding instrumental parameters though the agreement was higher for the color space. It is important to take into account that, since GPA treats identically both data sets (sensory and instrumental), it cannot be used to predict one set from the other one but only to analyze the relationships between data sets (Dijksterhuis 1995). Partial least squares (PLS) regression has also been applied for relating FCP sensory data and instrumental parameters. O'Riordan et al. (1998) successfully applied PLS regression for predicting the perceived differences on cheeses' flavor using aroma release data measured during consumption. More recently, Tarancón et al. (in press) studied the relationship between consumers' perception of biscuits texture and instrumental parameters, by applying PLS regression to the individual data provided by each consumer. Results proved that this approach is useful to gather information about the stimuli responsible for the different sensory characteristics perceived by consumers and to establish instrumental indices for predicting sensory differences among products.

Compared with sensory descriptive procedures that use a consensus vocabulary, FCP has several advantages: An extensive training is not required, vocabulary development is economical in terms of time and effort, and the generated vocabulary comprise more *consumer-recognizable* terms useful for marketing. The main drawbacks are the complexity of the analysis and the interpretation of the sensory vocabulary, which is not always easy due to the large number of individual terms and their nature. Compared with other methods for evaluating differences perceived by consumers like check all that apply (CATA), flash profile, and ultraflash profile combined with Napping® FCP produces intensity values that give more powerful information. Nevertheless, these other methods, such as FCP, also produce products' perceptual maps but in a very shorter time. For that reason, they are preferred to FCP when a rapid access to the relative sensory positioning of products is required.

6.2   EXPERIMENTAL DESIGN AND PROCEDURE

The procedure used for FCP evaluation varies considerably among the different studies. FCP procedure can include the same steps as conventional profile: recruitment, selection and training of assessors, vocabulary generation and sample evaluation. However, in FCP, the training step is considerably shorter than in a conventional profile, or it is simply not included, especially when studying consumer perception.

6.2.1   SELECTION OF ASSESSORS

There is not much consensus regarding the criteria for selecting assessors or their minimum number required for evaluating samples. As stated by Williams and Langron (1984), assessors for FCP must simply be objective, capable of using scales, and able to consistently use the vocabulary developed. In many studies, assessors are selected according to their availability or willingness to participate, their likes and dislikes, or their frequency of consumption of the product under evaluation. However, there are some few studies in which the selection is made on the basis of assessor's performance after completing acuity tests. Regarding the number of assessors, the studies focused on product differences involve from 8 to 24 assessors. For studies focused on consumers' perception, the number of subjects required should be larger. Guy et al. (1989) recruited 100 subjects to evaluate the suitability of FCP to consumers' data. Nevertheless, most of the recent studies were carried out with 25–35 consumers because a larger number implies difficulties, either in interpreting the data or in finding consumers that can attend many sessions.

6.2.2 VOCABULARY GENERATION

Conventional FCP uses a natural eliciting vocabulary procedure in which the samples or a representative set of samples is presented to each assessor, who is asked to describe them. The type of description required varies depending on the study: "to write down any descriptions or associations…," "to describe the samples using all the descriptors needed," "to describe differences among products," or "to choose descriptive terms, under the headings of appearance, texture and flavor, which caused the products to differ." In most cases, the assessors evaluate and describe samples in isolated sensory testing booths. In other cases, personal interviews are performed in which interviewers can help the assessors in the process of generating descriptors. Alternatively and as previously stated, many studies combine FCP with RGM, as it can facilitate the vocabulary development task. The procedure basically consists of individual interviews in which consumers are presented with a series of triads of samples. The number of triads and the samples included in each triad has to be selected to ensure as much as possible that most of the sensory differences among samples are covered. The number of triads is determined according to the number of samples (n) as (n − 1)/2. The usual procedure for selecting the triads is the variation of Kelly's RGM (Kelly 1955) proposed by Thomson and McEwan (1988). Samples are arranged into triads, by firstly selecting three of them at random from the initial pool. The second triad is constructed by randomly selecting one of the samples from triad 1 and two samples from the remaining pool. Triad 3 is constructed by randomly selecting one of the two *new* samples from triad 2, plus two other samples from the remaining pool. The procedure is repeated until all of the samples are included in triads. Different triads and order of triads are made to minimize order effects. Each subject is presented with the first triad that comprised three samples (a–c). The subject is asked in what ways two selected products (a and b) from the triad are similar to each other and different from the third (c). The interviewer records responses as they are elicited, and when no new constructs are forthcoming, the other two combinations (a and c vs. b; b and c vs. a) are presented. This procedure is repeated for the remaining triads. The generated terms are finally listed and the individual score sheets are made. In some cases, assessors are also asked to check their list of descriptive terms, define the meaning of each one, and eliminate those not deemed necessary.

6.2.3 EVALUATION OF SAMPLES

Each assessor, using her/his own scorecard, evaluates in a scale the intensity of the attributes she or he elicited for each of the evaluated samples. Although there is no standard type of scale, most studies use unstructured line scales with lengths varying from 6 to 15 cm, with anchors corresponding to minimum and maximum intensities.

Similarly to conventional profile, sample presentation and testing should be carried out following standard procedures and under controlled conditions (lighting, temperature, humidity, odors, and rounds) to avoid possible biases. Products are presented one at a time to the assessors, following a balanced (e.g., Williams Latin square) or randomized design. To avoid fatigue, the number of samples evaluated per session is limited to four to six, depending on the product category under evaluation. If the number of samples is large, pauses within the evaluation can be considered.

6.3 DATA ANALYSIS

As mentioned, GPA (Gower 1975) is a multivariate technique used for analyzing different types of sensory data, but its main application has been for FCP. GPA allows matching the different configurations obtained from the assessors, by transforming the individual spaces in three steps (translation, scaling, and rotation or reflection) and averaging them to produce a common or average space. GPA then eliminates the three sources of variation generated by the assessors: the level of scoring, the idiosyncratic use of descriptors, and the range of scoring (Arnold and Williams 1986). The result is a consensus (across subjects) representation of the products in a multidimensional space that can be interpreted at the individual level as it includes the descriptors of each assessor. For GPA, analysis of FCP data is arranged in K (one per assessor) matrices including each one the evaluations given by the assessor to M attributes (usually columns) in N products (usually rows). If replicate judgments of samples exist, supplementary rows are added to the matrices. Data corresponding to each assessor (each one of the K matrices) can be viewed as a configuration of N points (samples) in a space of M (attributes) dimensions. Figure 6.1a shows, as a theoretical example, the configurations corresponding to two assessors (K = 2) evaluating two attributes (M = 2) for three samples (N = 3). Observe that a representation of different configurations in the same figure makes full sense only when both assessors evaluated the same attributes, as in the case of

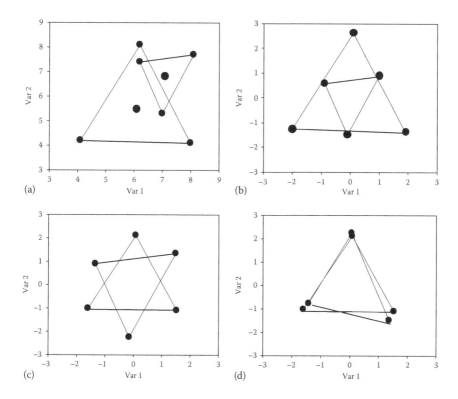

FIGURE 6.1 GPA transformation steps. Example for theoretical configurations of two assessors evaluating three samples: (a) initial configurations, (b) translation, (c) scaling, and (d) rotation.

data from a conventional profile. With FCP data, the first dimension can be, for instance, sweetness for one assessor and odor intensity for the other one, and thus, only the type of scale coincides between assessors. By using geometrical transformations, GPA adjusts the configurations and matches them to each other as closely as possible. The first step is a translation to superimpose assessors' configurations in a common center. This step that remains valid even when dimensions represent different attributes from assessor to assessor (Dijksterhuis and Gower 1991/1992). In our example, Figure 6.1b illustrates this step, which removes the variations due to the use of different parts of the scales by the assessors. The second step is an isotropic scaling where configurations are shrunk or stretched to obtain new configurations that are about the same size but keeping constant the relative distance between samples (Figure 6.1c). In this step, the effects of using different ranges of scales by the assessors are removed. Finally, the third step consists of rotations to match the configurations as much as possible (Figure 6.1d). After these steps, a common configuration is computed as the mean of all transformed individual configurations. GPA, through a series of iterative steps, repeats the steps until the distance between the transformed configurations and the mean (consensus) configuration is minimized.

The final consensus configuration obtained expands in a group average space with the same number of dimensions (M) than the original space. To facilitate representations, PCA is usually applied to select a lower number of dimensions, explaining most of variability. This final space presents an orientation generally different from that of the original space, making it necessary to reorient the individual final configurations obtained by GPA to facilitate their comparisons with the group average (Dijksterhuis 1996).

An analysis of the variance applied to the distances between samples before and after transformations allows estimating the decrease of the variability involved with the steps of rescaling, rotation, and translation. This analysis (Procrustes analysis of variance [PANOVA]) is included in the programs running GPA that also give as output the scaling factor applied to the sample configuration of each assessor during the matching process. A scaling factor higher than 1 indicates that the configuration has been stretched to match the consensus, and a scaling factor lower than 1 indicates that the configuration has been shrunk.

From the multidimensional plot of the consensus configuration, the analysis of variance allows to determine the percentage of variability explained by each dimension and the residual variances of samples and assessors. The number of PCA dimensions retained for analyzing and plotting the results is decided according to the amount of variance explained. The first dimensions representing a sufficient percentage of the total variability are retained while dimensions explaining <10% are usually discarded. In practice, two to six dimensions are considered adequate to represent the samples. In the consensus sample space, samples are represented by their coordinates in the selected dimensions, and according to their position, differences and similarities among samples can be established. Since these dimensions are linear combinations of all the terms/attributes used by the assessors, each dimension can be interpreted through its correlation coefficients with the attributes. In general, the higher the coefficient (negative or positive), the higher the influence that the attribute has in explaining the separation of the samples along the dimension. However, the critical value of the correlation coefficient for an attribute being considered as significant has differed among studies: 0.5 (Delahunty et al. 1997), 0.7 (Guerrero et al. 1997; Michel et al. 2011; Tarancón et al. 2013), and 0.8 (González-Tomás and Costell 2006b; Arancibia et al. 2011). Samples' residuals (squared distances from the position of the sample in the consensus configuration to the position of the sample on each assessor configuration) allow identifying the samples represented farther away from the consensus configuration, for which GPA has been less efficient. If all samples' residuals have similar values, the agreement among consumers about the position of a sample is similar for all samples. When a sample shows a higher residual variance than the rest, more disagreement among assessors exists about the position of that sample compared to the rest.

On the other hand, assessor's residuals (squared distances between the configuration of a particular assessor and the consensus configuration) indicate the agreement of a consumer with the rest of the group. Lack of fit of an assessor involves a large residual.

There are several programs available that can be used to perform GPA. Initial works used macros in GENSTAT and SAS language or PROCRUSTES—PC software (OP&P, Utrecht, the Netherlands). Nowadays, XLSTAT (Addinsoft, Paris, France), Senstools (OP&P, Utrecht, the Netherlands), and R are the most commonly used programs to run GPA. It should be also noticed that although GPA is the most used, other statistical procedures like multiple factor analysis (MFA, Escofier and Pages 1990) or multiblock analysis as STATIS or DISTATIS (Abdi et al. 2012) can also be used to analyze FCP data.

6.4 CASE STUDY: SENSORY PROFILE OF COMMERCIAL CHOCOLATE SOY DRINKS

6.4.1 Products and Assessors

Six commercial chocolate soy drinks from the Spanish market were studied. Ten assessors from Valencia (Spain) that declared to regularly consume soy drinks (at least once a week) participated in the study.

6.4.2 Vocabulary Generation: Repertory Grid Method

RGM was used to generate the vocabulary to describe the samples. In individual interviews, consumers were asked to describe the similarities and differences among the six samples concerning their texture and flavor.

In this case, the six samples were presented in four triads. From the six samples, three triads were initially obtained according to Thomson and McEwan (1988). Triad 1 comprised three samples from the initial pool of samples. Triad 2 was formed by combining one sample from the first triad with two from the remaining pool and the third triad was similarly defined. In this case, a fourth triad was added for assuring that most differences among samples were tested. Triad 4 comprised those samples that appeared only once in the previous three triads. For avoiding order effects, different triads and presentation orders of the triads were used (examples of four triads of samples for one assessor are ABC, CDE, EFA, and BDF). The first triad was presented to each assessor, who was asked to describe—using his or her own terms—how two of the samples (A and B) were alike and different from the third (C). This procedure was repeated for all the possible combinations within the triad. Once all possible descriptors were elicited from that triad, the assessor was presented with the next triad and the process was repeated until completing the four triads. The generated terms were listed by the interviewer on individual score sheets.

6.4.3   EVALUATION OF SAMPLES: FREE-CHOICE PROFILING

To evaluate the sensory characteristics of the six chocolate soya drinks, each assessor evaluated the samples with their individual score sheets. They rated the intensity of each attribute using a 10 cm unstructured line scale anchored with "not perceived" and "intense." The assessors evaluated the six samples in a unique session in a standardized test room with separate booths. Samples were presented in transparent glasses coded with three-digit random numbers and according to a balanced design to avoid serving order effects.

The data set is available for download from the CRC Web site: http://www.crcpress.com/product/isbn/9781466566293.

6.4.4   DATA ANALYSIS

GPA was applied to the FCP data using Senstools Version 3.3. PANOVA was used to assess the differences between individual and consensus configuration, as well as between samples.

6.4.5   RESULTS

6.4.5.1   Descriptors

A total of 82 descriptors (57 for flavor and 25 for texture) were generated by the assessors to describe differences among the chocolate soya beverages. The number of descriptors generated by the assessors ranged from 6 to 11.

6.4.5.2   Assessors' Scaling Factors, Distribution Plot, and Residual Variances

Evaluation of samples by FCP provided 10 individual matrices. Each matrix included the intensity scores for the six samples (rows) for the number of attributes generated by each assessor (columns). GPA was applied to these data to obtain the consensus/average configuration. The individual scaling factors that indicate to what extent the initial configurations of the assessors have been stretched or shrunk are indicated in Table 6.1. Assessor 3 had the greater scaling factor (1.29) indicating that the configuration has

TABLE 6.1
Scaling Factors for Each
Assessor Obtained with GPA

Assessor	Scaling Factor
1	1.04
2	1.12
3	1.29
4	0.68
5	1.12
6	1.25
7	0.78
8	1.12
9	1.15
10	0.98

been stretched because he or she used a narrower range of the scale, whereas assessor 4 had the lowest scaling factor (0.68) indicating that the configuration has been shrunk because he or she used a much wider range of the scale.

The assessor's plot obtained from principal coordinate analysis of the matrix of distances between each assessor's sample configuration is also provided. This plot (Figure 6.2) shows the distribution of assessors according to their similarity or difference in the assessment of samples. In this case, assessors were broadly distributed along the first dimension rather

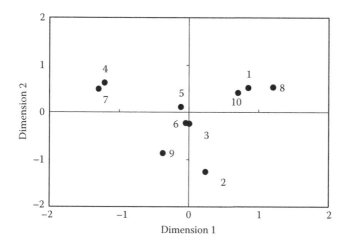

FIGURE 6.2 Assessor's plot obtained from the analysis of distances between configurations.

TABLE 6.2
Residual and Total
Variance of Assessors

Assessor	Residual Variance	Total Variance
1	1.21	10.5
2	2.14	9.15
3	0.95	10.88
4	2.01	9.34
5	1.55	10
6	1.07	10.71
7	2.01	9.34
8	2.2	9.06
9	1.31	10.36
10	1.1	10.66

than in the second. Assessors 4 and 7 lay close and separated from the rest, indicating that their configurations coincided and were different from the rest. Thus, these two consumers coincided between them in how they perceived samples but not with the rest, and they could be considered outliers. However, as stated by McEwan and Hallett (1990), differences in the assessors' positions do not necessarily lead to significant differences between them. Complementary information, such as the percentage variation explained or residual variance, should also be considered. In Table 6.2, residual variance values of assessors are listed. Assessors 4 and 7 had a high residual variance (2.0) in the group, although not very different from other assessors' residuals (assessors 2 and 8, with residuals of 2.2). In this case, since the number of assessors is low, all the assessors have been retained. In other studies, in the same situation, the final decision could have been taken by comparing the results obtained with and without considering the *outliers*. If, after the elimination of the data of those consumers, GPA provides a different sample average space, the assessors can be considered outliers and can be removed before performing data analysis. Also, if there are several groups of assessors clearly differentiated in distribution plot, assessors clustering should be considered and perceptual maps can be obtained for the different groups (González-Tomás and Costell 2006b).

6.4.5.3   GPA Consensus Space

The distribution of explained variance among the dimensions of the consensus configuration is shown in Table 6.3. The first three dimensions

---

TABLE 6.3

Distribution of Explained Variance among the Dimensions of the Consensus Configuration

Dimension	Real	Residual	Total
1	42.25	5.38	47.63
2	22.44	3.86	26.3
3	10.39	3.45	13.84
4	6.07	1.73	7.81
5	3.3	1.13	4.43
Total	84.45	15.55	100

---

explained 87.8% of total variance, and further dimensions that explained less than 10% of variance were therefore not considered in this study. The consensus configuration of samples is represented in two biplots (dimensions 1–2 and dimensions 1–3) that show the six samples distributed according to differences and similarities in flavor and texture (Figure 6.3). Results of PANOVA also give information for interpreting differences among samples. The distribution of the total variance among samples is shown in Figure 6.4. A large amount of variance can be attributed to samples B and C, which consequently should appear located in the extreme of the samples' space, whereas a very low amount of variance attributed to sample A indicates that it would remain in the center of the sample space. The distribution of the residual variance among samples can also be observed in Figure 6.4. The best agreement among consumers was found for sample D with the lowest value of residual variance (1.47), while for sample E, the residual was maximum (3.92), indicating more disagreement among assessors in the position of this sample in the consensus space. The distribution of samples' variance among dimensions can be useful when interpreting the GPA consensus plot to determine differences in sensory characteristics among samples (Figure 6.5). For samples A, B, and C, most of the variability is mainly explained by the first dimension, indicating that these samples are mainly characterized by the attributes constituting dimension 1. In the case of samples D and F, the variability is mainly explained by the second dimension, and for sample E, dimension 3 explained most of its variability.

Finally, for interpreting samples' space, it is necessary to know which attributes are related to each dimension. For that, correlations of each one of the attributes with each dimension of the average space are studied. For each dimension, descriptors with correlation coefficients >0.60

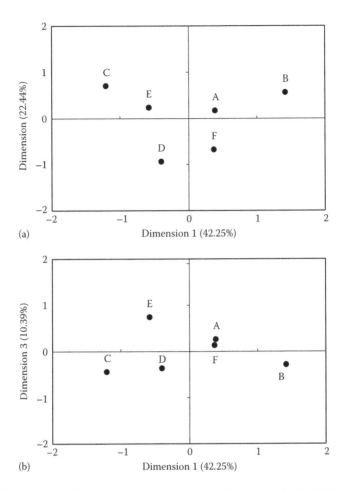

(a)

(b)

FIGURE 6.3   Sample average space (a) for the first two principal dimensions and (b) for the first and third dimensions.

were considered and are shown in Table 6.4. Considering the attributes related to dimension 1, the most important sensory differences were observed between sample B, which was perceived watery and with chocolate flavor, and sample C, which was described as thick and with artificial, perfume, vegetal, or sweetener flavor. In this case, the different terms used for describing flavor that characterized sample C appeared correlated, which may indicate that these terms were used by different assessors to describe the same sensation. Dimension 2 separated samples D and F in the bottom side of the space. These samples were considered thicker than the rest, with a "milk texture" and with chocolate, roasted, milk, or bean flavors. In this case, assessors agreed in the use of the descriptor "thick" for describing the texture of these samples, while

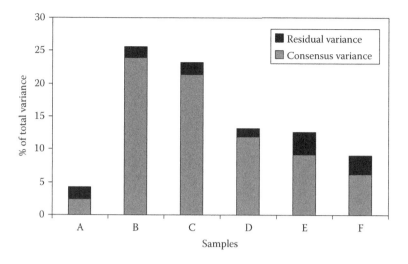

FIGURE 6.4   Distribution of consensus and residual variances among samples.

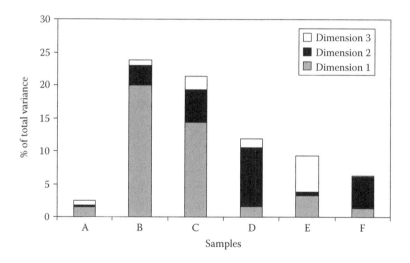

FIGURE 6.5   Distribution of the real variance explained by each sample among dimensions.

flavor was described again with different terms (chocolate, roasted, or bean) that could reflect the same sensation. In dimension 3, sample E was separated from samples C and D because it was perceived as having black chocolate and anise flavor, while artificial flavor and after taste were less intense than in samples C and D.

TABLE 6.4
Descriptors Correlated with Each of the Three Dimensions of the Average Space Generated by GPA (Correlations Higher than 0.6)

Dimension	Correlation (+/–)	Descriptors
1	Positive	Watery texture (4), chocolate flavor (2)
	Negative	Artificial flavor (6), sweetener flavor (3), perfume flavor, thickness (4), sweetness (3), vegetal flavor (2)
2	Positive	Watery texture (4), persistent flavor (2)
	Negative	Thickness (8), milk texture (3), chocolate flavor (3), roasted flavor (2), bean flavor (2), milk flavor (2)
3	Positive	Black chocolate flavor (2), anise flavor
	Negative	Artificial flavor (2), after taste (2), off flavor, and perfume flavor

Note: In parentheses, the number of times that a descriptor had a correlation greater than 0.60.

ACKNOWLEDGMENT

The authors wish to thank Luis Izquierdo for his valuable suggestions and comments on the manuscript.

REFERENCES

Abdi, H., Williams, L.J., Valentin, D., and Bennani-Dosse, M. 2012. STATIS and DISTATIS: Optimum multi-table principal component analysis and three way metric multidimensional scaling. *Computational Statistics* 4: 124–167.
Arancibia, C., Costell, E., and Bayarri, S. 2011. Fat replacers in low-fat carboxy-methyl cellulose dairy beverages: Color, rheology, and consumer perception. *Journal of Dairy Science* 94: 2245–2258.
Arnold, G.M. and Williams, A.A. 1986. The use of generalized Procrustes technique in sensory analysis. In *Statistical Procedures in Food Research*, ed. J.R. Piggott, pp. 233–255. London, U.K.: Elsevier Applied Science.
Bárcenas, P., Pérez Elortondo, F.J., and Albisu, M. 2003. Comparison of free choice profiling, direct similarity measurements and hedonic data for ewe's milk cheeses sensory evaluation. *International Dairy Journal* 13: 67–77.
Beilken, S.L., Eadie, L.M., Griffiths, I., Jones, P.N., and Harris, P.V. 1991. Assessment of the sensory characteristics of meat patties. *Journal of Food Science* 56: 1470–1475.
Costell, E., Trujillo, C., Damasio, M.H., and Duran, L. 1995. Texture of sweet orange gels by free-choice profiling. *Journal of Sensory Studies* 10: 163–179.

Delahunty, C.M., Mccord, A., O'neill, E.E.O., and Morrissey, P.A. 1997. Sensory characterization of cooked hams by untrained consumers using free choice profiling. *Food Quality and Preference* 8: 381–388.

Deliza, R., Macfie, H., and Hedderley, D. 2005. The consumer sensory perception of passion-fruit juice using free-choice profiling. *Journal of Sensory Studies* 20: 17–27.

Dijksterhuis, G. 1995. Multivariate data analysis in sensory and consumer science: An overview of developments. *Trends in Food Science and Technology* 6: 206–211.

Dijksterhuis, G.B. 1996. Procrustes analysis in sensory research. In *Multivariate Analysis of Data in Sensory Science*, eds. T. Næs and E. Risvik, pp. 185–220. Amsterdam, the Netherlands: Elsevier Science.

Dijksterhuis, G.B. and Gower, J.C. 1991/1992. The interpretation of generalized procrustes analysis and allied methods. *Food Quality and Preference* 3: 67–87.

Escofier, B. and Pages, J. 1990. Multiple factor analysis. *Computational Statistics & Data Analysis* 18: 121–140.

Gains, N. 1994. The repertory grid approach. In *Measurement of Food Preferences*, eds. H.J.H. Macfie and D.M.H. Thomson, pp. 51–76. Glasgow, Scotland: Blackie Academic & Professional.

González-Tomás, L. and Costell, E. 2006a. Relation between consumers' perceptions of color and texture of dairy desserts and instrumental measurements using a generalized Procrustes analysis. *Journal of Dairy Science* 89: 4511–4519.

González-Tomás, L. and Costell, E. 2006b. Sensory evaluation of vanilla dairy desserts by free choice profile. *Journal of Sensory Studies* 21: 20–33.

González Viñas, M.A., Garrico, N., and Wittig De Penna, E. 2001. Free choice profiling of Chilean goat cheese. *Journal of Sensory Studies* 16: 239–248.

González Viñas, M.A., Moya, A., and Cabezudo, M.D. 2003. Description of the sensory characteristics of Spanish unifloral honeys by free-choice profiling. *Journal of Sensory Studies* 18: 103–113.

Gower, J.C. 1975. Generalized Procrustes analysis. *Psychometrika* 4: 33–50.

Guerrero, L., Gou, P., and Arnau, J. 1997. Descriptive analysis of toasted almonds: A comparison between expert and semi-trained assessors. *Journal of Sensory Studies* 12: 39–54.

Guy, C., Piggott, J.R., and Marie, S. 1989. Consumer profiling of Scotch whisky. *Food Quality and Preference* 1: 69–73.

Kelly, G.A. 1955. *The Psychology of Personal Constructs*. New York: Norton.

Lachnit, M., Busch-Stockfisch, M., Kunert, J., and Krahl, T. 2003. Suitability of free choice profiling for assessment of orange-based carbonated soft-drinks. *Food Quality and Preference* 14: 257–263.

McEwan, J. and Hallett, E.M. 1990. *A Guide to the Use and Interpretation of Generalised Procrustes Analysis: Technical Manual No. 30*, 76pp. Campden, U.K.: Campden & Chorleywood Research Association Group.

McEwan, J.A., Colwill, J.S., and Thomson, D.M.H. 1989. The application of two free-choice profile methods to investigate the sensory characteristics of chocolate. *Journal of Sensory Studies* 3: 271–286.

McEwan, J.A. and Thomson, D.M.H. 1989. The repertory grid method and preference mapping in market research: A case study on chocolate. *Food Quality and Preference* 2: 59–68.

Michel, L.M., Punter, P.H., and Wismer, W.V. 2011. Perceptual attributes of poultry and other meat products: A repertory grid application. *Meat Science* 87: 349–355.

Oreskovich, D.C., Klein, B.P., and Sutherland, J.W. 1991. Procrustes analysis and its application to free choice and other sensory profiling. In *Sensory Science Theory and Application in Food*, eds. H.T. Lawless and B.P. Klein, pp. 353–394. New York: Marcel Dekker.

O'Riordan, P.J., Delahunty, C.M., Sheehan, E.M., and Morrissey, P.A. 1998. Comparisons of volatile compounds released during consumption of a complex food by different assessors with expressions of perceive flavour determined by free-choice profiling. *Journal of Sensory Studies* 13: 435–459.

Parolari, G., Virgili, R., Panari, G., and Zannoni, M. 1994. Development of a vocabulary of terms for sensory evaluation of Parmigiano Reggiano cheese by free choice profiling. *Italian Journal of Food Science* 3: 317–324.

Pérez-Aparicio, J., Toledano-Medina, M.A., and Lafuente-Rosales, V. 2007. Descriptive sensory analysis in different classes of orange juice by a robust free-choice profile method. *Analytica Chimica Acta* 595: 238–247.

Piggott, J.R., Sheen, M.R., and Apostolidou, S.G. 1990. Consumers' perceptions of whiskies and other alcoholic beverages. *Food Quality and Preference* 2: 177–185.

Tang, C. and Heymann, H. 2002. Multidimensional sorting, similarity scaling and free-choice profiling of grape jellies. *Journal of Sensory Studies* 17: 493–509.

Tarancón, P., Fiszman, S.M., Salvador, A., and Tárrega, A. 2013. Formulating biscuits with healthier fats. Consumer profiling of textural and flavour sensations during consumption. *Food Research International* 53: 134–140.

Tarancón, P., Sanz, T., Salvador, A., and Tárrega, A. In press. Effect of fat on mechanical and acoustical properties of biscuits related to texture properties perceived by consumers. *Food and Bioprocess Technology* DOI: 10.1007/s11947-013-1155-z.

Thomson, D.M.H. and McEwan, J.A. 1988. An application of the repertory grid method to investigate consumer perceptions of foods. *Appetite* 10: 181–193.

Williams, A.A., Baines, C.B., Langron, S.P., and Collins, A.J. 1981. Evaluating tasters' performance in the profiling of foods and beverages. In *Flavour'81*, pp. 83–92. Berlin, Germany: Walter de Gruyter.

Williams, A.A. and Langron, S.P. 1983. A new approach to sensory profile analysis. In *Flavour of Distilled Beverages: Origins & Development*, ed. J.R. Piggott, pp. 219–224, Chichester, U.K.: Ellis Horwood Ltd.

Williams, A.A. and Langron, S.P. 1984. The use of free-choice profiling for the evaluation of commercial ports. *Journal of the Science of Food and Agriculture* 35: 558–568.

7 Flash Profile

Julien Delarue

CONTENTS

7.1 ORIGINS AND CONTEXT

7.1.1 LIMITATIONS OF CONVENTIONAL APPROACHES

Conventional descriptive approaches are widely used and are indisputable references for sensory profiling. However, they come up against various limitations that may sometimes discourage or disappoint their users. Some of these limitations are intrinsic to descriptive analysis. They are indeed *low-level* limitations in that they relate to the difficulty of measuring perception, especially when subjects actually differ in their perceptions. In such cases, consensus-based methods may not be relevant.

Other limitations are of a practical nature. They derive from constraints such as the time and the resources needed to set up and manage a fully operational descriptive panel. Properly setting up a descriptive analysis study and training a panel indeed takes time (at least several weeks and up to several months). Unfortunately, when resources are limited, this would prevent the use of sensory descriptive analysis, even though there is a real need for sensory information. Such situations occur, for example, when a company cannot afford maintaining a panel dedicated to every product category they make. Also, in the frame of research or R&D projects, the products of interest are frequently prototypes that are made selectively and could not be produced for the sole sake of training a panel. This is a major constraint for running a descriptive analysis, especially for short-shelf-life products. In addition to this, recognized professional experts such as oenologists, flavorists, developers, and perfumers are usually reluctant to use conventional descriptive methods. The main reason for this, beside time constraints, is the restrictive nature of consensus-based descriptive techniques that limit the use of attributes to a *lowest-common denominator*.

7.1.2 Move toward Faster Methods

These considerations have led sensory scientists to develop faster methods that are more adapted to such situations where conventional approaches were limited. The flash profile is in line with this trend toward faster and more flexible methods. It was imagined by Sieffermann (2000) as a combination of free-choice profiling (FCP), in which the subjects are not imposed on the use of a common vocabulary, and comparative assessment of the product set. Although FCP was developed about 15 years earlier, the real novelty of flash profile lies in the fact that it was designed as a *one-shot* measurement method with emphasis on the relative sensory positioning of the products evaluated. Even though the subjects evaluate the products on separate attributes, more attention is then paid to the relative positioning of the objects rather than to product scores on the separate attributes. As noticed by Moskowitz (2003), this tends to create a more holistic picture of response to the stimuli. Interestingly, other methods such as projective mapping/napping have pushed the logic even further by directly asking the subjects to position products according to their sensory similarities. But contrary to direct nonverbal methods such as free sorting, or projective mapping/napping, the flash profile does primarily rely on quantitative description. In this perspective, attributes become only a mean to collect sensory data. Their definition and precise meaning is no longer central. The downside is that the interpretation of sensory dimensions is less straightforward since the experimenter may not rely on the definition of the attributes.

7.2 PRINCIPLE

7.2.1 Way to Rapidly Position Products According to Their Major Sensory Differences

The flash profile aims at providing a quick access to the relative sensory positioning of a set of products. The assessment task simply combines two well-known procedures:

1. FCP (Williams and Langron 1984)
2. Comparative evaluation with an immediate ranking

This method thus consists in asking the assessors to use their own descriptive terms in order to rank the evaluated products for each of these terms.

7.2.2   WAY TO OVERCOME LIMITATIONS DUE TO THE
NEED OF AN EXTENSIVE TRAINING

First, with this methodology, there is no need for seeking a consensus vocabulary since an FCP procedure is applied. In order to make the flash profile even faster, it was proposed not to train panelists specifically for the evaluation of the product set under consideration but to use experienced subjects instead. Besides, presenting the whole product set simultaneously allows a direct comparative evaluation, which makes the quantification easier. The fact that assessors have a simultaneous access to the whole sample set indeed forces them to focus on the differences they perceive and lead them to generate discriminant attributes only. The usual phases of familiarization with the product space, attribute generation, and rating have thus been integrated into a single step.

7.2.3   TO ADOPT A WORK PATTERN THAT IS COMPATIBLE
WITH STANDARD PROFESSIONAL CONSTRAINTS

Instead of multiple tasting sessions that usually come along with constraining schedules in conventional descriptive analysis, the product evaluation session in flash profile typically consists of a single session of 40 min to 2 h length. The duration of the flash profile session may of course vary depending on the number and the type of products. At first look, such long sessions may seem discouraging. However, in practice, they fit much better to standard professional work rhythm. Repetition is in fact much more prohibitive. In addition to this, only individual evaluations are needed (no group session). This opens ways for a flexible organization based on individual appointments. All these make the flash profile particularly well adapted for working with internal panelists.

7.2.4   CONSUMER-ORIENTED METHOD

Although flash profile was initially designed to be conducted with experienced panelists, it can also be seen as a consumer-oriented method. Conducting descriptive analysis with consumers is extremely appealing as it is a potential way to assess consumers' diversity in their perception of products (Jack and Piggott 1992; Faye et al. 2006; Thamke et al. 2009). A major limitation to performing descriptive analysis with consumers has long been the use of a common vocabulary and the time needed to train the panelists and to align concepts accordingly. Rapid techniques such as the flash profile overcome these difficulties. Even though the ability of "untrained consumers" to give reliable descriptive results is often questioned, it seems to be largely underestimated (Husson et al. 2001;

Worch et al. 2010). A number of researchers have thus started using flash profile as a way to capture consumers' perception.

7.3 FLASH PROFILE IN PRACTICE

7.3.1 TYPICAL SESSION

Flash profile sessions are individual and usually take place in sensory booths. The panelists are presented with the whole set of products after the experimenter has briefly explained the principle of the method.

7.3.1.1 Instructions to the Panelists

Instructions are very simple. The panelists are asked to observe, to manipulate, and (in case of food products) to taste the samples in order to describe them. The panelists are told to use any nonhedonic attributes they would consider appropriate to describe the samples, providing that they are sufficiently discriminative to allow a ranking of the samples (or even a partial ranking, since ties are allowed). They are asked to describe all the differences they perceive and the number of attributes is generally left open. In some cases, the experimenter may decide to restrain the study to some aspects only (e.g., texture or flavor) and instructions are thus adapted accordingly.

The panelists must then immediately rank all the samples for each elicited attribute, from the weakest to the strongest sensation, with ties allowed. Alternatively, the panelist may choose anchors that he or she finds appropriate (e.g., from *yellow* to *brown*). A printed example is usually given as represented in Figure 7.1. Care must then be taken to choose an example that is not directly related to the product set under investigation in order not to influence the description.

The panelists are free to compare and to retaste the samples as many times as necessary to perform the task. The ranks for each attribute are to be reported on a form (Figure 7.2) that can be either printed or presented on a digital screen (computer or portable device). When working with paper and pencil, the panelists are asked to make sure that for each attribute, they have reported the codes of all samples and that no sample is reported twice.

We generally recommend that these two steps (attribute elicitation and ranking) are performed conjointly. However, some panelists find it more convenient to first list all the attributes they wish to evaluate and then to rank the samples.

7.3.1.2 Attribute Elicitation

Eliciting attributes is a key stage in descriptive techniques that rely on free choice of attributes. Contrary to consensus-based techniques where attributes

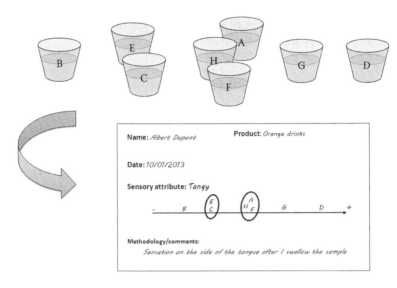

FIGURE 7.1   Example of the physical ranking of the samples along one sensory attribute and of its transcription onto a form.

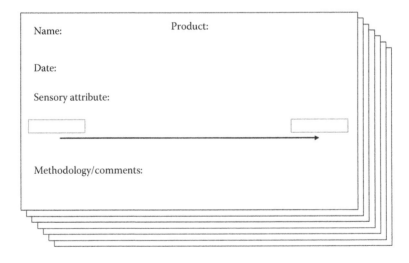

FIGURE 7.2   A series of blank forms is usually given to the assessors in order to report their rankings for each attribute they have used. Paper forms are simple and convenient but computers/digital screens may also be used.

are taught to the panelists (e.g., Spectrum method) or elaborated with the panel thanks to several dedicated sessions (e.g., quantitative descriptive analysis), the panelists here are left to themselves. In the flash profile procedure, the subjects are told to focus on the differences between the products. This implies that the subjects directly prioritize on major differences between the products. The number of attributes evaluated by a panelist in a flash profile is thus highly dependent on the actual differences between the products. This point widely differs from attribute elicitation stages of QDA for instance, where panelists are asked to cite all possible descriptive terms that would apply to the products. This methodological point certainly makes the attribute elicitation step more efficient (Dairou and Sieffermann 2002) and may explain a higher discrimination ability sometimes observed when compared to conventional profile (Delarue and Sieffermann 2004a).

Although there is no general rule regarding the number of attributes that are elicited by each panelist, in most studies that we have conducted, this number typically varied between 4 and 15 attributes depending on the product set and on the panelist.

In our experience with flash profile, attribute elicitation is usually not a problem when working with a panel of product experts or with experienced subjects. Such subjects are indeed used to describe their sensations using descriptive terms. Flavor description notwithstanding might be sometimes difficult, even with such subjects. It is therefore possible to provide help in the form of preexisting lists of attributes, from the literature (e.g., the wine aroma wheel) or from previous studies. These lists may indeed help the panelists to find words that match their sensations.

When working with consumers, the perspective is rather different. Depending on the objectives of the study, one might want to let the participants use their own terms without any suggestions. This may result in the elicitation of rather obvious attributes, although in past studies, the vocabulary used by consumers was found to be rich and diverse (Veinand et al. 2011). Providing a list may facilitate description (as in the check-all-that-apply technique), but it may also bias consumers' response. To my knowledge, these two options have not been formally compared yet. However, the choice really has to be made with respect to the goals of the study.

7.3.1.3 Preliminary Session

The initial flash profile procedure (Sieffermann 2000; Dairou and Sieffermann 2002) included a preliminary session during which the panelists are asked to individually generate descriptive terms based on a first evaluation of the product set. Following this session, the experimenter collects all the terms generated by the panelists and gathers them into a common list. Terms on this list are organized according to their corresponding sensory modality (aspect, visuotactile attributes, smell, texture in mouth, taste, flavor/aroma, other attributes [trigeminal sensations, lingering, etc.]) and sorted alphabetically. The compiled list is then provided to the panelists during the main evaluation session, together with their own initial list, in order to help them in finding attributes they might have either neglected or named with difficulty.

In our experience with flash profile, however, we observed that after the preliminary session, panelists tend to stick to their initial list, with very few exceptions.

These exceptions occurred mostly in the case of flavor description. Also, the list once led panelists to pay attention to a sound (perceived when consuming the product) that turned out to be very important for consumers. The preliminary session might thus be useful, notably when the differences to be described are subtle. Nevertheless, we now most often skip this step, considering that its added value is relatively limited. In addition to this, applying two sessions considerably affects the quickness and flexibility of the method.

7.3.1.4 Evaluation Task

As described earlier, the panelists must rank the samples for each attribute, with ties allowed. It is more intuitive and also more convenient for panelists to first physically rank the samples and then to report the corresponding codes on a form rather than directly ranking codes on the score sheet or on a computer screen. Manipulating the samples indeed allows the panelists performing more efficient comparisons because they can progressively adjust their rankings and they can work sequentially if desired.

The participants perform the task at their own pace. The evaluation sessions typically last between half an hour and an hour and a half depending on the subjects. This duration may however vary depending on the number of samples and the type of products that are evaluated.

Because the evaluation session is relatively long, the participants are free to take a break at any moment. If they do so, they are advised to take breaks between two attributes. Breaks help prevent sensory fatigue that is more likely to occur in comparative tasks. This is especially important when working with persistent products or product with strong characteristics.

Besides, water is usually provided, and in the case of the evaluation of food products, the panelists are asked to rinse their mouth. Additional cleansers (bread, crackers) may be provided depending on the product type. Unlike the methods based on monadic sequential evaluation, the mouth-rinsing procedure can hardly be controlled here. Although guidelines may be given to the panelists, the method essentially relies on their expertise and they are generally free to work the way they find most appropriate.

7.3.2 Samples

As explained, the flash profile method relies on the simultaneous presentation of the product set to the panelists who thus evaluate the samples comparatively. This has important practical consequences, notably in terms of sample preparation.

The first and most important consequence is that all the products must be available at the same time. If some products are not available together with the rest of the product set, they just cannot be included in the test. The experimenter and the stakeholder must thus pay attention to production constraints, availability on the market, etc. This also requires that all the products are presented in the same conditions of age (for short-shelf-life products), quantity, temperature, etc. For some products like hot beverages or ice creams, presenting all the samples simultaneously at the same temperature can be quite challenging. For instance, it is important to make sure that the samples remain stable while the subjects perform the evaluation task. Notwithstanding this difficulty, Albert et al. (2011) have also argued that a comparative assessment could actually represent an advantage. As they point out, slight temperature differences in a sample set would indeed not affect much the consistency of an assessor's description since it remains comparative. On the contrary, in QDA, a variation in a couple of degrees in temperature could cause a significant change in the ratings of a series of attributes (e.g., texture, flavor), which would ultimately change the final averages or the panel performance-related parameters.

In terms of quantity, the experimenter must provide as much product as necessary to perform rankings on several attributes. Additional samples or refill should thus be provided upon panelist' request. For products that may not be stable during the course of the session, it is probably better to renew the whole set of samples at once.

Naturally, the samples must be coded and when possible presented blind, as in any sensory evaluation. However, instead of three-digit random numbers that are customary in sensory analysis, it is better to use a simpler coding system (e.g., letters) in order to prevent mistakes and omissions when reporting the ranks on the score sheet.

It is also important that the subjects can manipulate the samples, change their positions in the booth, and yet always identify their codes. In the case of food products, the use of individual coded containers (glass, cups, plates, vials, etc., depending on the product type) is thus necessary. This will allow panelists to materially rank the samples in front of them.

7.3.3 PANELISTS

7.3.3.1 Number of Panelists

Although there is no absolute rule regarding the number of panelists that should participate to a flash profile, we generally consider four or five to be a minimum. This allows yielding relatively stable and complementary sensory information about the products. Naturally, recruiting more panelists is worthwhile but not strictly necessary.

In order to better define the number of panelists to be recruited for a study, it is important to understand the logic behind flash profiling in relation with the objectives of the study. In conventional profiling techniques, working with several assessors allows stabilizing the panel's outcome. In this perspective, resampling studies could help defining the adequate number of panelists (Monrozier and Danzart 2001; Cadoret and Husson 2013). This logic accounts for interindividual differences in sensitivity and performances. This also gives power to statistical tests that are applied to sensory data. For this purpose, panels of 10–12 subjects are recommended (King et al. 1995; Gacula and Rutenbeck 2006; Mammasse and Schlich 2010) even though these numbers vary from one product category to the other. In practice however, a larger number of panelists is usually trained for panel maintenance reasons and in order to ensure a minimal number of responses for each evaluation session.

In case of flash profile, the logic behind recruiting several panelists is slightly different. The panelists' responses are indeed seen as complementary measures instead of repeated measures, as it is (nearly) the case in conventional descriptive analysis. Expressly, the use of an FCP procedure accounts for the fact that subjects' perceptions may differ in nature, not only in terms of sensitivity. Consequently, the selection criteria are perhaps more important to the quality of the description than the number of recruited subjects. In addition to this, a flash profile is based on a single evaluation, and in most cases, there is no need for maintaining a larger panel.

7.3.3.2 Selection Criteria

Along this line of thinking, recruiting panelists who have a complementary expertise is usually sought. For instance, in our lab, we find it useful to recruit in the same study colleagues from the biophysics lab who are specially interested in food texture as well as colleagues from the chemistry lab who are specially interested in flavor analysis.

7.3.3.2.1 Experienced Sensory Panelists

In the original flash profile procedure, sensory evaluation experts—not necessarily product experts—are selected to participate in the panel. Such sensory evaluation experts are meant to be experienced subjects who have previously participated in several descriptive evaluation tasks and who are able to understand panel leader's instructions and to generate discriminant and nonhedonic attributes. It was anticipated that their ability to focus on their own perceptions and to communicate them quantitatively would allow them completing the descriptive task more efficiently. In practice, we have often observed that this point was indeed determining in the quality of the results.

Note that these subjects do not need to be trained on a specific product set. In this respect, the selection of subjects can be compared to that in the quantitative flavor profiling (QFP) procedure, in which the assessors who are trained before participation in the test protocol actually become sensory experts (Stampanoni 1993).

7.3.3.2.2 Product Experts

Professionals like product developers or traditional experts like oenologists and flavorists may also participate fruitfully to a flash profile study. Thanks to their extensive experience, they indeed meet the requirements described earlier. In addition, they have a specific productwise sensitivity that is highly valuable in a sensory description study. In some cases, including such experts in a flash profile study may be sought as way of improving communications among stakeholders.

7.3.3.2.3 Consumers

On the complete opposite way, one may want to use flash profile to measure consumers' perception. Recruiting consumers to participate to a flash profile would thus make sense. In this perspective, it is important to recruit targeted and representative consumers. The most frequent panel size is of about 40–50 participants. Yet, depending on the objectives of the study, flash profiles with consumers have been conducted with panels ranging from 24 (Moussaoui and Varela 2010) to 200 participants (Ballay et al. 2006).

In our experience, it is possible for totally untrained consumers to perform the descriptive task. However, specific guidance must be provided with very didactic examples. It would also require extra attention from the part of the experimenter, as in face-to-face interviews, even though several participants could be managed simultaneously.

7.3.4 DATA ANALYSIS

7.3.4.1 Coding the Ranking Scores

The first step in data handling is to convert the ranking positions into scores. The sample with the lowest intensity (presumably ranked at the left-hand side of the scale) is arbitrary given a score of 1. The scores are then incremented according to the ranking position of each sample. In case of ties, the mean rank is given to all tied samples. The scores are then filled into a table that is used for subsequent data analysis. One table is thus obtained for each panelist (Figure 7.3). The experimenter may double-check that the sum of ranks is the same for all attributes and equals $p(p + 1)/2$ where p is the number of samples. This step is of course automated when using a software interface. Using this ascending coding system allows a straightforward interpretation of factorial analyses.

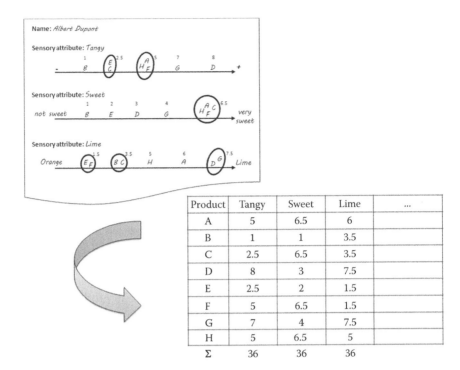

Product	Tangy	Sweet	Lime	...
A	5	6.5	6	
B	1	1	3.5	
C	2.5	6.5	3.5	
D	8	3	7.5	
E	2.5	2	1.5	
F	5	6.5	1.5	
G	7	4	7.5	
H	5	6.5	5	
Σ	36	36	36	

FIGURE 7.3 Example of the coding of one assessor's outcome (only the first three attributes are represented). The ranks are coded by the experimenter and entered into a spreadsheet. In case of ties, the mean rank is used. Note that for all attributes, the sum of ranks must always be the same. One table per assessor is obtained and individual tables may be then concatenated, depending on subsequent data analysis and on the statistical software that is used.

7.3.4.2 Statistical Analysis

Flash profile data analysis essentially relies on factorial or multidimensional analyses. Individual analysis can be applied to each panelist's data set, using, for instance, principal component analysis (PCA). However, in practice, the experimenters directly analyze the results at the panel level. Since the attributes evaluated by the panelists are presumably not the same, there is no way for aggregating data by computing mean scores or equivalent. Instead, multitable data analysis techniques are very useful. The generalized Procrustes analysis (GPA) (Gower 1975) that was initially used for analyzing FCP data by Williams and Langron (1984) is now traditionally applied to flash profile data. Other techniques such as multiple factor analysis (MFA) (Escofier and Pagès 2008) or STATIS (Lavit 1988; Schlich 1996) may also apply. In our experience, GPA and MFA provide very similar results.

These two techniques both yield factorial maps that can be used to assess the overall sensory positioning of the products as perceived by the panelists. They give access to the panel's *average* configuration in relationship with panelists' individual configurations. The quality of the mathematical consensus among panelists can be assessed using indices provided with each technique (e.g., Procrustes residual statistics in case of GPA, RV coefficients in cases of MFA and STATIS). Some data analysis routines also allow plotting/drawing confidence ellipses on the score plots (see, e.g., Husson et al. 2005).

In addition, the product discrimination can be tested in a multivariate way using, for instance, MANOVA or CVA (see Delarue and Sieffermann 2004a,b) applied to the consensus configuration that results of the GPA. Similarly, multiple discriminant analysis (MDA) (Jocteur Monrozier 2001) can be used instead of MFA.

The relative sensory positioning of the products may be considered as the main result of the flash profile and it may be sufficient in many cases. However, one may also want to interpret this positioning in terms of sensory attributes. This interpretation is made by means of the loading plots corresponding to each factorial map. A number of warnings must however be made at this stage. Attribute interpretation is indeed limited by the fact that the panelists do not define the attributes and furthermore they are not trained to evaluate these attributes. The identification of sensory dimensions is thus highly interpretative. In addition, some subjects may use the same attribute leading to the false idea that they evaluate the same characteristics, which may or may not be the case. Conversely, the same concept may be transcribed by different attributes depending on the subject. In the latter case, these attributes may be explicitly semantically related (e.g., *sour* and *tang*) or not (e.g., *smooth* and *fatty*). Interpretation regarding the meaning of the attributes thus remains hypothetical in the absence of external information. Accordingly, great care must be taken to interpret the correlation between attributes (Figure 7.4).

7.3.4.3   How to Evaluate Panel Performance?

Panel performance in flash profile can only be assessed by means of repetition. Other criteria such as quality of the consensus or accuracy are not pertinent given the nature of the method. A full repetition of the evaluation task implies repeated sessions. Delarue and Sieffermann (2004a), for instance, ran the flash profile evaluation sessions in triplicate and found a good repeatability of the panel's results over the three repetitions. Running multiple sessions is however seldom since it is quite contradictory to the objective of rapidity sought with the use of flash profile.

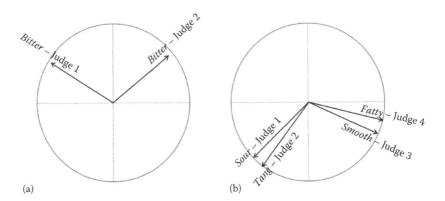

FIGURE 7.4 Illustration of the difficulty of interpreting the correlation between individual attributes. In some cases (a), the assessors may use the same attribute but rank the products differently on this attribute. In other cases (b), similar sensory notions may be described with different attributes.

Another option is to use duplicates. Including two samples of the same product in the set to be evaluated will indeed provide a way to probe repeatability. This can be done, for example, by analyzing the position of the duplicates on the sensory map. Note however that it is only an index and not an evaluation of the full repeatability. One may want to duplicate several products in order to get more confidence in the panels' outcome. A few cautionary notes are in order in this regard. First, it is wise not to multiply the number of duplicates up to the point where the product set will be distorted and where the subjects will implicitly be asked to rank too many identical stimuli. Second, the panelists must not, and at any price, suspect that duplicates are present. Their responses will otherwise be automatically biased (they will pay extra attention in *finding* the duplicates) and the experimenter will falsely conclude that the panelists are repeatable. At last, the interpretation of the position of the duplicates on a sensory map may be misleading. In effect, when two duplicated samples are not close on the score plot resulting from a GPA or MFA, the most obvious interpretation is that the panel is not repeatable. However, one should keep in mind that flash profile data are comparative. As a result, two replicated samples that are positioned apart on a sensory map could mean that all products in the product set are confusable and that interproduct distances are not larger than the difference between two samples of the same product. The experimenter should thus consider this ambiguity and question the interpretation of the results in each case and in the light of prior knowledge of the existing differences in the product set.

Apart from panel performance, strictly speaking, the validity of the results can be estimated by comparison with a reference technique such as

QDA (in the case of "expert profiling"). However, in the case of consumer profiling with flash profile, validity is a much less obvious notion, although comparison can still be made with other methods (Moussaoui and Varela 2010; Veinand et al. 2011).

7.3.5 FLASH PROFILE WITH CONSUMERS

As mentioned earlier in this chapter, flash profile can be seen as a consumer-oriented method although it was initially designed to be conducted with experienced panelists only. Techniques like flash profile indeed present a potential way to assess consumers' diversity in their perception of products because they leave a large degree of freedom to the subjects in their description of the products. In addition, their rapidity and relative simplicity make them applicable to the interview of consumers who cannot be trained.

Eladan et al. (2005) first used flash profile with consumers to describe fragrances. They obtained product configurations very similar to those obtained both with trained panelists and with professional perfumers. Since then, several studies have successfully applied flash profile with panels of consumers. In the same line of thinking, flash profile has been used in cross-cultural studies to compare the perceptions of products in different countries (Ballay et al. 2006; Blancher et al. 2007).

Moussaoui and Varela (2010) have found that the perceptual map obtained from a flash profile carried out with naïve consumers correlated well with those obtained from QDA for the description of hot beverages, with some disagreement for the description of "creamy/creaminess" that was less accurate when described by consumers.

In a study comparing the application of several techniques to get descriptive information from consumers, Veinand et al. (2011) have observed that flash profile could be used efficiently providing that specific guidance was given to participants. In that study, each participant was given a booklet with printed instructions, a detailed example and blank "score sheets." After the usual instructions were given orally, the task was split into two parts. Assessors were first told to find descriptive terms that differentiated the products. All the descriptive terms were then listed by the interviewer on a blank sheet of the assessor's booklet. Then, after a 5 min break, they were asked to rank the products according to each descriptive term they found previously and reported those ranks on the score sheets of their booklet.

Even though the studies mentioned earlier reported successful use of the flash profile with consumers, the reader should be warned that the semantic interpretation of the consumer descriptive terms can be very complex because of the number and the diversity of generated terms. These terms are usually not linked with any specific definitions or references because of the absence of training. Conversely, when a list of attributes is imposed to the consumers, there is no guarantee that they will understand these attributes in a similar way and maybe not the way the experimenter thinks they will do. Examples of such sensory misunderstandings are common in flavor development. It could also be noted that the comparative evaluation mode may induce contrast effects (especially for appearance). Likewise, it may lead the participants to focus on characteristics they would normally not pay much attention to in their consuming habits. Although to my knowledge such effects have not been formally tested, Albert et al. (2011) have, for example, noticed that flash profile tends to yield more detailed information about the product characteristics than QDA and Napping.

In addition to this, consumers tend to use composite sensory notions that may also refer to the benefits of consuming the product (e.g., thirst quenching, satiating) or past experiences. These *attributes* are often mixed up with hedonic considerations. Therefore, it is usually hard for food developers to know how these concepts can be translated into specific product characteristics (van Kleef et al. 2005). However, this can be turned into an advantage for consumer research, and flash profile could very well be used for assessing perceptual dimensions that go beyond sensory perception (Ballay et al. 2006; Delarue et al. 2008) (Figure 7.5).

7.4   APPLICATIONS OF FLASH PROFILE

7.4.1   When to Use Flash Profile

There are a number of situations where flash profile is particularly well suited. Such situations typically are when a defined set of products is to be analyzed at a given point in time. This would often be the case when the need for sensory information is linked to one project (either in R&D, in research or in market research) in the absence of a trained panel for the product category under investigation. In the following, the cases where flash profile would be most useful are listed:

- In any one-shot analysis, linked to one project with no preexisting descriptive tool and no need for a trained panel after the project is completed.
- In research projects, especially in first stages of the project in order to gain faster sensory knowledge and make informed decisions in project management. This may apply to NPD projects in R&D as well as to sensory research projects (see the specific section in the succeeding text).
- As a help for building more robust descriptive analysis tools.
- When the products to be analyzed have very short shelf lives (Lassoued et al. 2008).

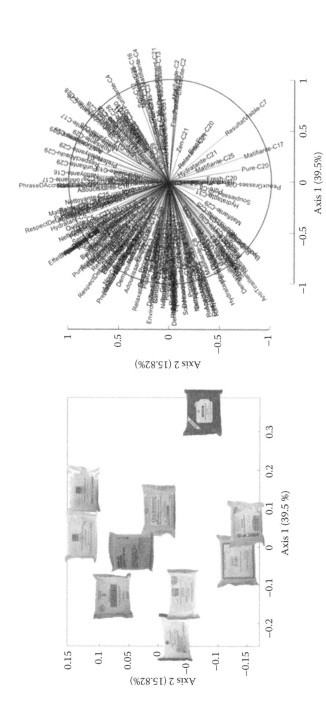

FIGURE 7.5 Example of the use of flash profile to assess the sensory expectations driven by the claims and packaging of makeup remover wipes. This study was achieved with 34 untrained women who were regular users of such products. (Adapted from Delarue, J. et al., Mapping claims according to consumers' sensory perception of cosmetics, Paper read at *25th IFSCC Congress*, Barcelona, Spain, October 6–9, 2008.)

- In market sensory analysis, when one wants to get rapid sensory insights into a market for a given product category.
- In sensory consumer research, in order to understand the diversity of consumers' perceptions.

7.4.2 EXAMPLES OF APPLICATIONS IN RESEARCH AND R&D

In many research and R&D projects, flash profile may prove to be very useful and particularly well fitted to project constraints. It is notably the case when a sensory positioning is sufficient, even when one's goal is to relate sensory to instrumental measurement data (Loescher 2003; Lassoued et al. 2008). Using such rapid methods as flash profile in this perspective represents a "psychophysical mind-set (functional relations between variables) applied to new types of data (locations of products using a multidimensional coordinate space)" (Moskowitz 2003). Shown in the following are few examples of such use of the flash profile (Delarue et al. 2004). Besides, a data set including results from a Flash Profile of six strawberry yogurts from the French market is available for download from the CRC Web site: http://www.crcpress.com/product/isbn/9781466566293.

7.4.2.1 Example 1: Using Flash Profile as a Tool for Formulation and Optimization

This study was conducted in the frame of Coustel's PhD thesis (Coustel 2005) and aimed at investigating the contribution of several distillate fractions to the odor of formulated rums. In this work, we used flash profile in combination with a screening design of experiment (DoE). Seven distillate fractions were thus used in order to blend eight rums (A–H) according to a Plackett–Burman design (Figure 7.6), where each fraction is either incorporated at the maximum level used at the distillery (+) or is not included in the blend (–). All blends were adjusted to 50% vol. of ethanol.

		Fractions					
	1	2	3	4	5	6	7
A	+	+	+	+	+	+	+
B	–	+	–	+	–	+	–
C	+	–	–	+	+	–	–
D	–	–	+	+	–	–	+
E	+	+	+	–	–	–	–
F	–	+	–	–	+	–	+
G	+	–	–	–	–	+	+
H	–	–	+	–	+	+	–

8 Blends

FIGURE 7.6 Formulation of eight rums using seven distillate fractions that were blended according to a Plackett–Burman DoE.

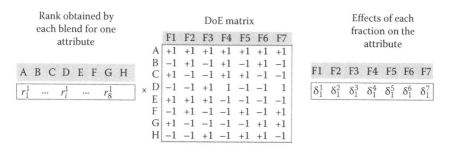

FIGURE 7.7 Calculation of the effect of each experimental variable on each sensory attribute thanks to the properties of the DoE.

The odor of these eight blends was then evaluated using flash profile by eight subjects experienced in olfactory evaluation. For this purpose, 15 mL of each blend was poured in standard wine tasting glasses and presented simultaneously to the panelists in standard individual sensory booths.

As a result, the subjects used between 4 and 8 attributes for a total of 38 attributes. However, instead of analyzing the data using a factorial analysis, we directly estimated the effect of distillate fractions on each attribute, thanks to the DoE properties (Figure 7.7). The outcome is a matrix of effects that can be plotted as shown in Figure 7.8. Here, it can be seen that the distillate fraction #3 has the strongest effects on nearly half of the attributes. In the light of this result, we suggested changes in the rum distillation process that led to significant improvement of the final product.

7.4.2.2 Example 2: Using Flash Profile in Psychophysics (Delarue and Sieffermann 2004b)

In this study, we used a sweetness inhibitor 2(-4-methoxy-phenoxy)propanoic acid (sodium salt) (Na-PMP) in the formulation of strawberry yogurts in order to investigate how sweetness affects overall flavor perception. Using the flash profile method to evaluate the samples allowed us to compare yogurts formulated with added inhibitor to a control set. Na-PMP is known to specifically inhibit sweet taste and it is efficient at low concentration (between 50 and 150 ppm), which makes it a potential tool for investigating flavor perception in real food products.

Seven strawberry yogurts (product set A) were formulated according to a Doehlert design with varying sugar content and titratable acidity (lactic acid concentration) as shown in Figure 7.9. Na-PMP (100 ppm) was added to a replicated set of products (product set B).

A descriptive analysis of each product set was performed separately using flash profile. Both product sets (A and B) were evaluated by the same panel of nine subjects. Product set B (with inhibitor) was evaluated first

Attributes	Fraction #3	Fraction #7	Fraction #17	Fraction #19	Fraction #35	Fraction 43	Heads
Banana_J04	10	6	4	1	7	3	11
Brandy_J07	13	7	2	7	2	4	1
Burst_J01	8	4	4	7	7	11	3
Clove_J06	10	10	0	6	8	4	2
Ethanol_J02	10	0	4	0	4	0	14
Floral_J05	10	1	7	1	7	10	4
Fruit brandy_J04	9	0	7	6	7	10	3
Fruit pit_J03	12	8	4	3	9	3	3
Fruit+brandy_J01	5	2	5	5	2	5	12
Grape_J02	2	1	1	12	2	9	9
Green almond_J05	5	10	5	2	7	8	7
Green walnut_J05	6	10	4	2	12	4	2
Honey_J04	12	4	8	0	4	4	8
Honey_J05	5	4	1	3	16	1	4
Lacton-butter_J08	12	10	2	0	4	6	6
Methylated alcohol_J03	16	3	1	5	3	4	4
Methylated alcohol_J06	0	2	0	11	13	3	5
Methylated alcohol_J07	10	3	3	3	3	10	4
Mushroom earthy_J06	8	3	3	12	4	9	3
Overall intensity_J05	16	2	6	1	3	3	1
Overall intensity_J06	4	9	11	2	4	9	3
Overall intensity_J07	13	7	2	7	2	4	1
Plum brandy_J08	16	4	4	4	4	0	4
Plum_J05	16	1	1	1	7	2	4
Plum_J07	13	7	2	7	2	4	1
Pungent_J03	8	8	4	0	8	8	8
Raisin_J04	9	2	11	7	6	1	6
Rubber_J03	16	6	2	4	4	2	2
Rum-woody_J08	6	6	12	4	2	10	0
Satur Odor_J01	8	2	0	7	9	7	9
Soaked grape_J01	14	2	8	4	2	6	4
Soaked grape_J07	16	0	0	0	0	0	0
Sourish-Kirsch_J08	14	2	4	2	4	0	10
Straw_J04	2	16	2	2	6	2	2
Straw_J06	3	5	2	12	5	3	10
Vanilla_J01	15	5	2	1	2	8	1
Vanilla_J04	6	3	7	10	8	3	7
White rum_J02	10	10	4	2	4	8	6

FIGURE 7.8 Representation of the matrix of effects of the distillate fractions on the sensory attributes. Fraction #3 has the strongest effect on a majority of attributes.

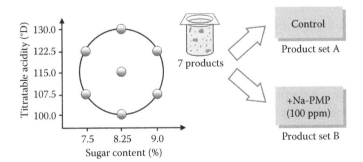

FIGURE 7.9 Formulation of strawberry yogurts with varying sugar content and titratable acidity. The Na-PMP sweetness inhibitor was then added to the products (product set A) versus the control products (product set B).

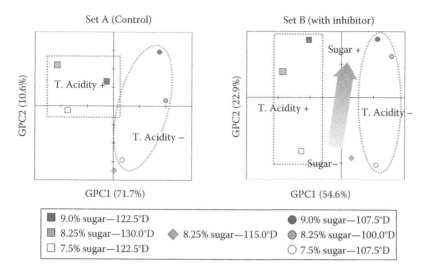

FIGURE 7.10 Score plots of flash profile results for the two sets of yogurts (consensus configurations obtained by GPA).

and product set A was evaluated 1 week later. Subjects had no information about the purpose of the study nor about the nature of the yogurt ingredients. The data obtained for each product set were compiled separately using GPA and the consensus configurations are plotted using PCA in order to represent the relative sensory positioning of the products within their respective product set.

The configurations obtained for the two product sets can be compared (Figure 7.10). As can be seen, the first sensory dimension was driven by titratable acidity in both cases. For the control (product set A), a regression analysis shows that the sugar content also had a significant effect on the first sensory dimension ($F = 23.6$, $p < 0.01$; $R^2 = 0.98$). Interestingly, when the inhibitor is present, the effect of sugar content transfers onto the second sensory dimension. This indicates that sugar acts differently on yogurt perception. The examination of the loadings of individual attributes gives further insights into what is happening when the inhibitor is present.

Overall the subjects used taste, flavor, odor, and texture attributes. In the control set, it can be seen in Figure 7.11 that the first dimension was a "sour to sweet" dimension, which is consistent with the significant effect of titratable acidity and sugar content. It can also be noticed that many aroma attributes were correlated with this main sensory dimension. When the inhibitor is present, only few attributes related to "sweetness" remained and they were loaded on the second axis. In this case, many aroma attributes were correlated with the second axis, which may indicate that the sugar content also drives aroma perception. This could be due to

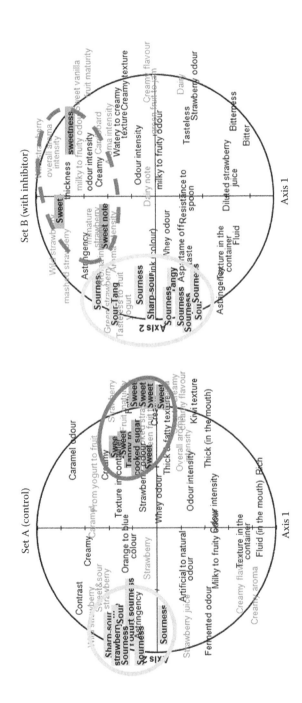

FIGURE 7.11   Loadings of individual attributes after GPA on flash profile data. Attributes related to *sourness* are highlighted in light gray, attributes related to *sweetness* are in dark gray, and flavor (aroma) attributes are printed in lighter gray.

residual sweetness perception or to indirect effect of sugars (e.g., effect on texture perception).

This work shows that the predominant sour/sweet dimension in strawberry yogurts can be dissociated into two independent sensory dimensions with the use of a sweetness inhibitor. In the latter case, it is not clear whether the second sensory dimension is related more to sweetness or to texture perception. However, we may reasonably assume that residual sweetness (maybe sub- or perithreshold) is enough to drive aroma perception on this dimension.

To conclude with this work, it must be noticed that getting access to the sensory perception of such specific product has been possible only because a fast and flexible descriptive method such as the flash profile was available. Here, the application of flash profile in combination with a specific design of the product set and control of formulation variables strives toward the psychophysical mind-set described by Moskowitz (2003) and allows relating product variables to sensory perception by means of the coordinates of the products in the multidimensional sensory space.

7.4.2.3 Example 3: Using Flash Profile (Combined with DoE) to Accelerate the First Stages of Research Projects/Sensory Projects

This third example is part of Taréa's PhD (Taréa 2005), who investigated the relationships between the structural properties of apple puree and the perception of its texture. In order to conduct this research, it was decided not to work on real apple purees but to mimic purees by the way of gelled particles dispersed in apple juice. This point was essential to accurately control the structure and the rheological properties of the food matrix. There are several ways to make such simulated purees and to change their properties. At the time of that study, it was possible to play with four physical variables in order to modify particle properties: their shape (when obtained either by emulsification or crushing), their size (by adjusting ultraturax speed), their firmness (by changing the Ca^{2+} concentration), and their *roughness* (with presence/absence of cellulose).

When starting such a research project, however, and in the absence of prior information on the effect of these physical variables (very few people studied these types of structure/texture relationships before), one may wonder if these variables are worthy of investigation and if a panel should be trained to evaluate these products. In this case, making the fake purees was long and tedious. It was thus decided first to run a sensory experiment based on a flash profile evaluation of a first series of prototypes. As in Example 1 described earlier, the products were prepared according to a Plackett–Burman DoE with – and + levels of the four physical variables.

The combinations of these variables in the DoE resulted in the formulation of eight products that were sensory evaluated by six experienced subjects according to the flash profile methodology. The subjects used between 2 and 5 attributes and a total of 21 attributes were generated and evaluated. The evaluation lasted between 30 min and 90 min depending on the subjects.

The data were compiled in a single *attributes × products* matrix that was used to calculate the corresponding matrix of effects following the same calculation as described in Figure 7.7 for Example 1. In order to tell whether the four variables of formulation were worthy of investigation, we simply submitted them to a PCA. The most striking result from that analysis was that three independent dimensions accounted for 92% of the total variance (Figure 7.12). These dimensions were strongly correlated, respectively, to the shape and the *roughness* of the particles (PC1, 52% of variance), to the size of the particles (PC2, 21% of variance), and to the firmness of the particles (PC3, 19% of variance).

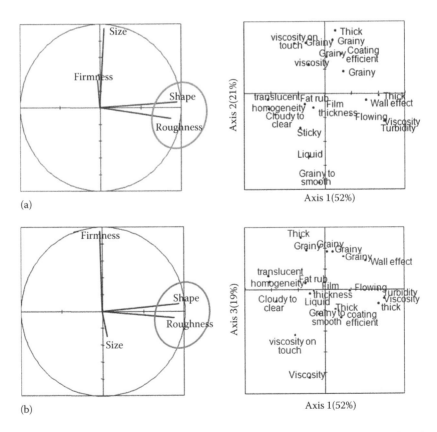

FIGURE 7.12 Effect of the physical variables on the sensory properties of apple purees as evaluated by flash profile. The first three axes from the PCA of the matrix of effects are represented: (a) first and second axis and (b) first and third axis.

Thanks to the DoE, these three dimensions could be separated and controlled. However, two of the initial variables (*shape* and *roughness*) were correlated together and thus contributed to the sensory perception of the purees in a similar way.

Following this first experiment, it was thus decided to set the *shape* of the particles for further experiments and to work on the *roughness* instead. Working with only one type of shape allowed us using the crushing process only and avoiding the emulsification process that was much more time consuming. Overall, when considering all the experiments that were further conducted within the frame of that PhD, we estimated that 2 months of product preparation were saved. This experience shows that using a rapid descriptive method such as flash profile in the early stages of a research project may help in guiding the research, by adjusting the product set and the domain of investigation. This can be done with minor expenses and costs, and in some cases, it may even allow the researcher to save time. Hopefully, this type of approaches may also prove to be useful in product development projects where efficiency is key.

7.5   LIMITS AND PERSPECTIVES

7.5.1   WHEN NOT TO USE FLASH PROFILE

There are also a number of situations where flash profile is to be avoided. Various reasons could lead to choose another descriptive method. They range from practical constraints to more fundamental considerations:

- When all the products are not available at the same time, in the same place.
- When the products to be evaluated are too numerous (although this point could be discussed, see succeeding text) or when products are difficult to compare. This could be because of the way they are usually consumed (cosmetics in self-application [makeup, shampoo, etc.], toothpaste, etc.) or because they have very strong, longing, or irritating characteristics (chili pepper, liquor, etc.).
- For quality control and in any situation where building a database is sought. The comparative nature of the evaluation in flash profile and the absence of calibrated scales indeed prevent data comparisons from one session to another.

7.5.2   WHEN TO USE FLASH PROFILE WITH CAUTION

In some situations, flash profile could be used but it would bear some limitations and thus may not be the best option. Also, in some cases, more research would be needed to test its applicability. In addition to this, of course, flash profile does not solve all problems of conventional descriptive profiling. Notably, the quantitative description of complex odors or flavors remains very challenging (Lawless 1999). For example, taking into account characteristics that only apply to one product is still problematic and takes us back to the issue of prototypal description. Other points that may be critical are listed in the following:

- Panelists who do not match selection criteria: when recruiting untrained or naïve subjects who are neither experts nor representative consumers. Most often, students would fall in this category. It is thus advised not to conduct a flash profile with students except for pedagogic purposes.
- A related question is as follows: "Is it possible to train a panel dedicated to flash profile?" To my knowledge, some companies have implemented such panels. However, to date, there is no standard method for training a flash profile panel. If one wants to do so, I would advise to train the subjects in the view of developing their attention, introspective sensory focus, and general quantification abilities. Ideally, they should also experience the description of a diversity of product types. Some sensory teams have experimented the use of flash profile with their own panelists previously trained for conventional sensory profile (either QDA or spectrum). Unfortunately, these attempts have failed, essentially because the panelists were so much used to their routine evaluation that they did not conform to the new instructions. At best, they performed the flash profile using the exact same list of attributes they were trained on.
- Measurement of time-related perceptions: We attempted to use flash profile to measure the dynamics of the perception of chewing gums (Delarue and Loescher 2004) with some success. The samples had to be provided in small screw-cap plastic containers so that the subjects could evaluate longer-term attributes out of the sensory booths. In that study, we encouraged the assessors to use time-dependent attributes and indeed they used both instant and longer-term attributes. However, the link with the dynamics of preferences was not straightforward. The flash profile could thus be used in this perspective, but methods like TDS that are specifically dedicated to the measurement of time-related perceptions would certainly be preferable.
- Important number of products: It is one of the most frequently asked questions by those sensory practitioners who are willing to conduct a flash profile. The comparative evaluation task is indeed

believed to limit the number of samples that can be evaluated in a single session. The answer of course depends on the nature of the products to be evaluated. In our experience though, it seems that this number is often underestimated by sensory practitioners. For instance, Tarea et al. (2007) have shown it was possible to get relevant sensory information from the flash profile of as many as 49 different apple and pear purees. Besides, one should keep in mind that the way a session is conducted in flash profile is flexible and can be designed so that the evaluation task stays feasible. For example, it is possible to plan large breaks between attributes or, if necessary, to set up separate sessions of two or three attributes only. Another option is to first apply a simpler task (mapping or sorting) and then run the flash profile on a representative subset (that would be selected based on the first task results).

7.5.3 PERSPECTIVES AND FURTHER DEVELOPMENTS

To conclude with this chapter, it must be noted that flash profile is not a completely fixed methodology and that, as shown through the various examples, it can be adapted to the needs of the study or even combined with other methods as long as basic principles are respected (individual evaluation, appropriate sample preparation, careful subject recruitment, respect for interindividual differences, recognition of expertise, etc.). A correct understanding of the method would allow further developments. For example, there certainly is an important potential in the use of flash profile with developers and project team members in R&D in order to make informal tasting during team meetings more efficient. Naturally, this would hold for other rapid methods as well. In our experience, we have found that flash profile is very close to what people in companies actually do when they confront their views on samples, should they be prototypes, competitors' products, or samples provided by suppliers. Product developers have a strong knowledge of their products and also a form of sensory expertise although it is often not formalized like that of trained panelists. As long as products are presented blind and that the evaluation is individual, using such methods as flash profile in a working context may thus facilitate dialog between team members and between services (e.g., R&D and marketing). If managed well, the development of this practice could be highly beneficial both to sensory teams and to their stakeholders.

Finally, Dairou and Sieffermann (2002) have suggested the use of flash profile as an additional tool to facilitate the setting up of a conventional profile. In the same line of thinking, Giboreau et al. (2001) have used a categorization task in a preliminary phase for more thorough analysis. Although this possibility sounds very promising, its added value has not been formally tested yet. This opens perspectives for research on the use of such rapid methods and of their combinations.

7.A APPENDIX: SCRIPT FOR ANALYZING DATA FROM FLASH PROFILE USING GPA IN R

GPA is performed in the package FactoMineR, which has to be loaded

```
library(FactoMineR)
```

The first step is to load data into R from the clipboard. The data matrix for data analysis should be constructed for all consumers by placing next to each other the individual matrixes shown in Figure 7.3. The global matrix should be copied from Excel into the clipboard and imported into R using the following command.

```
Dataset <- read.table("clipboard", header = TRUE, sep =
    "t", na.strings = "NA", dec = ".", strip.white = TRUE)
```

The following command indicates that the name of the samples is included in the column identified as *Product*:

```
row.names(Dataset) <- as.character(Dataset$Product)
Dataset$Product <- NULL
```

The next step is performing GPA on the imported data set. It is important to take into account that the GPA function needs the group structure of the data. The number of attributes used by each panelist for evaluating samples should be included within the argument *group*. In the following, the command for analyzing data from 20 panelists who used 9, 5, 4, 3, ..., 6 attributes for evaluating the samples is shown:

```
GPAresob<-GPA(Dataset, group = c(9,5,4,3,7,3,7,8,9,12,
    14,5,7,8,9,7,3,7,5,6))
```

The following commands provide the coordinates of the consensus and the partial configurations. The GPA function of FactoMineR does not provide the projection of the attributes.

```
GPAresob$consensus
GPAresob$Xfin
```

REFERENCES

Albert, A., P. Varela, A. Salvador, G. Hough, and S. Fiszman. 2011. Overcoming the issues in the sensory description of hot served food with a complex texture. Application of QDA®, flash profiling and projective mapping using panels with different degrees of training. *Food Quality and Preference* 22(5):463–473.

Ballay, S., S. Warrenburg, J.-M. Sieffermann, L. Glazman, and G. Gazano. 2006. A new fragrance language: Intercultural knowledge and emotions. Paper read at *24th IFSCC Congress—Integration of Cosmetic Sciences*, October 16–19, 2006, Osaka, Japan.

Blancher, G., S. Chollet, R. Kesteloot, D.N. Hoang, G. Cuvelier, and J.M. Sieffermann. 2007. French and Vietnamese: How do they describe texture characteristics of the same food? A case study with jellies. *Food Quality and Preference* 18(3):560–575.

Cadoret, M. and F. Husson. 2013. Construction and evaluation of confidence ellipses applied at sensory data. *Food Quality and Preference* 28(1):106–115.

Coustel, J. 2005. Mise au point du contre typage d'échantillons de rhum par voie organoleptique et analyses physico-chimiques. PhD thesis, Génie Industriel Alimentaire, ENSIA, Paris, France.

Dairou, V. and J.-M. Sieffermann. 2002. A comparison of 14 jams characterized by conventional profile and a quick original method, the Flash Profile. *Journal of Food Science* 67(2):826–834.

Delarue, J., F. Beurier, and J.M. Sieffermann. 2008. Mapping claims according to consumers' sensory perception of cosmetics. Paper read at *25th IFSCC Congress*, October 6–9, 2008, Barcelona, Spain.

Delarue, J., M. Danzart, and J.-M. Sieffermann. 2004. Flash profile gives insights into human sensory perception. Paper read at *Fifth International Multisensory Research Forum*, June 2–5, 2004, Sitges, Spain.

Delarue, J. and E. Loescher. 2004. Dynamics of food preferences: A case study with chewing gums. *Food Quality and Preference* 15(7–8):771–779.

Delarue, J. and J.-M. Sieffermann. 2004a. Sensory mapping using Flash Profile. Comparison with a conventional descriptive method for the evaluation of the flavour of fruit dairy products. *Food Quality and Preference* 15(4):383–392.

Delarue, J. and J.-M. Sieffermann. 2004b. Use of 2(-4-methoxyphenoxy)propanoic acid (Na-PMP) to investigate flavour interactions in real food products. Paper read at *International Symposium on Olfaction and Taste (ISOT/JASTS)*, July 5–9, 2004, Kyoto, Japan.

Eladan, N., G. Gazano, S. Ballay, and J.-M. Sieffermann. 2005. Flash profile and fragrance research: The world of perfume in the consumer's words. Paper read at *ESOMAR Fragrance Research Conference*, May 15–17, 2005, New York.

Escofier, B. and J. Pagès. 2008. *Analyses factorielles simples et multiples, objectifs méthodes et interprétation*, 4th edn. Paris, France: Dunod.

Faye, P., D. Bremaud, E. Teillet, P. Courcoux, A. Giboreau, and H. Nicod. 2006. An alternative to external preference mapping based on consumer perceptive mapping. *Food Quality and Preference* 17(7–8):604–614.

Gacula, M. and S. Rutenbeck. 2006. Sample size in consumer test and descriptive analysis. *Journal of Sensory Studies* 21(2):129–145.

Giboreau, A., S. Navarro, P. Faye, and J. Dumortier. 2001. Sensory evaluation of automotive fabrics: The contribution of categorization tasks and non verbal information to set-up a descriptive method of tactile properties. *Food Quality and Preference* 12:311–322.

Gower, J.C. 1975. Generalised Procrustes analysis. *Psychometrika* 40(1):33–51.

Husson, F., S. Le, and J. Pages. 2005. Confidence ellipse for the sensory profiles obtained by principal component analysis. *Food Quality and Preference* 16(3):245–250.

Husson, F., S. Le Dien, and J. Pagès. 2001. Which value can be granted to sensory profiles given by consumers? Methodology and results. *Food Quality and Preference* 12:291–296.

Jack, F.R. and J.R. Piggott. 1992. Free choice profiling in consumer research. *Food Quality and Preference* 3:129–134.

Jocteur Monrozier, R. 2001. Le profil sensoriel de la mesure a l'analyse discriminante multiple. PhD thesis, Sciences de l'Aliment, ENSIA, Paris, France.

King, B.M., P. Arents, and N. Moreau. 1995. Cost/efficiency evaluation of descriptive analysis panels—I. Panelsize. *Food Quality and Preference* 6:245–261.

Lassoued, N., J. Delarue, B. Launay, and C. Michon. 2008. Baked product texture: Correlations between instrumental and sensory characterization using Flash Profile. *Journal of Cereal Science* 48(1):133–143.

Lavit, C. 1988. *Analyse conjointe de tableaux quantitatifs*. Paris, France: Masson.

Lawless, H.T. 1999. Descriptive analysis of complex odors: Reality, model or illusion? *Food Quality and Preference* 10:325–332.

Loescher, E. 2003. Evaluations instrumentale et sensorielles de la texture de produits alimentaires de type semi-liquide. Applications au cas de fromages blancs et compotes. PhD thesis, Science de l'Aliment, ENSIA, Paris, France.

Mammasse, N. and P. Schlich. 2010. Number of assessors in descriptive sensory panels: A database approach. Paper read at *11e Congrès Agrostat*, February 23–28, 2010, Benevento, Italy.

Monrozier, R. and M. Danzart. 2001. A quality measurement for sensory profile analysis: The contribution of extended cross-validation and resampling techniques. *Food Quality and Preference* 12:393–406.

Moskowitz, H.R. 2003. The intertwining of psychophysics and sensory analysis: Historical perspectives and future opportunities—A personal view. *Food Quality and Preference* 14(2):87–98.

Moussaoui, K.A. and P. Varela. 2010. Exploring consumer product profiling techniques and their linkage to a quantitative descriptive analysis. *Food Quality and Preference* 21(8):1088–1099.

Schlich, P. 1996. Defining and validating assessor compromises about product distances and attribute correlations. In *Multivariate Analysis of Data in Sensory Science*, T. Noes and E. Risvik, eds. Amsterdam, the Netherlands: Elsevier Science.

Sieffermann, J.-M. 2000. Le profil flash—Un outil rapide et innovant d'évaluation sensorielle descriptive. Paper read at *AGORAL 2000—XIIèmes rencontres "L'innovation: de l'idée au succès"*, Montpellier, France.

Stampanoni, C.R. 1993. The "quantitative flavor profiling" technique. *Perfumer and Flavorist* 18:19–24.

Taréa, S. 2005. Etude de la texture de suspensions de particules molles concentrées. Relations entre la structure, la rhéologie et la perception sensorielle: Application aux purées de pommes et poires et mise au point de milieux modèles. Doctorat Sciences alimentaires, Laboratoire de Perception Sensorielle et Sensométrie, ENSIA, Paris, France.

Tarea, S., G. Cuvelier, and J.M. Sieffermann. 2007. Sensory evaluation of the texture of 49 commercial apple and pear purees. *Journal of Food Quality* 30(6):1121–1131.

Thamke, I., K. Dürrschmid, and H. Rohm. 2009. Sensory description of dark chocolates by consumers. *LWT—Food Science and Technology* 42(2):534–539.

van Kleef, E., H.C.M. van Trijp, and P. Luning. 2005. Consumer research in the early stages of new product development: A critical review of methods and techniques. *Food Quality and Preference* 16(3):181–201.

Veinand, B., C. Godefroy, C. Adam, and J. Delarue. 2011. Highlight of important product characteristics for consumers. Comparison of three sensory descriptive methods performed by consumers. *Food Quality and Preference* 22(5):474–485.

Williams, A.A. and S.P. Langron. 1984. The use of free-choice profiling for the evaluation of commercial ports. *Journal of the Science of Food and Agriculture* 35:558–568.

Worch, T., S. Lê, and P. Punter. 2010. How reliable are the consumers? Comparison of sensory profiles from consumers and experts. *Food Quality and Preference* 21:309–318.

8 Free Sorting Task

Sylvie Chollet, Dominique Valentin,
and Hervé Abdi

CONTENTS

8.1 THEORY BEHIND THE METHOD

The sorting task is a simple procedure for collecting similarity data in which each assessor groups together stimuli based on their perceived similarities. Sorting is based on categorization—a natural cognitive process routinely used in everyday life—and does not require a quantitative response. The final objective of the sorting task is to reveal—via statistical analyses—the structure of the product space and to interpret its underlying dimensions.

The sorting task originated in psychology (Hulin and Katz 1935) and this field has used it routinely since (see, e.g., Miller 1969; Imai 1966, for early applications; see also Coxon 1999, for a thorough review and historical perspectives). It was first used in the field of sensory evaluation in the early 1990s to investigate the perceptual structure of odors (Lawless 1989; Lawless and Glatter 1990; MacRae et al. 1992; Stevens and O'Connell 1996; Chrea et al. 2005). Lawless et al. (1995) were the first to use a sorting task with a food product.

8.2 DESIGN OF EXPERIMENTS

8.2.1 PROCEDURE

The free sorting task is the basic method, but different variations of the sorting task emerged according to the applications and the objectives of the study.

8.2.1.1 Free Sorting Task

The sorting task is performed in a single session. All products are presented simultaneously and randomly displayed on a table with a different order per assessor. Assessors are asked first to look at, smell, and/or taste (depending on the objectives of the study) all the products and then to sort them in mutually exclusive groups based on perceived product similarities. Assessors can use the criteria they want to sort the stimuli, and they are free to make as many groups as they want and to put as many products as they want in each group.

The sorting task can be stopped at this point or can be followed by a description step where assessors are asked to describe each group of products (Lawless et al. 1995; Tang and Heymann 1999; Faye et al. 2004, 2006; Saint-Eve et al. 2004; Lim and Lawless 2005; Cartier et al. 2006; Blancher et al. 2007; Lelièvre et al. 2008, 2009; Santosa et al. 2010). This procedure is called *labeled sorting* by Bécue-Bertaut and Lê (2011). To facilitate both the assessors' task and data analysis, a preestablished list can be provided during this step to help assessors in labeling their groups (Lelièvre et al. 2008).

8.2.1.2 Variations of the Sorting Task

Several variations of the sorting task have been developed. A first variation consists in providing information on either the number and/or the nature of the groups. This type of directed sorting is very useful to evaluate if assessors are able to discriminate between different categories of products (Ballester et al. 2009; Parr et al. 2010). For example, Lawless (1989), using citrus and woody odors, noted that when assessors had to sort the odors into only two categories, the multidimensional scaling (MDS) configuration showed only two clusters (woody and citrus), with ambiguous odors on the edges of each cluster, whereas when the assessors could use as many groups as they wished, the MDS configuration showed four groups, with the lime fragrances and ambiguous odors failing into a central cluster between woody and citrus groups.

Another variation—first proposed by Rao and Katz (1971)—is called hierarchical sorting task. In this variation, after the assessors have performed the sorting task, they are asked to successively merge the two groups that are most similar up to the time where a single group is formed (ascendant hierarchical sorting; see Coxon 1999), or inversely, the assessors are asked to separate each group into finer groups up to the time where no further separation is possible (descendant hierarchical sorting, Clark 1968), or both (Kirkland et al. 2000). Ascendant hierarchical sorting has been applied to milk chocolate recently (under the name of taxonomic free sorting [TFS]) by Courcoux et al. (2012). Descendant hierarchical sorting has been applied to olive oil by Santosa et al. (2010) and to cards by Cadoret et al. (2011). Hierarchical sorting could give more precise information than the free sorting task as it provides a more graduate measurement of the similarity between products than the 0/1 data provided by the free sorting task. According to Courcoux et al. (2012), the position of the products on the sensory map could be more stable than the one provided by the free sorting task, but further studies are needed to validate this conclusion.

8.2.2 ASSESSORS

Who can perform a sorting task? *A priori* everybody, but the obtained results might not be perhaps exactly the same! Some studies showed that untrained and trained assessors generate similar perceptual maps (breakfast cereals, Cartier et al. 2006; beers, Lelièvre et al. 2008; Chollet et al. 2011). But other studies reported some differences between the maps generated by assessors with different levels of expertise (wine, Solomon 1997; Ballester et al. 2008; beers, Chollet and Valentin 2001; Patris et al. 2007; and fabrics, Soufflet et al. 2004). It seems that these discrepancies depend upon the nature of the products and the nature of the differences between the products. In some cases, novices tend to categorize the products according to basic sensory features, whereas experts tend to use rather higher-level types of categorization (e.g., grape variety, Ballester et al. 2008). In other cases, it was shown that experts are more precise in their sorting (Beguin 1993; Soufflet et al. 2004; Patris et al. 2007). Overall, it seems that untrained panelists can provide a coarse map of the products.

Concerning the number of assessors needed in sorting tasks, it has been suggested that a large number of assessors is required (Faye et al. 2004). However, other studies carried out with beers indicated that stable results could be reached with 20 untrained assessors (Lelièvre et al. 2008; Chollet et al. 2011). So it is likely that the stability of the results may vary with some aspects of the task, and recently, Blancher et al. (2012) have suggested that the stability of sorting task results depends on the characteristics of the product sets and on the assessor expertise level. These authors proposed to evaluate these effects with a cross-validation procedure and, specifically, suggested to use bootstrapping techniques to draw large numbers of samples of different sizes from the original set and compute the average R_V coefficient (which they called R_{Vb}) to determine the number of assessors necessary to obtain stable results. They consider that an average R_{Vb} at least equal to 0.95 is a good indicator of stability. All in all, although the quality of sorting task results is clearly influenced by the nature of the products, it seems that a sorting task with about 20 assessors can provide relevant and interpretable results but that this recommendation, however, may depend on the specific characteristics of the data.

8.2.3 PRODUCTS

A recurrent question when using the sorting task is: Is there a limit to the number of products that can be evaluated? Two phenomena need to be taken into account when answering this question. The first one is linked to the product itself: Some products cannot be tasted in large number because of their intrinsic properties such as alcohol content or high taste persistence. It is obvious that a sorting task with (tasting) 15 whiskies is likely to be problematic! The second phenomenon is linked to memory. Because of the necessity of comparing products, performing a sorting task involves short-term memory that has a capacity limited to about seven *chunks* (Miller 1956). As a consequence, when the number of products to sort exceeds the assessor's memory span, then these products have to be tasted several times and this increases the risk of confusion between products. These memory problems have previously been highlighted in an experiment where verbal data, collected after the sorting task, were analyzed in conjunction with behavioral indicators (Patris et al. 2007). Results showed that trained and untrained participants expressed difficulties to memorize beer samples during the task. Besides, the number of times that the products have to be tasted increases with the number of products to sort but depends also on the resemblance between products: The more similar the products, the more difficult the task and the more often assessors have to taste the products.

The efficiency of the sorting task has also been shown to decrease when the number of products is too small. For example, Nava Guerra et al. (2004) report that a sorting task performed with eight beers gave poor results compared to a similar task carried out with 12 beers. So even though the optimal number of products to sort is product dependent, as a general rule of thumb, we can advise to carry out a sorting task using between 9 and 20 products with an optimum number being around 12 products. Finally, when the sorting task is carried out only with visual criteria, the number of stimuli can be considerably increased as the memory load and fatigue issues are alleviated.

But what to do when we have more than 20 products to sort? A first solution is to use incomplete block designs. But in this case, a larger number of assessors are needed (e.g., we would need 84 assessors for an entire set of 50 products with 12 products in one sort and each product being tested by at least 20 assessors). Another possibility is to split a large set of products into several smaller sets and to add in each smaller set the same product (called a prototype) and to compare the products to this prototype. Obviously, in this case, the choice of the prototype is crucial. Although these approaches have not yet given rise to any scientific publications and thus still need to be validated, comparing products to a set of references seems to be a promising idea. Some recent descriptive methods such as the pivot profile (Thuillier 2007) and the polarized sensory positioning (Teillet et al. 2010) relying on the comparison with reference products have, however, been successfully applied, respectively, to champagne and mineral water.

8.3 IMPLEMENTATION AND DATA COLLECTION

8.3.1 EXAMPLE OF QUESTIONNAIRE

An example of instructions is shown as follows:

You have 12 samples in front of you. Please, look at, smell, and taste these samples. Then make groups according to the similarity of these products. You are free to make the groups according to any criteria that you may choose, and you do not need to specify your criteria. You can make as many groups as you want and group together as many samples as you want but please make more than one group and fewer groups than the number of products. Take as much time as you want.

8.3.2   EXAMPLE OF SCORE SHEET

Examples of score sheets are shown in Figure 8.1.

8.3.2.1   Free Sorting Task

To analyze the sorting data with an MDS approach, the results of each assessor are encoded in an individual co-occurrence matrix where the rows and the columns are products. A value of 1 at the intersection of a row and a column indicates that the assessor sorted these two products together, whereas a value of 0 indicates that the products were not put together. All the individual matrices are then summed up to obtain a global similarity matrix (Figure 8.2).

The individual matrices could also be built automatically (see Appendix) starting from a table with the products in line, the assessors in the column, and, at the intersection, the group affiliation number (products in the same group have the same affiliation number). Of course, other equivalent schemes could be used; they will all lead to the same distance matrices.

An alternative approach to analyze the sorting task is to use multiple correspondence analysis (MCA; see Abdi and Valentin 2007b; Cadoret et al. 2009). In this case, the results of the task performed by each assessor are expressed by using a group coding (also called *complete disjunctive coding*). In this scheme, each assessor is represented by as many binary vectors as there were groups of objects, and all objects in a given group are given the value of 1 in the column representing their group and a value of 0 for the other groups. An equivalent approach is to compute between

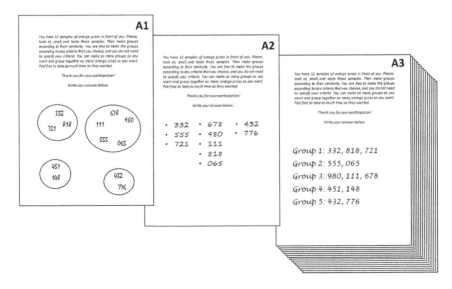

FIGURE 8.1   Example of score sheets obtained in a sorting task.

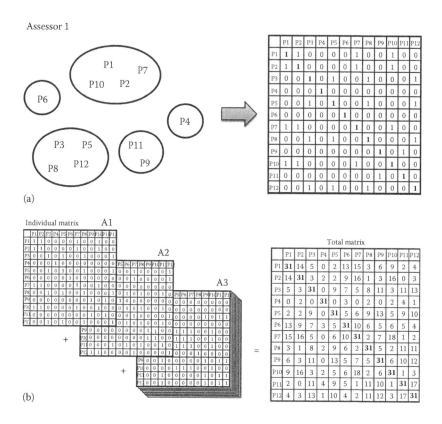

FIGURE 8.2 Example of data obtained in a free sorting task: (a) individual data and (b) group data.

product's χ^2 distance matrices for each subject and sum these matrices prior to obtain a global similarity matrix.

8.3.2.2 Descendant Hierarchical Sorting

In a descendant hierarchical sorting task, the similarity between products is coded as the last level at which they have been sorted together divided by the number of sorting levels. For example, two products sorted together at the third level will have a similarity score of 3/3 and two products sorted together at the second level will have a score of 2/3 (see Figure 8.3).

8.4 DATA ANALYSIS

8.4.1 PERCEPTUAL DATA ANALYSIS

The sorting similarity matrix is generally analyzed by MDS, a technique used to visualize proximities or distance between objects in a low dimensional space. In MDS, each object is represented by a point in a map.

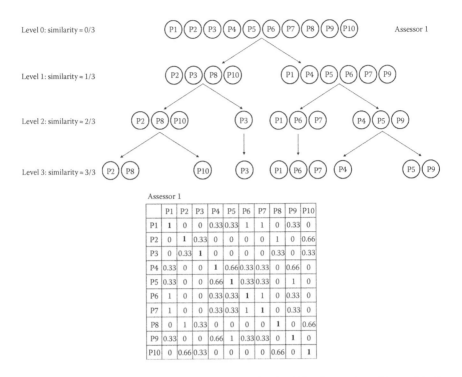

FIGURE 8.3 Example of individual data obtained in descendant hierarchical sorting task.

In this map, the points are arranged so that objects that are perceived to be similar to each other are placed near each other and objects that are perceived to be different are placed far away. Different algorithms can be used to obtain the visual representation of the objects. These algorithms can be classified into two main categories: metric MDS (also called classical MDS or principal coordinate analysis; see Abdi 2007a,b) and nonmetric MDS. In metric MDS, the proximities are treated directly as (squared Euclidean) distances. The input matrix is first transformed into a cross product matrix and then submitted to an eigendecomposition (a technique equivalent to principal component analysis (PCA); see Abdi and Williams 2010). In nonmetric MDS, the proximities are treated as ordinal data. An iterative stepwise algorithm is used to create a visual representation of the objects. This algorithm (1) starts by creating an arbitrary configuration of the objects, (2) computes distances among all pairs of points, (3) compares the input matrix and the distance matrix using a stress function (the smaller the value of the stress, the greater the correspondence between the two matrices), and (4) adjusts the position of the objects in the configuration in the direction that best decreases the stress. Steps 2 through 4 are repeated until the value of the stress is small enough or cannot be decreased any more. Different authors have different standards regarding the amount of stress to tolerate. The rule of thumb used in sensory evaluation is a stress value lower than 0.2 is acceptable.

In the sensory field, the similarity matrix is often analyzed with nonmetric MDS, but metric MDS is also frequently used because the sorting similarity matrix is equivalent to a squared Euclidean metric (see Abdi et al. 2007, for a proof); note that the sorting similarity matrix can also be replaced by a χ^2 distance matrix.

Multiblock analyses that take into account individual data such as DISTATIS (Abdi et al. 2007, 2012; Abdi and Valentin 2007c), MCA (Takane 1980; Abdi and Valentin 2007b; Cadoret et al. 2009), or common components and specific weights analysis (SORT CC, Qannari et al. 2009) have also been used recently. Multiple factor analysis ([MFA] Escofier and Pagès 1983; see also Abdi and Valentin 2007a; Abdi et al. 2013) can also be used (see Dehlholm et al. 2012a). All these techniques provide a common map (often called a compromise) and also show how each assessor positions the products in the common space. Some of these techniques also provide a map of the assessors. These techniques will generally lead to similar conclusions for the relative position of the products. However, the specific χ^2 distance metric used in MCA makes salient products that are rarely associated with other products, and therefore, these rare products may define dimensions by themselves (see also the section on analysis of the example).

8.4.2 Descriptive Data Analysis

The analysis of the descriptors associated to the groups of products depends upon the authors. Most analyses start by constructing a contingency table with descriptors in rows and products in columns. The values in the contingency table indicate the frequency at which each descriptor was employed for a stimulus. The descriptors given for a group of stimuli are assigned to each stimulus of the group and descriptors given by several assessors are assumed to have the same meaning. If the intensity of the descriptors is evaluated as suggested by Lelièvre et al. (2008), geometric means (see Dravineks 1982) can be used instead. The resulting contingency tables are quite large and so the number of descriptors is generally reduced by grouping together terms with similar meanings and by discarding descriptors used by fewer than a certain proportion of assessors (e.g., 10%). The frequency data can then be projected onto the similarity maps by computing the correlations between the occurrence of descriptors and the stimuli factor scores (Faye et al. 2004; Cartier et al. 2006; Abdi and Valentin 2007c). Alternatively, the contingency table can be submitted to a correspondence analysis (CA; see Abdi and Williams 2010) to position both stimuli and descriptors on a descriptor-based space (Picard et al. 2003; Soufflet et al. 2004), or to an MCA (Cadoret et al. 2009).

8.5 ADVANTAGES AND DISADVANTAGES

8.5.1 ADVANTAGES

One of the main issues with verbal-based methods such as conventional profile is that they rely heavily on an analytical perception of the products as well as on the ability to translate sensations into words. As a consequence, it is likely that product aspects difficult to verbalize will be overlooked by these methods. The free sorting task alleviates this problem by relying first on a global perceptual step in which the similitude between the products is evaluated. The verbalization of the differences between products occurs only in a second step or can even be omitted.

From a practical point of view, the sorting task associated with a verbalization step is a time-effective way of describing products as long as only a coarse description of the products is required. Moreover, the sorting task can be used with both consumers and trained panelists on a relatively large set of products.

Finally, the sorting task is well adapted to select a subset of products for conducting further descriptive analysis (Giboreau et al. 2001; Piombino et al. 2004). Despite a few differences, perceptual maps obtained with a free sorting task are globally comparable with those obtained from a conventional profile (Faye et al. 2004; Saint-Eve et al. 2004) and seem to be reproducible (Falahee and MacRae 1997; Cartier et al. 2006; Lelièvre et al. 2008; Chollet et al. 2011).

8.5.2 DISADVANTAGES

A common problem for the sorting task is that the whole set of products needs to be presented at the same time. Therefore, this method is not suitable, for example, for hot products nor for quality control.

Another aspect, rarely addressed in the literature, is the fact that vocabulary used by novices in a sorting task is often difficult to analyze and interpret. Because novices have not been trained with common references, their descriptions vary a lot from one assessor to the other, and it is often necessary to preprocess the attributes (e.g., categorization of similar terms, elimination of idiosyncratic terms) before projecting them onto the MDS maps or performing a CA. An additional problem is that assessors spontaneously qualify their attributes with various quantitative terms such as "very," "many," "slightly," "more than," and "less than," and this makes data interpretation rather cumbersome. To take into account this problem,

Lelièvre et al. (2008) suggested to provide the assessors with a predefined set of quantifiers to indicate the intensity of the perceived attributes (e.g., "not," "a little," "medium," and "very") and to analyze the data using geometric means.

Finally, authors using the sorting task generally report that this is a natural and easy task for consumers. However, even if the principle of the task is easy to understand, the task itself is not always so easy: for example, Patris et al. (2007) using a verbal report methodology showed that both trained and untrained assessors declared having memory and saturation problems when performing a sorting task on beers.

8.6 APPLICATIONS

8.6.1 PRODUCTS APPLICATIONS

The sorting task has been used on a large variety of food products including vanilla beans (Heymann 1994), cheese (Lawless et al. 1995), drinking waters (Falahee and MacRae 1995, 1997; Teillet et al. 2010), fruit jellies (Tang and Heyman 1999; Blancher et al. 2007), beers (Chollet and Valentin 2001; Abdi et al. 2007; Lelièvre et al. 2008, 2009), wines (Piombino et al. 2004; Ballester et al. 2005, 2008; Campo et al. 2008; Bécue-Bertaut and Lê 2011), yoghurts (Saint-Eve et al. 2004), spice aromas (Derndorfer and Baierl 2006), cucumbers and tomatoes (Deegan et al. 2010), olive oil (Santosa et al. 2010), and meat (Hoek et al. 2011). It has also been used in sensory evaluation of nonfood products such as fabrics or leathers (Giboreau et al. 2001; Picard et al. 2003; Soufflet et al. 2004; Faye et al. 2006), plastic cards (Faye et al. 2004), perfumes (Cadoret et al. 2009), or sounds (Gygi et al. 2007).

8.6.2 USEFULNESS OF SORTING TASK

The sorting task can be a very useful tool in various areas of industry including R&D, quality control, and marketing. Its role is particularly important in the development of products as well as in routine control to maintain product quality. In R&D, in addition to being a useful tool for selecting products, the sorting task could also be appropriate to determine the general characteristics of a product from a given family when we know *a priori* the relevant sensory characteristics of different members from this family. Indeed, based on the proximity structure of the members, we can deduce the membership of the studied products and thus derive their sensory characteristics. In quality control, the sorting task could be used in order to obtain an estimation of the variation of sensory characteristics according to the age of products or to the different batches. Finally, the sorting task could also help marketing research by providing map in which products are compared to their competitors.

8.6.3 EXAMPLE OF APPLICATION

As an illustration, we carried out a sorting task with 16 spices, 10 individual spices (cardamom, chili, cinnamon, cloves, coriander, ginger, nutmeg, pepper, star anise, and turmeric) and 6 blends of spices (chili + turmeric + coriander, chili + turmeric, cinnamon + cloves + cardamom, pepper + nutmeg, ginger + pepper, and ginger + cardamom). Twenty-one French assessors participated to this experiment.

Results were analyzed with two statistical methods: metric MDS and DISTATIS.

8.6.3.1 Metric MDS

Figure 8.4 presents the MDS representation. Figure 8.4 suggests that there are four groups of spices: the first one composed of Cinnamon and the blend Cin + Clo + Car; the second one of Chili, Turmeric, and the blends Chi + Tum and Chi + Cor + Tum; the third one of Pepper and the blend Pep + Nut; and the last one of the other spices.

8.6.3.2 DISTATIS

To take into account individual differences, multiblock analyses such as DISTATIS are particularly well adapted (Abdi et al. 2005, 2007, 2009). DISTATIS provides two types of MDS-like maps: (1) a map of the assessors that can be used, for example, to identify clusters among the assessors

FIGURE 8.4 A 2D metric MDS map.

(in this map, assessors far to the right of dimension 1 have a large commu-
nality with the other assessors, whereas assessors close to the origin would
be atypical; if a few assessors were outliers, their data may be eliminated
and the analysis rerun) and (2) a map of the products that reflects how
the group of assessors evaluated the products (in this map, the product
positions of each assessor can also be displayed and also some confidence
ellipsoids that represent the variability of the results over the assessors
[when confidence intervals do not overlap, the products are perceived as
different by the assessors; see Abdi et al. 2009; Dehlholm et al. 2012b;
Cadoret and Husson 2013]).

Figure 8.5 shows the map of the assessors. As most of the assessors are
positioned on the right, they mostly agree on their sorting. The second
dimension (which account for only 7% of the variance) shows that assessors 12, 14, 15, and 16 are slightly more distant from the others.

Figure 8.6 shows the spices along with their confidence ellipsoids. We
observe that for some spices, the ellipsoids are small (e.g., Pepper, Pep+Nut
and Cinnamon, Cin+Clo+Car), and for others, the ellipsoids are larger (e.g.,
Chili, Coriander, Clove). This pattern indicates that for specific spices, the
assessors are rather in agreement, whereas they tend to diverge for other
spices. The first dimension opposes the spices with Chili and Turmeric to
the spices with Pepper and Nutmeg. The second dimension opposes the
spices with Cinnamon to the spices with Pepper. Concerning the product
positioning, as can be seen by comparing Figures 8.4 and 8.6, the general
solution of DISTATIS is very close to the metric MDS map. This conclusion is confirmed by computing a R_V coefficient between the factor spaces
of the MDS and DISTATIS. Its value of 0.99 confirms that the two spaces
are almost identical.

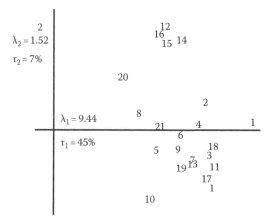

FIGURE 8.5 A 2D DISTATIS map of the assessors. The map suggests that the
assessors are rather homogeneous.

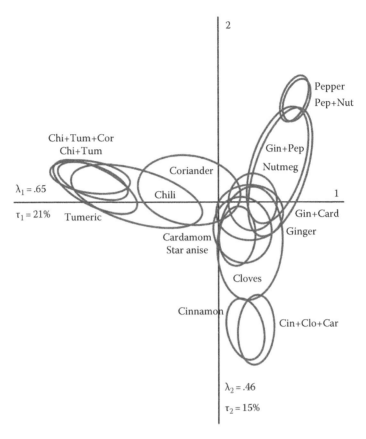

FIGURE 8.6 A 2D DISTATIS map showing the products with their 95% confidence ellipsoids. When the confidence ellipsoids of two products do not intersect, the products are perceived as significantly different by the group of assessors.

8.A APPENDIX: ANALYSES WITH R

The techniques described in this chapter can be performed with R using the packages DistatisR (developed by Abdi, Beaton, and Chin-Fatt and available from CRAN and the senior author's homepage). The data of the example are available as a data set from this package. The analysis of the results of this chapter can be obtained with the following R commands:

```
# install the packages
install.packages("prettyGraphs")
install.packages("car")
install.packages("DistatisR")
#----------------------------------------------------
```

```
# 1 Load the library DistatiR
library(DistatisR)
# The data to illustrate the paper are in the data set
# SortingSpice
# that is provided by the package DistatisR.
data(SortingSpice)
# This data set gives the data.frame called SortSpice
# that we will analyze
#-----------------------------------------------------

# 2 Create the set of distance matrices (one distance
# matrix per assessor)
#    (use the function DistanceFromSort) applied to
#    SortSpice
DistanceCube <- DistanceFromSort(SortSpice)
#    The results will be stored into the 3D array
#    called DistanceCube
#-----------------------------------------------------
# 3 For the first analysis, we will perform
#    a metric multidimensional analysis
#    on the distance matrix obtained by summing
#    all the assessors' distance matrices.
#    The sum is obtained with this instruction
TotalDistance = apply(DistanceCube,c(1,2),sum)
#    The sum is stored in TotalDistance
#
# 3.1  Analyze TotalDistance with metric mds.
#       Use the function mmds (from package DistatisR)
mdsRes <- mmds(TotalDistance)
# The results of mmds are in mdsRes
# 3.2 Now a pretty plot with the prettyPlot function
# from prettyGraphs
# For this plot we will use the factors scores (stored
# in mdsRes$FactorScore)
PlotMDS <- prettyPlot(mdsRes$FactorScore,
            display_names = TRUE,
            display_points = TRUE,
            contributionCircles = TRUE,
            contributions = mdsRes$Contributions)
#-----------------------------------------------------
# 4  For the second analysis, we use the distatis
#    method
#    This is performed by the distatis function
#    (with cube of distance "DistanceCube" as a
#    parameter)
testDistatis <- distatis(DistanceCube)
```

```
# The factor scores for the products (i.e., spices)
# are in
# testDistatis$res4Splus$F
# the factor scores for the assessors (i.e., analysis
# if the RV matrix)
# are in testDistatis$res4Cmat$G
##---------------------------------------------------
# 4.1    Inferences on the products obtained via
#        bootstrap
#        here we use two different bootstraps:
# 4.1.1 Bootstrap on factors (very fast but could be
#        too liberal
#        when the number of assessors is very large)
BootF <- BootFactorScores(testDistatis$res4Splus$Parti
  alF,niter=1000)
# 4.1.2 Complete bootstrap obtained by computing sets
#        of compromises
#        and projecting them (could be significantly
#        longer because a lot
#        of computations is required)
#
F_fullBoot <- BootFromCompromise(DistanceCube,
                     niter=1000)
#---------------------------------------------------
# 4.2    Create the Graphics
# Get the Factor Scores and Partial Factor Scores for
# the plot Routine
LeF        <- testDistatis$res4Splus$F
PartialFS <- testDistatis$res4Splus$PartialF
# 4.2.1 plot the Observations with the Bootstrapped
#        CI from the factor scores
PlotOfSplus <- GraphBootDistatisCpt(LeF,
                     BootF,PartialFS,ZeTitle='Bootstrap
                     on Factors')
# 4.2.2 Plot the Observations with the bootstrapped CI
#        from the Compromises
PlotOfSplusCpt <- GraphBootDistatisCpt
                        (LeF, F_fullBoot,
                        PartialFS,ZeTitle='Full
                        Bootstrap')
# 4.2.3 Plot the Bootstrap results with ellipses
#        instead of convex hulls
PlotOfSplusElli <- GraphBootDistatisEllipseCpt(Le
     F,F_fullBoot,PartialFS,
     ZeTitle='Full Bootstrap 95% CI', nude=TRUE,
     color = PlotOfSplus$col)
```

```
# 4.2.4 Plot the RV map
PlotOfRvMat <- GraphDistatisRv
                (testDistatis$res 4Cmat$G,
                ZeTitle='Rv Mat')
# 5  A MCA type of approach can be obtained by
#    replacing step 2 by
DistanceCube <- Chi2DistanceFromSort(SortSpice)
# This will compute a Chi-2 distance for each
# assessor,
# suffice to repeat steps 3 to 4 to obtain the
# analysis
```

REFERENCES

Abdi, H. 2007a. R_V coefficient and congruence coefficient. In N.J. Salkind (ed.): *Encyclopedia of Measurement and Statistics*. Thousand Oaks, CA: Sage. pp. 849–853.

Abdi, H. 2007b. Metric multidimensional scaling. In N.J. Salkind (ed.): *Encyclopedia of Measurement and Statistics*. Thousand Oaks, CA: Sage. pp. 598–605.

Abdi, H., Dunlop, J.P., and Williams, L.J. 2009. How to compute reliability estimates and display confidence and tolerance intervals for pattern classifiers using the bootstrap and 3-way multidimensional scaling (DISTATIS). *NeuroImage* 45: 89–95.

Abdi, H. and Valentin, D. 2007a. Multiple factor analysis. In N.J. Salkind (ed.): *Encyclopedia of Measurement and Statistics*. Thousand Oaks, CA: Sage. pp. 657–663.

Abdi, H. and Valentin, D. 2007b. Multiple correspondence analysis. In N.J. Salkind (ed.): *Encyclopedia of Measurement and Statistics*. Thousand Oaks, CA: Sage. pp. 651–657.

Abdi, H. and Valentin, D. 2007c. Some new and easy ways to describe, compare, and evaluate products and assessors. In D. Valentin, D.Z. Nguyen, and L. Pelletier (eds.): *New Trends in Sensory Evaluation of Food and Non-food Products*. Ho Chi Minh, Vietnam: Vietnam National University—Ho Chi Minh City Publishing House. pp. 5–18.

Abdi, H., Valentin, D., Chollet, S., and Chrea, C. 2007. Analyzing assessors and products in sorting tasks: DISTATIS, theory and application. *Food Quality and Preference* 18: 627–640.

Abdi, H., Valentin, D., O'Toole, A.J., and Edelman, B. 2005. DISTATIS: The analysis of multiple distance matrices. *Proceedings of the IEEE Computer Society: International Conference on Computer Vision and Pattern Recognition*, San Diego, CA, pp. 42–47.

Abdi, H. and Williams, L.J. 2010. Principal component analysis. *Wiley Interdisciplinary Reviews: Computational Statistics* 2: 433–459.

Abdi, H., Williams, L.J., and Valentin, D. 2013. Multiple factor analysis. *Wiley Interdisciplinary Reviews: Computational Statistics* 5: 149–179.

Abdi, H., Williams, L.J., Valentin, D., and Bennani-Dosse, M. 2012. STATIS and DISTATIS: Optimum multi-table principal component analysis and three way metric multidimensional scaling. *Wiley Interdisciplinary Reviews: Computational Statistics* 4: 124–167.

Ballester, J., Abdi, H., Langlois, J., Peyron, D., and Valentin, D. 2009. The odors of colors: Can wine expert or novices distinguish the odors of white, red, and rosé wines? *Chemosensory Perception* 2: 203–213.

Ballester, J., Dacremont, C., Le Fur, Y., and Etievant, P. 2005. The role of olfaction in the elaboration and use of the Chardonnay wine concept. *Food Quality and Preference* 16: 351–359.

Ballester, J., Patris, B., Symoneaux, R., and Valentin, D. 2008. Conceptual vs. perceptual wine spaces: Does expertise matter? *Food Quality and Preference* 19: 267–276.

Bécue-Bertaut, M. and Lê, S. 2011. Analysis of multilingual labeled sorting tasks: Application to a cross-cultural study in wine industry. *Journal of Sensory Studies* 26: 299–310.

Beguin, P. 1993. *La classification et la dénomination de parfums chez des experts et des novices*, Thèse de Doctorat, Université de Louvain, Louvain-la-Neuve, Belgique.

Blancher, G., Chollet, S., Kesteloot, R., Nguyen, D., Cuvelier, G., and Sieffermann, J.-M. 2007. French and Vietnamese: How do they describe texture characteristics of the same food? A case study with jellies. *Food Quality and Preference* 18: 560–575.

Blancher, G., Clavier, B., Egoroff, C., Duineveld, K., and Parcon, J. 2012. A method to investigate the stability of a sorting map. *Food Quality and Preference* 23: 36–43.

Cadoret, M. and Husson, F. 2013. Construction and evaluation of confidence ellipses applied at sensory data. *Food Quality and Preference* 28: 106–115.

Cadoret, M., Lê, S., and Pagès, J. 2009. A factorial approach for sorting task data (FAST). *Food Quality and Preference* 20: 410–417.

Cadoret, M., Lê, S., and Pagès, J. 2011. Statistical analysis of hierarchical sorting data. *Journal of Sensory Studies* 26: 96–105.

Campo, E., Do, V.B., Ferreira, V., and Valentin, D. 2008. Sensory properties of a set of commercial Spanish monovarietal young white wines. A study using sorting tasks, lists of terms and frequency of citation. *Australian Journal of Grape and Wine Research* 14: 104–115.

Cartier, R., Rytz, A., Lecomte, A., Poblete, F., Krystlik, J., Belin, E., and Martin, N. 2006. Sorting procedure as an alternative to quantitative descriptive analysis to obtain a product sensory map. *Food Quality and Preference* 17: 562–571.

Chollet, S., Lelièvre, M., Abdi, H., and Valentin, D. 2011. Sort and beer: Everything you wanted to know about the sorting task but did not dare to ask. *Food Quality and Preference* 22: 507–520.

Chollet, S. and Valentin, D. 2001. Impact of training on beer flavor perception and description: Are trained and untrained assessors really different? *Journal of Sensory Studies* 16: 601–618.

Chrea, C., Valentin, D., Sulmont-Rossé, C., Ly, M.H., Nguyen, D., and Abdi, H. 2005. Semantic, typicality and odor representation: A cross-cultural study. *Chemical Senses* 30: 37–49.

Clark, H.H. 1968. On the use and meaning of propositions. *Journal of Verbal Learning and Verbal Behavior* 7: 421–431.

Courcoux, P., Qannari, E.M., Taylor, Y., Buck, D., and Greenhoff, K. 2012. Taxonomic free sorting. *Food Quality and Preference* 23: 30–35.

Coxon, A.P.M. 1999. *Sorting Data: Collection and Analysis.* Thousand Oaks, CA: Sage.

Deegan, K.C., Koivisto, L., Näkkilä, L.J., Hyvönen, L., and Tuorila, H. 2010. Application of a sorting procedure to greenhouse-grown cucumbers and tomatoes. *Food Science and Technology* 43: 393–400.

Dehlholm, C., Brockhoff, P.B., and Bredie, W.L.P. 2012a. Confidence ellipses: A variation based on parametric bootstrapping applicable on multiple factor analysis results for rapid graphical evaluation. *Food Quality and Preference* 26: 278–280.

Dehlholm, C., Brockhoff, P.B., Meinert, L., Aaslyng, M.D., and Bredie, W.L.P. 2012b. Rapid descriptive sensory methods—Comparison of free multiple sorting, partial napping, napping, flash profiling and conventional profiling. *Food Quality and Preference* 26: 267–277.

Derndorfer, E. and Baierl, A. 2006. Development of an aroma map of spices by multidimensional scaling. *Journal of Herbs, Spices & Medicinal Plants* 12: 39–50.

Dravineks, A. 1982. Odor quality: Semantically generated multidimensional profiles are stable. *Science* 218: 799–801.

Escofier, B. and Pagès, J. 1983. Méthode pour l'analyse de plusieurs groupes de variables: Application à la caractérisation des vins rouges du Val de Loire. *Revue de Statistique Appliquée* 31: 43–59.

Falahee, M. and MacRae, A.W. 1995. Consumer appraisal of drinking water: Multidimensional scaling analysis. *Food Quality and Preference* 6: 327–332.

Falahee, M. and MacRae, A.W. 1997. Perceptual variation among drinking waters: The reliability of sorting and ranking data for multidimensional scaling. *Food Quality and Preference* 8: 389–394.

Faye, P., Brémaud, D., Durand-Daubin, D., Courcoux, P., Giboreau, A., and Nicod, A. 2004. Perceptive free sorting and verbalization tasks with naive subjects: An alternative to descriptive mappings. *Food Quality and Preference* 15: 781–791.

Faye, P., Brémaud, D., Teillet, E., Courcoux, P., Giboreau, A., and Nicod, H. 2006. An alternative to external preference mapping based on consumer perceptive mapping. *Food Quality and Preference* 17: 604–614.

Giboreau, A., Navarro, S., Faye, P., and Dumortier, J. 2001. Sensory evaluation of automotive fabrics: The contribution of categorization tasks and non verbal information to set-up a descriptive method of tactile properties. *Food Quality and Preference* 12: 311–322.

Gygi, B., Kidd, G.R., and Watson, C.S. 2007. Similarity and categorization of environmental sounds. *Perception and Psychophysics* 69: 839–855.

Heymann, H. 1994. A comparison of free choice profiling and multidimensional scaling of vanilla samples. *Journal of Sensory Studies* 9: 445–453.

Hoek, A.C., Van Boekel, M., Voordouw, J., and Luning, P.A. 2011. Identification of new food alternatives: How do consumers categorize meat and meat substitutes? *Food Quality and Preference* 22: 371–383.

Hulin, W.S. and Katz, D. 1935. The Frois-Wittmann pictures of facial expression. *Journal of Experimental Psychology* 18: 482–498.

Imai, S. 1966. Classification of sets of stimuli with different stimulus characteristics and numerical properties. *Perception and Psychophysics* 1: 48–54.

Kirkland, J., Bimler, D., Drawneek, A., McKim, M., and Schölmerich, A. 2000. A quantum leap in the analyses and interpretation of attachment sort items. In B. Vaughn, E. Waters, G. Posada, and D. Teti (eds.): *Patterns of Secure Base Behavior: Q-Sort Perspectives on Attachment and Caregiving in Infancy and Childhood.* Hillsdale, NJ: Lawrence Erlbaum.

Lawless, H.T. 1989. Exploration of fragrances categories and ambiguous odors using multidimensional scaling and cluster analysis. *Chemical Senses* 14: 349–360.

Lawless, H.T. and Glatter, S. 1990. Consistency of multidimensional scaling models derived from odor sorting. *Journal of Sensory Studies* 5: 217–230.

Lawless, H.T., Sheng, N., and Knoops, S.S.C.P. 1995. Multidimensional scaling of sorting data applied to cheese perception. *Food Quality and Preference* 6: 91–98.

Lelièvre, M., Chollet, S., Abdi, H., and Valentin, D. 2008. What is the validity of the sorting task for describing beers? A study using trained and untrained assessors. *Food Quality and Preference* 19: 697–703.

Lelièvre, M., Chollet, S., Abdi, H., and Valentin, D. 2009. Beer trained and untrained assessors rely more on vision than on taste when they categorize beers. *Chemosensory Perception* 2: 143–153.

Lim, J. and Lawless, H.T. 2005. Qualitative differences of divalent salts: Multidimensional scaling and cluster analysis. *Chemical Senses* 30: 719–726.

MacRae, A.W., Rawcliffe, T., Howgate, P., and Geelhoed, E.N. 1992. Patterns of odour similarity among carbonyls and their mixtures. *Chemical Senses* 17: 119–125.

Miller, G.A. 1956. The magic number seven plus or minus two: Some limits of our capacity for processing information. *Psychological Review* 63: 81–97.

Miller, G.A. 1969. A psychological method to investigate verbal concepts. *Journal of Mathematical Psychology* 6: 169–191.

Nava Guerra, M.T., Chollet, S., Gufoni, V., Patris, B., and Valentin, D. 2004. *Chemo-Sensorial Expertise and Categorization: The Case of Beers.* Dijon, France: European Chemoreception Research Organisation.

Parr, W.V., Valentin, D., Green, J.A., and Dacremont, C. 2010. Evaluation of French and New Zealand Sauvignon wines by experienced French wine assessors. *Food Quality and Preference* 21: 56–64.

Patris, B., Gufoni, V., Chollet, S., and Valentin, D. 2007. Impact of training on strategies to realize a beer sorting task: Behavioral and verbal assessments. In D. Valentin, D.Z. Nguyen, and L. Pelletier (eds.): *New Trends in Sensory Evaluation of Food and Non-Food Products.* Ho Chi Minh, Vietnam: Vietnam National University—Ho Chi Minh City Publishing House.

Picard, D., Dacremont, C., Valentin, D., and Giboreau, A. 2003. Perceptual dimensions of tactile textures. *Acta Psychologica* 114 : 165–184.

Piombino, P., Nicklaus, S., LeFur, Y., Moio, L., and Le Quéré, J. 2004. Selection of products presenting given flavor characteristics: An application to wine. *American Journal of Enology and Viticulture* 55: 27–34.

Qannari, E.M., Cariou, V., Teillet, E., and Schlich, P. 2009. SORT-CC: A procedure for the statistical treatment of free sorting data. *Food Quality and Preference* 21: 302–308.

Rao, V.R. and Katz, R. 1971. Alternative multidimensional scaling methods for large stimulus sets. *Journal of Marketing Research* 8: 488–494.

Saint-Eve, A., Paçi Kora, E., and Martin, N. 2004. Impact of the olfactory quality and chemical complexity of the flavouring agent on the texture of low fat stirred yogurts assessed by three different sensory methodologies. *Food Quality and Preference* 15: 655–668.

Santosa, M., Abdi, H., and Guinard, J.X. 2010. A modified sorting task to investigate consumer perceptions of extra virgin olive oils. *Food Quality and Preference* 21: 881–892.

Solomon, G.E.A. 1997. Conceptual change and wine expertise. *Journal of the Learning Sciences* 6: 41–60.

Soufflet, I., Calonnier, M., and Dacremont, C. 2004. A comparison between industrial experts' and novices' haptic perception organization: A tool to identify descriptors of handle of fabrics. *Food Quality and Preference* 15: 689–699.

Stevens, D.A. and O'Connell, R.J. 1996. Semantic-free scaling of odor quality. *Physiological Behavior* 60: 211–215.

Takane, Y. 1980. Analysis of categorizing behavior by a quantification method. *Behaviometrika* 8: 75–86.

Tang, C. and Heymann, H. 1999. Multidimensional sorting, similarity scaling and free choice profiling of grape jellies. *Journal of Sensory Studies* 17: 493–509.

Teillet, E., Schlich, P., Urbano, C., Cordelle, S., and Guichard, E. 2010. Sensory methodologies and the taste of water. *Food Quality and Preference* 21: 967–976.

Thuillier, B. 2007. Rôle du CO_2 dans l'appréciation organoleptique des champagnes—Expérimentation et apports méthodologiques. Thèse de Doctorat, Université de Reims, Reims, France.

9 Projective Mapping and Napping

Christian Dehlholm

CONTENTS

9.1   METHOD OVERVIEW

- Projective mapping is a sensory evaluation technique in which samples are—as the name implies—projected onto a 2D surface, for example, a piece of paper or as icons on a screen. Napping is a commonly applied variant of projective mapping.
- Evaluation instructions are as follows: to place samples perceived as similar closer to each other, while samples perceived less similar are placed further apart. In this way, all samples are placed according to one another. Normally, assessors decide on their own discrimination criteria.
- It is common to attain descriptive sample information by adding on a few describing words to each sample.

- The projections' Cartesian coordinates (x, y) and the frequencies of similar sample descriptions constitute the data of an evaluation. Data are treated with multivariate multiblock analysis.
- This evaluation type is commonly regarded as being a rapid as well as a low-resource approach. Studies show that projective mapping methods may not give as detailed sample information as conventional profiling, and that results show rather large confidence intervals. Nevertheless, the information derived has proven both multidimensional and well applied for numerous purposes. Hence, the method has gained in popularity during the last decade.

9.2   THEORY BEHIND THE METHOD

9.2.1   INTRODUCTION

Projective mapping was introduced to the sensory science field by 1994. It was presented as a rapid approach to collect (dis)similarities within a set of products, with a more holistic sample approach. Thompson and MacFie (1983) highlighted the important challenges of conventional descriptive evaluation: the employment of analytic profiles, when perception is in fact thought to be holistic. The projective mapping technique in various forms applied the idea of "placing" psychology (Dun-Rankin 1983; Risvik et al. 1994), as simultaneously presented samples had to be projected within a 2D area framed by a sheet of paper. Assessors were introduced to the projective mapping technique, but had no formal training. Risvik et al. (1994) originally supplied assessors with a rectangular paper sheet and the sample set and instructed them to place samples perceived as similar close to each other and samples perceived different further apart. By 2003, the projective mapping variant Napping was introduced, which applied the same evaluation instructions as projective mapping (Pagès 2003, 2005). However, Napping assumed that a set of data analysis instructions were followed and hence could be seen as a more defined case of projective mapping (Dehlholm 2012; Dehlholm et al. 2012b). The ultra-flash profiling method (Pagès 2003) was normally considered to constitute the collection of the assessors' semantic descriptions of products as add-on methodology for Napping. With inspiration from sorting, the sorted Napping variation (Pagès et al. 2010) suggested to replace ultra-flash profile by a description of groupings of projected products. As a recent method variation, the "global" (normal) Napping was seen in opposition to the "partial" Napping, where assessors were guided by, for example, sensory modalities (Pfeiffer and Gilbert 2008; Dehlholm 2012; Dehlholm et al. 2012b). An even faster evaluation approach has been implemented as the consensus Napping for group judgments, but it did not show reliable results for assessors untrained in the methodology (Dehlholm 2012). Several authors have recently compared projective mapping with other rapid descriptive methods and highlighted the technique as successfully applied for various purposes (Ares et al. 2010, 2011; Nestrud and Lawless 2010; Albert et al. 2011; Dehlholm 2012; Dehlholm et al. 2012a,b; Valentin et al. 2012; Varela and Ares 2012). Today, the technique has been broadly included into sensory software enabling computer screens, tablets, and other mobile devices to constitute the projective response area.

The easy-to-apply practical aspect of projective mapping evaluations opens up for multiple method variations. They have been applied with various geometries of the evaluation frame, with or without the use of scales, in different couplings to semantic-collecting techniques and with all types of assessors. This chapter will guide the reader and the practical user of projective mapping in understanding the implications of some of these variations and at the same time work as a practical manual.

9.2.2 FRAMING HOLISTIC PERCEPTION

While no prior consideration has been given to the geometric frame in which sample projections are made, recent studies indicate that the choice of geometry affected the results (Dehlholm 2012; Dehlholm et al. 2014a; Hopfer and Heymann 2013). At present, projective mapping has been performed on 2D spaces. This makes it possible to vary either the size of the frame or the shape of the frame. In addition, scales have been applied in various ways. Risvik et al. (1994) applied unstructured line scales in a study with trained assessors. In a later study with consumers, Risvik et al. (1997) applied structured line scales. King et al. (1998) compared the structured and unstructured approach, in a study where untrained assessors were asked for hedonic comparisons, and concluded the unstructured approach provided the best product representation. The sensory professional's choice of scales has received little attention since. The choice of adding scales to projective mapping is discussed later.

Regarding frame size, Risvik et al. used both the metric formats A4 and A3 paper (Risvik et al. 1994, 1997). Pagès (2003) used 60 cm × 40 cm size of paper (approximately A2) for Napping. Others have used 60 cm × 60 cm paper sheets (King et al. 1998; Kennedy and Heymann 2009; Nestrud and Lawless 2010), and a 40 cm in diameter round frame was tried (Dehlholm 2012). None has examined if a relationship existed between varying sizes of a similar geometry. And until now, variations of projective mapping geometry seem to have been applied without further considerations. But, there has been proven perceptual relations between a distance, in the form of a line, and its frame (Künnapas 1955) and the geometry of the frame influences the directionally perceived distances (Künnapas 1957b).

Both Dehlholm (2012) and Hopfer and Heymann (2013) have recently confirmed that the geometry of the frame impact projective mapping results. Although significantly different results were found, Dehlholm (2012) concluded that it is important to understand how method variations affect assessor responses.

In Napping, data are collected in the rectangular geometry under the assumption that a wider horizontal direction is perceived as being more important. Since projective mapping methods are generally regarded as being holistic procedures, it is surprising that assessors are indirectly encouraged to be more unidimensional in their product projection. The rectangular geometry encourages the assessor to apply a small number of global main attributes to describe main differences along imagined x- and y-axes of the frame. The 60×60 cm geometry is thought to generate a projection strategy like that of the rectangular geometry but without guidance for the more important dimension as with the rectangular geometry. The application of the round geometry was based on the idea of removing this potential simplification and splitting up the frame into axes. But the fact that our eyes cover a horizontal elliptical field of view (Künnapas 1957a) with the horizontal to vertical aspect ratio of 1.5 (Egge 1984) might actually ease the task of projecting in a rectangular frame. So far, no sensory studies have reported the application of other sizes of frames, for example, the elliptical frame shaped according to our field of view.

9.2.3 PROJECTION STRATEGIES

The distances between samples in projective mapping are currently handled by model algorithms as Euclidean distances. Hence, the differences between samples are handled according to the evaluation instructions, namely, to place more similar products closer and more distant products further apart. But where the assessor instructions initially may seem clear and straightforward, it now appears that assessors interpret them differently. Several authors have suggested that projection strategies may vary from assessor to assessor (Kennedy 2010; Dehlholm 2012; Hopfer and Heymann 2013). Variation in projection strategies depends much on the assessor type and level of training. A recent study shows that 83 naive assessors who performed three various projective mapping tasks clearly apply different projection strategies for placing the samples inside the projection frame (Dehlholm 2012). Six types of strategies were identified in the study (Figure 9.1):

1. One-way linear projection patterns, sometimes observed to be horizontal or vertical, but most often diagonal. The diagonal arrangement might be the naive assessors' reaction to indulge experiment

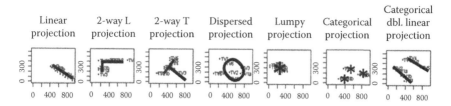

FIGURE 9.1   Observed projection strategies by naive assessors in 249 projective mappings of sound quality recorded from the loudspeakers of 7 TV set. Observed strategies were (1) one-way linear projections, (2) two-way L or T projections, (3) dispersed projections, (4) lumpy projections, (5) categorical projections, and (6) categorical double linear projections. (From Dehlholm, C., *Descriptive Sensory Evaluations: Comparison and Applicability of Novel Rapid Methodologies*, SL grafik, Copenhagen, Denmark, 2012.)

instructions, as the use of the whole projective space was recommended. With this projection pattern, the given responses are primarily one-dimensional at an individual assessor level.

2. Two-way L- or T-shaped projection patterns, reminding of the linear case but with added dimensionality. This was seen when a few products were placed in an orthogonal dimension to the main construct. It seems to be a reasonable assumption that products in some cases are placed in the diverging dimension according to the product that binds together the diverging dimensions only rather than according to the whole product space.

3. Dispersed projections, which occupy a greater area of the projective frame, and sometimes seen in an O or S shape pattern. Here, two or more perceptual dimensions have probably been into play or the assessor has been able to place the products without substantial considerations and thus promoting a more conceptual placement.

4. Lumpy projections, where all products are placed closely together in a small area of the projection frame and where it is difficult to evaluate or recognize a clear projection pattern. This is possibly a consequence of the naive assessor being unfamiliar with the methodology and afraid of making *judgment errors* rather than the case where the assessor did not perceive any differences between products.

5. Categorical projections, showing clear clusters of products and seeming like the intracluster distances are not of any importance to the assessor.

6. Categorical double linear projections, observed as a combination of two linear patterns clearly separated from each other in a categorical manner. It is uncertain whether the interlinear distance can be matched to the intralinear ones.

According to the mentioned strategies, the category of dispersed projections with full use of the projection area is the wish of the sensory professional. However, this implies that the assessor have some methodological experience. Therefore, the sensory professional may choose to spend more time to introduce the method for complete naive assessors, while it is less important for the more experienced assessors. In fact, projection strategies might relate to assessors' general knowledge and experience of working with sensory spaces.

9.2.4 Semantics in Projective Mapping

Projective mapping in its single form consists only of the assessor carrying out the projections with no further actions. However, as results become much more valuable with added product descriptions. Various techniques exist to add on a descriptive step to the projected products. The names of these add-on techniques are often omitted and hence assumed to be an integrated part of projective mapping. As of today, the most applied add-on description step is called ultra-flash profile. The name relates to flash profiling, a rapid version of the earlier free choice profiling, as being an ultrarapid independent vocabulary generation. In other words, the assessor *just* supplies a free number of adjectives for each sample after having finalized the projections. Nevertheless, an important difference exists between the ultra-flash profiling and flash profiling. Flash profiling is similar to conventional profiling, an attribute-based approach in which assessors evaluate a vocabulary determined a priori. Attributes are fixed and the evaluation approach becomes more reductionistic, or instrumental. On the contrary, projective mapping techniques are thought to be of a more holistic nature where the perceived differences are expressed via the projections. The sensory professional will not force the assessor to note or explain why a sample is placed as it is. Hence, the placements of the samples are based more on intuition. If the assessor knows why the samples were placed as they were, he or she is encouraged to add adjectives for each sample. This would be the ultra-flash profile step. As this very open procedure often results in a large variation of reported semantics, various semantic-guiding options are applicable to let assessors stay within similar semantic boundaries, for example, a flavor wheel, prior established vocabulary or partial Napping, as explained in the following section.

9.2.5 Partial Napping

It has been proposed to distinguish between two different forms of Napping: the conventional version without restrictions, alternatively called *global Napping*, versus the *partial Napping*, where the assessor is restricted in his or her response (Dehlholm 2012; Dehlholm et al. 2012b). Bearing in mind that the panel leader cannot restrict the assessor's cognitive response to a stimulus, the panel leader can still guide the focus of the assessor by only asking for specific terms of interest. This will result in a partial Napping on mouthfeel, a partial Napping on appearance, a partial Napping on emotions, etc. Dehlholm demonstrated that the three consecutive partial Nappings by the same panel of expert assessors based on the evaluation of appearance, flavor, and mouthfeel could be aggregated into one analysis. And the result of that analysis showed a product space very similar to that obtained from conventional profiling by an expert panel. However, only few attributes were similar between methods. As a rapid profiling technique, projective mapping variations rarely include training.

9.2.6 CONSENSUS NAPPING

Dehlholm et al. (2014b) examined the potentials of carrying out descriptive profiling in the fastest possible form, namely, as a group session where the group provided one consensus result only. This was attempted in order to evaluate the process, as it actually takes place in several environments. One example would be a group of marketing professionals that want to gain a fast overview of a product landscape. In many cases, they would not trouble themselves with recording and analyzing individual measures. Decisions would then be taken immediately and on the basis of the agreed consensus placements. Another example would be the product development process (where sensory professionals were not involved) where decisions had to be taken immediately on the basis of a single rapidly obtained consensus result. These group evaluations were based on various procedures, one of them projective mapping, specifically named as consensus Napping. In such an evaluation, the absolute number of assessors present in the group might vary. As consensus Napping only provides one set of coordinates per product, having less assessors in a consensus group would correspond to having fewer measurements from individual assessors. Hence, the value of the input from a 4-person group evaluation would have the weight 0.25 per person compared with 0.1 from a 10-person group evaluation. This leaves the result from a smaller group more vulnerable to the variation and skill within the individual assessor.

The amount of semantic attributes derived from a consensus evaluation based on a free-choice vocabulary is lower than those from a similar individual-based evaluation. As explained later, a traditional projective mapping data set could consist of one set of coordinates per sample, followed by a contingency table with the semantic attributes and their frequencies. A general multiple factor analysis (MFA) on Napping data suggests leaving the semantic information as supplementary and letting the model configuration rely on projective information only. But, for consensus data, the reported semantics accounts for a substantial amount of the product information. It should generally be considered to include the semantic information in the product configuration modeling.

9.3   DESIGN OF EXPERIMENTS

Projective mapping is a rapid evaluation technique. Hence, the planning of a projective mapping evaluation is a relative simple task for the sensory professional who is familiar with conventional profiling. Since projective mapping and variations belong to the newer methodological developments in sensory evaluation, conventions and standards are not yet established for their use. Various studies have applied projective mapping in various ways and compared those with each other and with other methods, for example, conventional profiling. But a golden path needs yet to be found. As a result, there are some factors to consider for the sensory professional who has to prepare a projective mapping evaluation. These are listed in Table 9.1 together with their known variations and notes on their possible implications.

For each specific factor, a choice must be made, to which the following notes are meant as a practical guidance:

1. *Choice of frame geometry.* Both projective mapping and the later Napping used the rectangular piece of paper to collect projective measurements. The rationale behind (for Napping) was to guide the assessor to span the more important space in the horizontal direction.

    In specialized sensory software, you will often get the choice between the rectangular and the square frame. Even though studies prove frames to have a possible influence on the placement of a sample, the practical significance might not be evident.
2. *Scales.* In the first sensory projective mapping study by Risvik (1994), crossed axes were applied within the frame. But since, studies seem to avoid axes. As the results are based on Euclidean distances between samples, it is only the distances between samples that matters for the results. As soon as scales and lines are put within the evaluation frame, they will define some area or dimension. If the wish is to let the assessor be more spontaneous or holistic, axes should be omitted. If it for some reason is a wish to probe two main dimensions within the frame, the result will be stronger, but not based on free basic Napping principles. In that case, one could consider examining the same variation by a simple two independent line scale evaluation.

TABLE 9.1
Factors to Consider for a Projective Mapping (Napping) Evaluation

Factor	Variation (Not Exhaustive)	Implication (Theoretical)
Frame geometry	Rectangular	Conventional—assessors generally consider the longer direction as more important.
	Square	Applied in several studies—dimensionality is not taken into consideration.
	Round	Applied in few studies—aims at removing dimensionality thinking from the assessor.
	Ellipse	Suggested (not yet applied)—aims at mimicking the visual field.
Added scales	None	Conventional—more holistic perceptual approach.
	Grid or scales	Not conventional—adds dimensionality to the projective space and promotes a constructual sample approach.
Type of assessor	Naïve	Consumer study—more holistic thinking, but might apply various projection strategies.
	Professional panelists	Assessors might apply semantics from a priory trained vocabulary and apply a more constructual sample approach.
Collection of semantics	For each product (Napping with ultra-flash profiling)	Conventional—semantics are noted for each sample after placement.
	For groups of products (sorted Napping)	Semantics are noted for each group of similar samples after placement.
Free or restricted/ guided responses	Global Napping	Assessors are allowed to place samples according to a "perceptual" system free of choice.
	Partial Napping	Evaluation instructions tell assessors to focus their projections on a pregiven subject, e.g., a sensory modality or emotional states.
Repetitions	None	Supports the idea of the spontaneous evaluation.
	Some, e.g., three	Provides a larger data set for each assessor.

3. *Assessors.* As in any study, the assessors must be chosen according to the aim of the analysis. The more trained or expert an assessor is, the more they would tend to evaluate with the use of trained or learned constructs. So, if it is a wish to obtain more holistic and spontaneous product differentiations, use the more naive consumer. The number of assessors varies greatly according to the type of product to be evaluated and to the type of assessor. Studies have been made with a number of assessors varying from 8 (Risvik et al. 1997; Barcenas et al. 2004; Pagès 2005; Kennedy and Heymann 2009) to 83 (Dehlholm 2012). As it is most likely dependent on the sample set, no research has yet concluded on the minimum number of assessors necessary to achieve stable product spaces.

4. *Semantics.* Projective mapping techniques are not descriptive unless supplied with the additional collection of sample descriptors. The most widespread technique has been the simple ultra-flash profile. Other techniques exist but have only been applied to a low extent.

5. *Free or restricted/guided responses.* Conventional or global Napping is the projective mapping, where the assessors design their own system for product separation. For various reasons, the sensory professional might be interested in a separation of products according to, for example, mouthfeel only. Then, assessors would be instructed to separate the products according to mouthfeel. This is a partial Napping, a directed projective mapping, where assessor focus is guided by the instructions. A not insignificant advantage with the partial Napping is that it can be repeated with different foci, with the same assessors and still within a short time frame. For instance, the sensory professional could ask assessors to perform first a partial Napping on appearance, then a partial Napping on flavor, and finally, a partial Napping on mouthfeel.

6. *Repetitions.* There are no clear standards of whether to use repetitions in projective mapping evaluations. Some authors do use replicates and some do not. To make a competent choice of whether to use replicates, spare a thought on the purpose of the replicate. If the evaluation is performed with trained panelists that use predefined attributes, maybe you want repetitions to look at individual performances. But as described later, there are other ways to evaluate individual performances. Furthermore, studies show that individually repeated projective mappings are seldom alike (Kennedy 2010; Dehlholm 2012), even though the average panel

configurations are similar. If one assessor is asked to perform repetitions in an evaluation that is supposed to retrieve measures of spontaneous perception, the second and third repetition must be somewhat less spontaneous and might involve some memory biases. If the purpose of the evaluation is to measure spontaneous perception, it seems more rational to perform one evaluation per assessor only. If it is a wish to retrieve more data from each assessor, one can consider performing several partial Nappings, where each of them has a different focus.

The appropriate amount of samples to include in a projective mapping evaluation varies according to the type of product under investigation. As the assessor projects the samples on a 2D surface, the minimum number of samples to include must be four, which are needed to span the dimensions, although more samples are typically involved. Five (Risvik et al. 1994; Pagès 2005) to eighteen (King et al. 1998) samples have been reported in various projective mapping studies. The sensory professional must choose the maximum number of samples according to the nature of the sample and with respect to sensory fatigue of the assessors.

A balanced and randomized evaluation order is ideal. However, projective mapping is a rapid evaluation technique and is most often used to highlight main differences between products. The level of noise is often higher for projective mapping than for conventional profiling. Hence, small perceptional differences normally found due to a minimization of carryover effects and other biases might fade away in evaluation uncertainty. For this reason, and based on the often rather practical applications of projective mapping, the methodology can be applied without the implementation of a balanced and randomized evaluation design. In a laboratory study, one should aim toward the more ideal randomization of samples.

It is a good idea to use duplicates in the evaluation. Duplicates are one or more of the samples blindly repeated within the set. The duplicates, treated as separate samples in results, can be used as a measure of validation, for example, graphically or as the people performance index explained in Section 9.5.

Samples for the evaluation will in most food product cases have to be served blinded in such a way that the assessor cannot recognize the original product brand or origin. Projective mapping of nonfood products are also immediate. Nonmaterial samples like sound clips can be handed by appropriate software. If overall product concepts are to be evaluated, a product fact sheet or the actual product packaging can be applied. Samples are best identified by three-digit random numbers.

9.4 IMPLEMENTATION AND DATA COLLECTION

The samples are to be placed in front of the assessor who has been instructed to evaluate samples, for example, from left to right. The instructions are (for a conventional projective mapping) as follows:

> You will have to evaluate the samples in front of you one by one. You have to place them on the piece of paper according to the strategy that two samples placed closer to each other are more alike than two samples placed further apart. The criteria for how to separate the samples just have to make sense to you. In this way, there are only right solutions.

In some situations, it is more suitable to move stickers with the sample identifier around within the evaluations frame rather than the samples.

After evaluation of the first sample, it can be place in anywhere within the projection frame. Typically an assessor would place it in the middle of the frame. It is allowed to move the projections around and the positions are not fixed until the end of the evaluation. Next, the second sample is placed according to the first sample, closer or apart according to their (dis)similarity. As the assessor proceeds, more and more samples are placed within the projection frame. As the projections are comparative placements, the assessor should be allowed to go back and forth and reevaluate samples pairwise. When the final sample is placed, the projective mapping is done and the assessor has now built an individual system illustrating the differences among samples (as illustrated in Figure 9.2).

If semantics are wished for, they can now be written down directly on the paper sheet next to the individual sample. This task is easier, if assessors have been taking notes during their projective mapping.

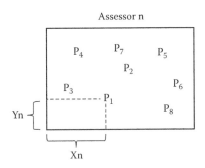

FIGURE 9.2 Illustration of a projective mapping frame after projection of eight products (P). The placement of assessor n's product coordinates (Xn, Yn) are normally measured from the lower left corner.

	Coordinates from assessors 1 to n									Total frequency of attributes 1 to n				
Sample	X1	Y1	X2	Y2	X3	Y3	...	...	Xn	Yn	Attribute 1	Attribute 2	Attribute 3	...	Attribute n
P_1															
P_2															
P_3															
...															
P_n															

Sample n

FIGURE 9.3   An optimal data table structure of projective mapping data before analysis with MFA in R.

Data collection is done for each assessor by measuring the coordinates for each sample placement. The zero point (0, 0) is the lower left corner of the paper. Data can be structured like illustrated in Figure 9.3. The data table is constructed so that one row corresponds to one sample. The first column consists of the sample names. The following part of the data table consists of all assessors' coordinates. Two adjacent columns correspond to one assessor's x- and y-coordinates, respectively. The first row corresponds to assessor identification, for example, X1 and Y1 for the coordinates from assessor 1. Following the assessor coordinate part of the data table is the attribute part, if the ultra-flash profile was performed. This is the conventional data table structure for data analysis with MFA as implemented in the FactoMineR package (Lê et al. 2008; Husson et al. 2010) in the free statistical software R (Ihaka and Gentleman 1996; R Development Core Team 2010). The part of the data table consisting of the applied attributes is a contingency table (or cross tabulation). Hence, each of the columns represents one attribute, for example, *sweet*, as applied in the evaluation. The first row corresponds to the name of the attribute, and the following rows correspond to the total number of occurrences of that attribute for all assessors. It can be cumbersome to construct this table manually, but the FactoMineR package contains the function (*textual*) that constructs this contingency table from a list of ultra-flash profiling results. Such a list consists of three columns, one for the assessor identification, one for the sample name, and one with all the attributes (see the function help file in R for more information). Accordingly, it is only needed to write down the assessors notes once.

Even though projective mapping techniques were developed for use with standard-size paper sheets in mind, most commercial sensory software now offers the option of running a projective mapping session. Then, the projective mapping frame size will be according to the computer screen size rather than the paper size. What are projected according to each other in this case will actually be the sample codes, icons, or images. For some applications, for example, where assessors are naive consumers or elderly, the step might be a point of extra difficulty.

In the spirit of free user-license programs like R, the author wish to advocate a free categorization task program originally developed for psychoacoustical experiments. The program is called TCL-LabX (Gaillard 2009) and is easily searchable online. This program offers sensory professionals and students a free way to administer a projective mapping session. It has the option of adding descriptive terms to each sample, hence carrying out the ultra-flash profile. In addition, it supplies a number of interesting features such as a time log of all movements done by the assessor. The assessor's final projective mapping coordinates are easily copied from the result file to a data table.

9.5 DATA ANALYSIS

There is yet no clear consensus on how to evaluate projective mapping data. For this reason, studies have so far applied principal component analysis variations, generalized Procrustes analysis, or multidimensional scaling method variations. Only few studies compare these statistical approaches (Nestrud and Lawless 2011). As projective mapping is a rapid approach, it tends to highlight major differences among products. Hence, statistical approaches that are overall comparable will tend to generate results that are overall comparable.

There has been a general increasing popularity in using R and the special developed sensory function packages. There are several packages developed for various sensory purposes, for example, *sensR* for discrimination data (Christensen and Brockhoff 2011). The FactoMineR package contains relevant algorithms for the professional that work with multivariate exploratory data analysis. The SensoMineR package (Lê and Husson 2008; Husson et al. 2011) is more directed toward problem-based tasks as encountered by the sensory professional. Both FactoMineR and SensoMineR have drop-down menu packages for the R graphical user interface RCommander, for the less programming experienced professional.

The projective mapping data consist, conventionally, of two variables per assessor, one being the x-coordinate and one being the y-coordinate. As the assessors may have applied different criteria to distinguish between the samples, the sets of variables cannot be assumed to be of similar nature. Hence, a data analysis taking account for such varying variable sets (k-sets) is desirable. For projective mapping data, SensoMineR suggest to apply an Indscal model or an MFA. MFA is a two-step analysis where the first step is a principal component analysis of each set of variables (unless data are categorical in which case it is a multiple correspondence analysis). This is performed to normalize each set of variables. The second step is a new

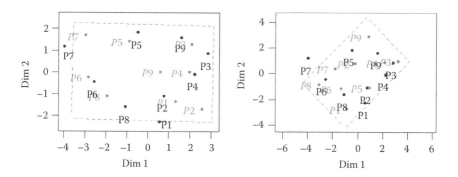

FIGURE 9.4 Two individual assessors' projective mappings (dotted line and names in italic) superimposed on the global analysis.

principal component analysis on the concatenated results, which end up with the global configuration (Escofier and Pagès 1994; Abdi et al. 2013).

If it is wished to control as many model parameters as possible and get full control of the model output, perform the MFA in FactorMineR. SensoMineR offers the function Procrustes MFA (PMFA). This is not an MFA performed with pretreated data. This is the standard MFA, but with the individual assessor's projective result rotated and superimposed on the global result (Figure 9.4). This is presented as a validation of the individual assessor's data and is somewhat a graphical parallel to the numerical RV coefficient (explained in the following). A selection of the standard graphical output is shown in Figure 9.5. The individual weights plot illustrates how each group of variables (each assessor) is weighted on the global factor map. The data used in the illustrated example include results from the ultra-flash profile. It is possible in MFA to include groups of variables as supplementary groups in the analysis. That means that the group will not have any impact on the global configuration, but it will be included as superimposed variables. In the individual weights plot (Figure 9.5b), the shaded colored group 10 means that this group is supplementary. Group 10 is the semantic variables of the ultra-flash profile. The correlations of the variables from the ultra-flash profile are also presented (Figure 9.5d). Analogous to the result plot from flash profiling, the plot may be so over-weighed by the semantics that any graphical evaluation becomes unmanageable. As a possible remedy, FactoMineR holds a function for category description (*catdes*), which will highlight the semantics that is applied significantly more for any given sample than for the rest of the samples.

The often wished for bi-plot is not implemented in SensoMineR or FactorMineR.

Except for the graphics, the RV coefficient (Robert and Escoufier 1976) can be used for individual assessor validation. The RV coefficient is a

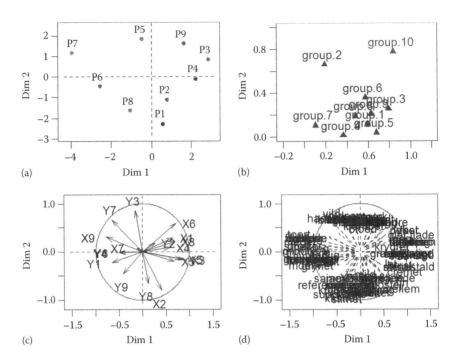

FIGURE 9.5 MFA of projective mapping data. (a) Individual factor map/score plot. (b) Individual weights of the groups of variables. (c) Correlation circle/loading plot of individual variables. (d) Correlation circle/loading plot of individual variables (semantics) from the ultra-flash profile.

multivariate correlation coefficient that lies between 0 and 1, where 1 is a complete match. It is found in FactoMineR as the function *coeffRV* but is also implemented as a part of several of the multivariate analyses. The PMFA approach provides the RV coefficient calculated between the global configuration and each individual assessor's configuration. Compared relatively to each other, one can gain an overview of which assessors that are closer to, or farther from, the global configuration. The *coeffRV* function also provides significance testing of the RV coefficient according to Josse et al. (2008). It is important to remember that a significant difference here would mean that the coefficient is different from zero.

The people performance index is another approach worth mentioning to evaluate assessor consistency, reported in use by Hopfer and Heymann (2013). The index is the ratio between the Euclidean distance of a sample and its duplicate (blind repeated sample) and the Euclidean distance between the two samples farthest apart. Opposite the RV coefficient, a low people performance index is preferred. It will be lower, if the assessor places the duplicates closer to each other or if the assessor uses the projective area to a large extend. The index has not yet been implemented as a function in R.

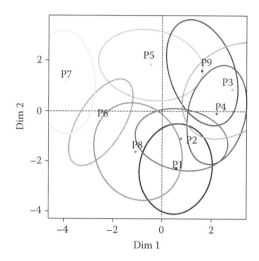

FIGURE 9.6 Individual factor map/score plot with confidence ellipses based on parametric bootstrapping.

It is possible to graphically display uncertainty on the score plot by the use of confidence ellipses around each sample mean (Figure 9.6). There are currently no confidence ellipse functions in R directed toward projective mapping evaluations. For this reason, a solution has been proposed that can easily be implemented into R (Dehlholm et al. 2012a). The procedure bootstraps the standardized groups of variables generated by the initial part of the MFA (see Appendix for function). Hence, the function input has to be the MFA result object as generated by the *MFA* function rather than the *pmfa* function. The more general applicability of confidence ellipses for sensory tasks have been discussed by Cadoret and Husson (2013).

As mentioned earlier, the semantic group of variables is treated as supplementary data by default by the *pmfa* function. If it is instead chosen to work with the more general *MFA* function, more options are available. It should then be taken into consideration, whether the semantic information should contribute to the model or if the model should be based on the projections only. One approach would be to include the compilation of all assessors' semantic information, as represented by the contingency table. Another approach would be to keep the semantic and the projective information separated by each individual assessor. Data would then be structured as illustrated in Figure 9.7 and a hierarchical MFA would suit this task.

The case of consensus Napping is shown in Figure 9.8. Semantic variables are included in the model and provide stronger data for modeling the product configuration. Contributions from both the semantic variables and

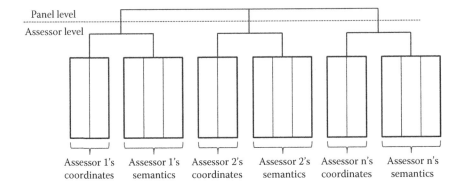

FIGURE 9.7   Illustration of the groups of variables and hierarchies arranged for a hierarchical MFA of projective mapping data including semantic information at an individual assessor level.

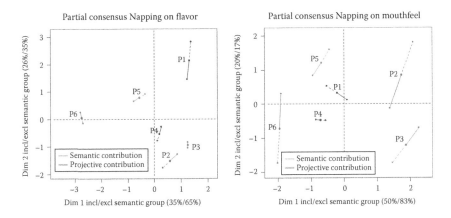

FIGURE 9.8   Consensus Napping score plots showing model contributions from the semantic groups of variables as well as product coordinates.

coordinates are highlighted in the figure. Clearly, the weights within some semantic variable groups contribute to draw some products in new directions. This is mainly in the second model dimension.

9.6   ADVANTAGES AND DISADVANTAGES

Projective mapping performed as global Napping and partial Napping, respectively, has recently been compared with conventional descriptive profiling, flash profiling, and free multiple sorting (Dehlholm et al. 2012a,b). Some main conclusions are graphically illustrated in Figure 9.9. It shows the first two model dimensions in sensory profiles of the same products. The plots are ordered according to the overall time spent on

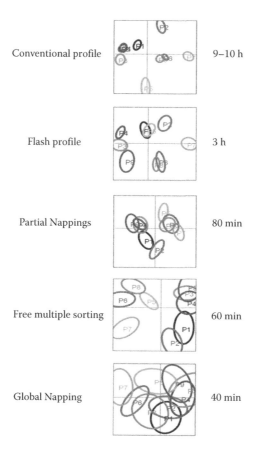

Conventional profile — 9–10 h

Flash profile — 3 h

Partial Nappings — 80 min

Free multiple sorting — 60 min

Global Napping — 40 min

FIGURE 9.9 Comparison of descriptive sensory methodologies with regard to precision and overall time spend on training/introduction and evaluation. (Modified reprint from Dehlholm, C. et al., *Food Qual. Preference*, 26, 278, 2012a; Dehlholm, C. et al., *Food Qual. Preference*, 26, 267, 2012b.)

the training/introduction and evaluation and it seems somewhat related to the area covering the confidence ellipses. This illustrates both the most important advantage as well as the most important disadvantage of projective mapping.

Time-wise, projective mapping is a rapid alternative to conventional descriptive profiling. It is fast to perform and hence saves resources such as assessor and panel leader man-hours. Its most important disadvantage is its poor precision. That is why projective mapping is also seen referred to as a "quick-and-dirty" method (Valentin et al. 2012). As illustrated, it is more important to consider reliability when rapid approaches are chosen. For instance, the sensory professional might be used to study modeled data on third and fourth model dimension. This is often not possible with projective mapping results, as uncertainty dominates the picture. Regarding the first two model dimensions, Figure 9.9 illustrates the main dimensionality among the products. It means that one single product cannot be independently described and that all product overlap at least one other product with regard to their confidence ellipses. Accordingly, one can say that "this" group of similar products differentiates significantly from "that" other group of products, or similar. This is of course product dependent and hence, relative. The products included in the study illustrated were commercially available and based on similar recipes. Pilot tests had shown that some of them varied weakly in flavor and some in mouthfeel, while some were very similar. If products are with such similarity that pilot tests repeatedly do not highlight any differences among products, they should not be considered for a projective mapping approach. If products have larger differences, the sensory professional can be more confident that products can be significantly differentiated.

Although categorized as a rapid approach, projective mapping techniques should not be chosen before conventional profiling with rapidity as the incentive. The two approaches differ from a perceptual point of view and, on that basis, suit various purposes. The continuum of sensory assessor types as described by Dehlholm (2012) can be used to understand the situations in which projective mapping is better applied than other descriptive alternatives (Figure 9.10). If no concept alignment (training) has been performed, projective mapping techniques without added scales are suited better for a more general overview of products, where semantics would be more conceptual. This is also referred to as a more holistic sample approach. A holistic sample approach is not the same as working with a holistic vocabulary, but it is the assessment of the product in its whole instead of measuring of selected attributes.

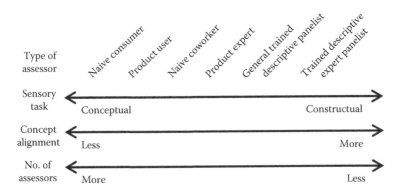

FIGURE 9.10 The continuum of sensory assessor types. (Reprinted from Dehlholm, C., *Descriptive Sensory Evaluations: Comparison and Applicability of Novel Rapid Methodologies*, SL grafik, Copenhagen, Denmark, 2012. With permission.)

9.7 APPLICATION

A recent study by Dehlholm et al. (2014b) compared consensus Napping results from group evaluations between groups of assessors. The groups consisted of experienced panelists or product producers/experts. Results showed large variations, largest among the product experts, but large enough between the groups of experienced panelists. Thus, the authors would not recommend consensus evaluations to test reliability between groups. Nevertheless, they noted that if assessors were both skilled in the methodology and also had in-depth product knowledge, results were influenced positively. Pilot tests had shown that assessors in a group consensus evaluation would typically start to discuss which of the sensory modalities was the more dominant. That led to difficulties in reaching a global group consensus. It was easier to create group consensus by restricting the discussions to partial Nappings (specific sensory modalities). The choice of having an experienced facilitator was important to reach consensus, although a group of assessors with high experience would most probably be able to perform the consensus profile without external facilitation. The RV coefficient can be used to compare a global model configuration with the individual assessors' configuration, and to compare various global model configurations with each other.

Within a company, projective mapping would be advantageously applied for exploration and idea generation, prototype assessment and design, and pilot in-house "market" tests of new products. For instance, this could be in the new product development exploration phase where many products with large differences have to be examined, or in the design phase where product characteristics must be outlined. These situations call for an easy applicable approach that can give a swift overview. In the company–market sphere, projective mapping would also be applicable for exploration and idea generation as well as for product comparisons within or between product categories. Although seldom applied, it would also add value in pure market-related evaluations, for example, a chef competition or product reviews in popular science and daily papers.

Hopfer and Heymann (2013) provide a comprehensive overview of specific projective mapping studies. They list variations according to the number of products involved in the study, the exact method variation, the geometric frame, the possible use of repetitions, duplicates and semantic descriptions, the statistical approach, and the major findings. Besides, they compare in total 14 previous studies that apply a projective mapping technique. Of those studies, three were published during the 1990s, while the rest are newer. This shows the novelty of the method and, hence, explains why projective mapping is still explored in all possible variations.

Even though Napping has been proposed as a specified projective mapping approach, a general projective mapping convention has not yet been adopted by the sensory community. There is no doubt that projective mapping will continue to be a focus area within sensory method research in the years to come.

9.A APPENDIX

A data set of projective mapping data and the script for drawing confidence ellipses around sample configurations are available for download from the CRC Web site: http://www.crcpress.com/product/isbn/9781466566293.

```
# — — — — — — — — Beginning of function — — — — — — —
##########
### Function to draw confidence ellipses on any MFA
### result object obtained via the MFA function
### in FactoMineR.
### To save the function to your hard drive, copy all
### these lines of text and save it in a text file with
### the name 'conf.R' (and not 'conf.txt').
### Load this function in R by writing
### 'source(file.choose())' in R command promt and
### choose this file (named 'conf.R').
### Run this function in R by writing 'MFAconf(X)'
### in R command promt where X defines your MFA
### result object.
### The function gives you the option of plotting
### specific model axes.
##########

# MFA result object and axes to draw inside the
# function.
MFAconf = function(MFAresob, axes = c(1, 2)){

if (!require("FactoMineR")) install.packages(
   "FactoMineR"); library("FactoMineR")

# The number of samples (n)
n = dim(MFAresob$ind$coord)[1]

# The number of groups of variables (m)
m = dim(MFAresob$group$coord)[1]

# Creating a new data frame with one row for
# each sample's associated MFA group of variables
CATnames <- vector(mode = "character",
   length = n*m)
```

```
for (j in 1:n){CATnames[(((j-1)*m)+1):(j*m)] <-
   dimnames(MFAresob$ind$coord[order(
   row.names(MFAresob$ind$coord)),])[[1]][j]}
PartielDim <-cbind.data.frame(names = CATnames,
   MFAresob$ind$coord.partiel)

# Bootstrapping the new data frame
Boot <- simule(PartielDim, nb.simul = 500)

# Creating ellipses around 95% of the
# bootstrapped means
EllipCoord <- coord.ellipse(Boot$simul,
   level.conf = 0.95, bary = FALSE, axes = axes)

# Plotting the ellipses
plot.MFA(MFAresob, title = " ", axes = axes,
   choix = "ind", ellipse = EllipCoord,
   ellipse.par = NULL)
}
# — — — — — — — — End of function — — — — — — — — —
```

REFERENCES

Abdi, H., Williams, L. J., and Valentin, D. 2013. Multiple factor analysis: Principal component analysis for multitable and multiblock data sets. *Wiley Interdisciplinary Reviews: Computational Statistics* 5: 149–179.

Albert, A., Varela, P., Salvador, A., Hough, G., and Fiszman, S. 2011. Overcoming the issues in the sensory description of hot served food with a complex texture. Application of QDA (R), flash profiling and projective mapping using panels with different degrees of training. *Food Quality and Preference* 22: 463–473.

Ares, G., Deliza, R., Barreiro, C., Giménez, A., and Gámbaro, A. 2010. Comparison of two sensory profiling techniques based on consumer perception. *Food Quality and Preference* 21: 417–426.

Ares, G., Varela, P., Rado, G., and Giménez, A. 2011. Identifying ideal products using three different consumer profiling methodologies. Comparison with external preference mapping. *Food Quality and Preference* 22: 581–591.

Barcenas, P., Elortondo, F. J. P., and Albisu, M. 2004. Projective mapping in sensory analysis of ewes milk cheeses: A study on consumers and trained panel performance. *Food Research International* 37: 723–729.

Cadoret, M. and Husson, F. 2013. Construction and evaluation of confidence ellipses applied at sensory data. *Food Quality and Preference* 28: 106–115.

Christensen, R. H. B. and Brockhoff, P. B. 2011. sensR—An R-package for sensory discrimination package version 1.2.10. http://www.cran.r-project.org/package=sensR/ (accessed May 16, 2011).

Dehlholm, C. 2012. *Descriptive Sensory Evaluations: Comparison and Applicability of Novel Rapid Methodologies.* Copenhagen, Denmark: SL grafik.

Dehlholm, C., Brockhoff, P. B., and Bredie, W. L. P. 2012a. Confidence ellipses: A variation based on parametric bootstrapping applicable on Multiple Factor Analysis results for rapid graphical evaluation. *Food Quality and Preference* 26: 278–280.

Dehlholm, C., Brockhoff, P. B., Meinert, L., Aaslyng, M. D., and Bredie, W. L. P. 2012b. Rapid descriptive sensory methods—Comparison of free multiple sorting, partial napping, napping, flash profiling and conventional profiling. *Food Quality and Preference* 26: 267–277.

Dehlholm, C., Lê, S., and Bredie, W. L. P. 2014a. Projective mapping: Consequences of variations in projective frame geometry and semantic restrictions. *Food Quality and Preference*, submitted for publication.

Dehlholm, C., Meinert, L., and Bredie, W. L. P. 2014b. Consensus group product assessments: A valid approach for rapid sensory profiling? *Journal of Sensory Studies*, submitted for publication.

Dun-Rankin, P. 1983. *Scaling Methods.* Hillsdale, NJ: L. Erlbaum.

Egge, K. 1984. The visual field in normal subjects. *Acta Ophthalmologica Supplementum* 169: 1–64.

Escofier, B. and Pagès, J. 1994. Multiple factor-analysis (AFMULT package). *Computational Statistics & Data Analysis* 18: 121–140.

Gaillard, P. 2009. Laissez-nous trier! TCL-LabX et les tâches de catégorisation libre de sons. In *Le Sentir et le Dire: Concepts et méthodes en psychologie et linguistique cognitives*, ed. D. Dubois, pp. 189–210. Paris, France: L'harmattan.

Hopfer, H. and Heymann, H. 2013. A summary of projective mapping observations—The effect of replicates and shape, and individual performance measurements. *Food Quality and Preference* 28: 164–181.

Husson, F., Josse, J., Lê, S., and Mazet, J. 2010. FactoMineR: Multivariate exploratory data analysis and data mining with R. R package version 1.14. http://CRAN.R-project.org/package=FactoMineR (accessed May 10, 2010).

Husson, F., Lê, S., and Cadoret, M. 2011. SensoMineR: Sensory data analysis with R. R package version 1.14. http://CRAN.R-project.org/package=SensoMineR (accessed May 11, 2011).

Ihaka, R. and Gentleman, R. 1996. R: A language for data analysis and graphic. *Journal of Computational and Graphical Statistics* 5: 299–314.

Josse, J., Pagès, J., and Husson, F. 2008. Testing the significance of the RV coefficient. *Computational Statistics & Data Analysis* 53: 82–91.

Kennedy, J. 2010. Evaluation of replicated projective mapping of granola bars. *Journal of Sensory Studies* 25: 672–684.

Kennedy, J. and Heymann, H. 2009. Projective mapping and descriptive analysis of milk and dark chocolates. *Journal of Sensory Studies* 24: 220–233.

King, M. C., Cliff, M. A., and Hall, J. W. 1998. Comparison of projective mapping and sorting data collection and multivariate methodologies for identification of similarity-of-use of snack bars. *Journal of Sensory Studies* 13: 347–358.

Künnapas, T. M. 1955. Influence of frame size on apparent length of a line. *Journal of Experimental Psychology* 50: 168–170.

Künnapas, T. M. 1957a. The vertical–horizontal illusion and the visual field. *Journal of Experimental Psychology* 53: 405–407.

Künnapas, T. M. 1957b. Vertical–horizontal illusion and surrounding field. *Acta Psychologica* 13: 35–42.

Lê, S. and Husson, F. 2008. Sensominer: A package for sensory data analysis. *Journal of Sensory Studies* 23: 14–25.

Lê, S., Josse, J., and Husson, F. 2008. FactoMineR: An R package for multivariate analysis. *Journal of Statistical Software* 25: 1–18.

Nestrud, M. A. and Lawless, H. T. 2010. Perceptual mapping of apples and cheeses using projective mapping and sorting. *Journal of Sensory Studies* 25: 390–405.

Nestrud, M. A. and Lawless, H. T. 2011. Recovery of subsampled dimensions and configurations derived from napping data by MFA and MDS. *Attention Perception and Psychophysics* 73: 1266–1278.

Pagès, J. 2003. Direct collection of sensory distances: Application to the evaluation of ten white wines of the Loire Valley. *Sciences des Aliments* 23: 679–688.

Pagès, J. 2005. Collection and analysis of perceived product inter-distances using multiple factor analysis: Application to the study of 10 white wines from the Loire Valley. *Food Quality and Preference* 16: 642–649.

Pagès, J., Cadoret, M., and Lê, S. 2010. The sorted napping: A new holistic approach in sensory evaluation. *Journal of Sensory Studies* 25: 637–658.

Pfeiffer, J. C. and Gilbert, C. C. 2008. Napping by modality: A happy medium between analytic and holistic approaches. *Ninth Sensometrics Meeting*, July 21–23, 2008, St. Catherines, Ontario, Canada.

R Development Core Team. 2010. R: A language and environment for statistical computing. *R Foundation for Statistical Computing*, Vienna, Austria. URL: http://www.R-project.org/ (accessed February 1, 2010).

Risvik, E., McEwan, J. A., Colwill, J. S., Rogers, R., and Lyon, D. H. 1994. Projective mapping: A tool for sensory analysis and consumer research. *Food Quality and Preference* 5: 263–269.

Risvik, E., McEwan, J. A., and Rodbotten, M. 1997. Evaluation of sensory profiling and projective mapping data. *Food Quality and Preference* 8: 63–71.

Robert, P. and Escoufier, Y. 1976. Unifying tool for linear multivariate statistical-methods—Rv-coefficient. *Journal of the Royal Statistical Society Series C-Applied Statistics* 25: 257–265.

Thompson, D. M. H. and MacFie, H. J. H. 1983. Is there an alternative to descriptive sensory assessment? In *Sensory Quality in Foods and Beverages: Definition, Measurement and Control*, eds. A. A. Williams and R. K. Atkin, pp. 96–107. Chichester, U.K.: Ellis Horwood Ltd.

Valentin, D., Chollet, S., Lelievre, M., and Abdi, H. 2012. Quick and dirty but still pretty good: A review of new descriptive methods in food science. *International Journal of Food Science and Technology* 47: 1563–1578.

Varela, P. and Ares, G. 2012. Sensory profiling, the blurred line between sensory and consumer science. A review of novel methods for product characterization. *Food Research International* 48: 893–908.

10 Polarized Sensory Positioning Methodologies

Eric Teillet

CONTENTS

10.1   CONTEXT AND POLARIZED SENSORY POSITIONING PHILOSOPHY

10.1.1   WHERE DOES PSP COME FROM?

Polarized sensory positioning (PSP) has been developed by Eric Teillet (SensoStat, Dijon, France) and Pascal Schlich (INRA, Dijon, France) during a PhD, which one of the purposes was to determine the taste of water (partnership CNRS/Lyonnaise des Eaux; Teillet 2009). The aim of Lyonnaise des Eaux (French supplier of tap water) was to routinely determine the taste of one or more water samples. This example of the taste of water will be used in this chapter in order to present the PSP methodology. Nevertheless, adaptation of PSP to other product spaces is obviously possible and will be discussed.

Water, if not contaminated by pollutants, is only composed of H_2O molecules and minerals (K^+, Na^+, Cl^-, Ca^{2+}, SO_4^{2-}, Mg^{2+}, etc.). At first glance, water is a simple and easy product for sensory characterization since sensory modalities such as odor or texture are not evaluated. However, sensory stimuli induced by water are generally so low that sensory characterization of this product represents a real challenge for sensory analysis. Best results (in terms of discrimination among samples and sensory positioning) have been obtained using comparative methods, such as the free-sorting task (Teillet et al. 2010a). Nevertheless, free-sorting task doesn't enable data aggregation, that is, to aggregate data from samples evaluated in different sessions. In order to analyze a new sample, the entire set of products should be presented again. Therefore, comparative methodologies that do not require the presentation of an excessive number of samples in each new analysis and that enable data aggregation from several studies are necessary. Besides, it is desirable that this comparative methodology could be performed by both trained, semitrained, and untrained assessors.

10.1.2   WHAT IS PSP?

PSP is not strictly a sensory methodology for which protocols are well defined and standardized but an approach for evaluating the sensory characteristics of products. PSP simply relies on a philosophy: using well-known standard products (ideally stable over time) as "poles" in order to position products in a sensory space.

PSP is based on the comparison of samples to reference products or "poles." The principle is actually to evaluate the overall similarity (or dissimilarity) between samples and each of the poles, wisely chosen in the product space. The poles should be different prototypes from the whole sensory product space, which represent the main sensory characteristics that can be encountered in the product category of interest.

In the case of the taste of water, poles have been chosen among French natural mineral waters in order to best represent the three main tastes of water: the metallic and bitter taste for low mineral content waters (e.g., Volvic), the neutral taste and the sensation of freshness for medium mineral content waters (e.g., Evian), and finally the more salty and astringent taste for highest mineral content waters (e.g., Vittel) (Teillet et al. 2010b). An interesting feature of natural mineral waters is that their mineral composition and sensory characteristics are legally required to be stable over time, which enables data aggregation from several PSP studies.

There is no clear rule about the number of poles needed in a PSP trial. Nevertheless, in a geometric point of view, a minimum of three "point to point" distances are necessary to locate a sample in a 2D map. Thus, we recommend a minimum of three poles. The issue of the choice of the poles will be further discussed later in the chapter.

10.2 PSP METHODOLOGIES

In the example, Volvic, Evian, and Vittel have been chosen as poles and have been named "A", "B," and "C", respectively, during the trials.

Different approaches can be used to collect information about the similarities (or dissimilarities) between samples and poles. The first one directly collects dissimilarities to poles by asking assessors to score overall dissimilarity on a continuous scale. The second one, which can be regarded as a *polarized triadic* test (MacRae et al. 1990), proposes to deduct similarities and dissimilarities to poles from co-occurrences, by asking assessors to indicate to which of the poles each sample is most similar to. A third alternative, called *napping PSP*, consists of merging projective mapping or Napping with PSP. However, considering that the author has tried this approach and found it quite difficult to understand by assessors, it won't be developed here.

These approaches have been the first to be applied but consist of only particular proposals. Other alternatives can be used to gather information about similarity between samples and each of the poles. Feel free to imagine your own protocols. The principle will remain the same provided that you have similarities or dissimilarities between the product and poles.

10.2.1 Continuous Scale

The first method allows getting scores, given by assessors, which reflect the distance between samples and each of the poles. This score is given on an unstructured linear scale. In the example, the scale was anchored from

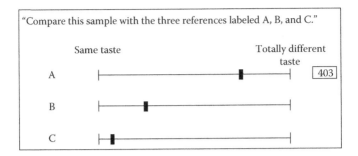

FIGURE 10.1   Example of continuous scale used for PSP.

"same taste" to "totally different taste" (Figure 10.1), but in a more general approach, it can be anchored from "exactly the same" to "totally different."

In the example given in Figure 10.1, a product coded 403 would have been evaluated as very different from A and similar to C than to B.

10.2.2   TRIADIC

In the case of *triadic* PSP (Figure 10.2), a panelist is simply asked to indicate to which pole a sample resembles the most and to which pole it resembles the least. In the example given in Figure 10.2, the water sample coded 403 resembles the most to C and resembles the least to A.

10.3   DATA ANALYSES

10.3.1   CONTINUOUS SCALE

In a first step, data from the continuous scale can be coded from 0 for "same taste" to 10 for "totally different taste." By analogy with

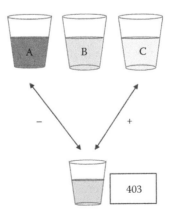

FIGURE 10.2   Principle of *triadic* PSP.

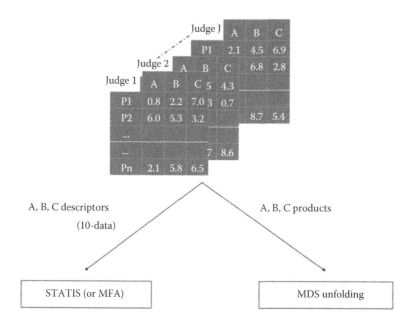

FIGURE 10.3 Two ways to process PSP data: STATIS (or MFA) and MDS unfolding.

the free-sorting task, data from PSP can be arranged as a dissimilarity matrix between samples and poles (Figure 10.3). Thus, data can be analyzed by multidimensional scaling (MDS) techniques. However, considering that dissimilarity matrices from PSP are rectangular, they cannot be processed with classical MDS and therefore MDS unfolding algorithms must be used (Green et al. 1989). These methods are an extension of MDS to matrices where products in rows can be different from products in columns but the purpose still remains to locate products, of which 2-by-2 distances are known, in a space. Like with classical algorithms of MDS, unfolding algorithms can incorporate nonmetric transformations.

Until recently, unfolding algorithms presented convergence problems to trivial solutions (equidistant products), but this problem seems to have been solved (Busing et al. 2005). Although unfolding is rarely used in sensory analysis, it is the most natural method to process PSP data from a continuous scale.

Nevertheless, data can be seen in a different way. If poles are considered as *meta-descriptors*, data can be encoded from 0 for "totally different taste" to 10 for "same taste." In the example of the taste of water, Volvic could be a "metallic and bitter" descriptor, Evian "tasteless and cool," and Vittel "salty and astringent." Therefore, using this approach intensities of each sample on several descriptors are considered and

Sample	Pole A	Pole B	Pole C	Pole A	Pole B	Pole C	...	Pole A	Pole B	Pole C
1	1.4	9.6	7.8	0	7.5	6.4	...	3.5	1.4	8.9
2	5.6	5.4	7.6	3.4	5.6	7.1	...	5.2	4.1	7.8
...	...	...	...	...	...	...	...	...	...	...
X	9.8	5.6	0.5	5.3	9	0.6	...	9.8	7.8	1.5

Assessor 1   Assessor 2   Assessor n

FIGURE 10.4 Example of the data matrix used for analyzing data from continuous-scale PSP using multiple factor analysis.

classical factor analyses such as PCA, MFA, STATIS, or GPA can be used (Figure 10.3). These methods can easily be implemented using the freely available R statistical software.

The use of the scales can be very different from one assessor to another since the criterion for deciding what is a "totally different taste" is not homogenous. For this reason, three-way analyses, such as STATIS, MFA, and GPA, are highly recommended. The matrix for analyzing PSP data using MFA is shown in Figure 10.4, while the script for data analysis in R is shown in Appendix 10.A.

PSP on continuous scales also enables hypothesis testing. Classical analysis of variance (ANOVA) applied to descriptive analysis can be used to identify significant differences among products. Ongoing studies are also considering the application of the mixed assessor model (MAM) (Brockhoff et al. 2011) to take into account the expected high scaling differences among assessors (especially when working with untrained subjects).

10.3.2 TRIADIC PSP

When working with *triadic* PSP, similarity information is qualitative, for example, "the product 403 resembles the most to C" and "the product 403 resembles the least to A." This information can be regarded as an occurrence between the product 403 and the poles and transformed using qualitative variables. For example, it can be considered that product 403 corresponds to the variables "C+" and "A–." It is therefore possible to create a global co-occurrence matrix product × variables (A+, B+, C+, A–, B–, C–) where occurrences are summed up over assessors, as shown in Table 10.1. This matrix can be analyzed by correspondence analysis (CA). The script for analyzing data from triadic PSP using CA is shown in the Appendix.

TABLE 10.1

Example of the Data Matrix Used for Analyzing Data from Triadic PSP Using Correspondence Analysis

Sample	A+	B+	C+	A–	B–	C–
1	14	1	5	0	15	5
2	0	20	0	18	0	2
...	...	...	...	...	...	...
x	7	7	6	5	8	7

Note: The table shows the number of assessors that considered that a sample was the most similar (+) and most different to each of the poles (A, B, and C).

10.4 PSP AND THE TASTE OF WATER

10.4.1 FIRST EXAMPLE

A first study, dedicated to validate the application of PSP for evaluating the taste of water has been conducted with 32 assessors and 10 water samples.

Waters were chosen in order to span the range of total mineralization found in French mineral waters. The poles Volvic, Evian, and Vittel were also blind tested in order to verify the consistency of PSP data. A product with off-flavor (Evian with chlorine) was added to the product space.

The unfolding analysis determined a three-dimensional space. The first two axes were driven by the degree of mineralization (Figure 10.5). Poles were located close to their respective blind tested samples, suggesting that PSP data were consistent.

On this first plane, the off-flavor "Evian with chlorine" was closer to Evian than the other poles. However, this product was isolated on dimension 3 (Figure 10.6).

Even though the results obtained were consistent, validation of this PSP method is currently based only on the logic of mineral content observed in prior studies, suggesting that other performance criteria should be defined. However, it is important to highlight that these results are equivalent to those obtained with different classical sensory methods in terms of interpretation. In particular, PSP has been found to discriminate better the taste of low and high mineralized water than classical descriptive analysis (Teillet et al. 2010b).

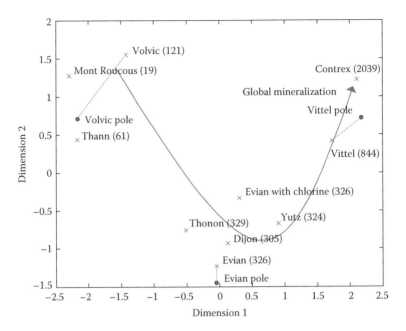

FIGURE 10.5 Product space, dimensions 1 and 2, for the evaluation of water samples. Global mineralization (mg/L) of the samples is presented between brackets.

Data have been also analyzed using the STATIS approach, providing equivalent conclusions (Teillet et al. 2010b).

10.4.2 Comparison between Continuous-Scale and Triadic PSP

A second study, conducted with 88 naïve consumers, is presented comparing the 2 PSP approaches developed (continuous scale and *triadic*). The same consumers evaluated eight water samples using both approaches. Half of them began with the continuous scale, and the other half began with the *triadic*.

Bootstrap sampling (Efron 1979) was used to evaluate the variability of the results. Bootstrap methods (random sampling with replacement) allow the estimation of the sampling distribution of almost any statistic and are often useful for sensory methodologies where no inference can be done (e.g., the location of products in a sensory map, for free-sorting task, Napping, or PSP). We won't develop these techniques here, but they lead to the construction of confidence ellipses in the sensory map (90% confidence ellipses in the present case). One thousand bootstrap samples were simulated for 88, 80, 70, …, 20, 10 consumers, which enabled to compare the variability of continuous-scale and triadic method. Using this approach, a minimum number of consumers could also be suggested for both methods.

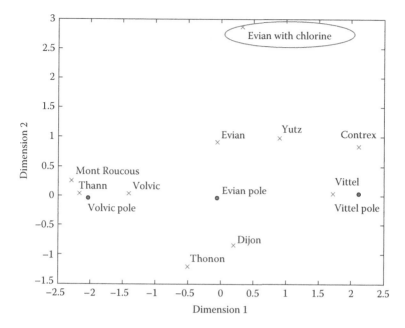

FIGURE 10.6 Product space, dimensions 1 and 3, for the evaluation of water samples.

Figure 10.7 presents the results of the bootstrap approach. Cartographies obtained by continuous scale and *triadic* were very similar and mainly driven by the global mineralization of waters. Besides, variability (the size of ellipses) was comparable for both methods, even if *triadic* seemed to better discriminate some waters (less overlapping of ellipses).

Variability remains pretty much the same from 88 to 40 consumers. However, variability largely increases when less than 40 consumers are considered, suggesting that for this particular example, a minimum of 40 untrained panelists seems necessary for a PSP study.

It would be very interesting to study the effect of training on this variability. One could think that training would improve this variability.

Considering these results, the two methodologies seem to be equivalent in terms of conclusions about water samples. However, consumers declared that the triadic PSP was easier to perform than scale-based PSP.

10.4.3 Example of Data Aggregation

A last study was conducted in order to assess the feasibility of PSP data aggregation. Three hundred and fifty-four consumers tasted 9 out of 18 water samples (from all over the world) in a single trial (at a water

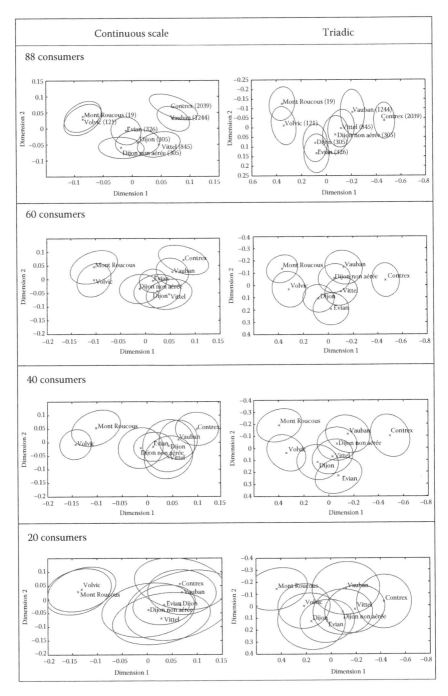

FIGURE 10.7 Cartographies obtained with continuous-scale and *triadic* PSP for 88, 60, 40, and 20 consumers (1000 bootstrap samples).

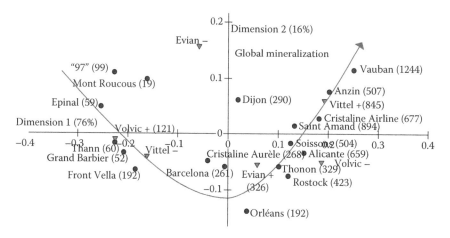

FIGURE 10.8 CA of aggregated data of water samples. On the graph, a water sample is close to a pole (Evian, Volvic, Vittel) if the co-occurrence sample/pole is high. Pole "+" means "resembles the most" and pole "−" means "resembles the least."

congress). The purpose was to aggregate this incomplete data and to draw a common map for the whole participants. Results are shown in Figure 10.8.

The resulting map is still driven by the global mineralization of water samples. Nevertheless, this kind of aggregation from incomplete blocks is quite common in the sensory field. It would be more interesting to aggregate data from several PSP studies.

10.5 DISCUSSION, CONCLUSION, AND PERSPECTIVES

10.5.1 ABOUT THE CHOICE OF THE POLES

We have seen that a minimum of three poles seems necessary for PSP 2D representations. And the choice of the poles seems to be a critical point of PSP. However, the main question that remains is how to choose them. Unfortunately, there is no real rule by now, only some pieces of advice.

In the first place, if data from several studies is going to be aggregated, the poles must be stable over time. If you are not sure these poles will be available for future trials, you must at least know their *formulation* in order to reproduce them.

A first easy case is when you want to compare your products to several *references* of the market. Choose them as poles.

If there is available knowledge or expertise about your product space, use it. For example, you can use previous classification of products based on cluster analysis performed on sample configurations from quantitative descriptive analysis or free-sorting task. Considering these data, you can define the sensory space by choosing the prototype that best represents each cluster (the closest to the barycenter, for example).

If you don't have any a priori information about product space, you will feel uncomfortable with the choice of your poles. Nevertheless, recent unpublished studies have shown that if several types of products are considered in the sensory space, there is no real importance in the choice of the product in each type. Besides, an iterative sequence of the choice of poles can be considered. If samples are placed outside the sensory space (very dissimilar of the poles), it is possible that a pole will be missed. On the contrary, if two of the poles remain very close to each other in all the studies, maybe one of them can be removed.

10.5.2 CONCLUSION

PSP seems to be a promising methodology. It has offered bright perspectives for routine sensory analysis of water samples. PSP methodology have provided good results and should ultimately enable the sensory characteristics of a given product to be determined, while enabling data aggregation and the evaluation of samples in multiple sessions.

The consistency criterion adopted for water samples only relies on the structure of the sensory space. Other criteria should be defined, such as repeatability, discrimination power of products (via ANOVA and confidence ellipses), or others. These considerations will be the topic of further work. There still remain a substantial number of questions to improve and generalize the use of PSP philosophy, such as studying the way data are collected (structure of the continuous scale), the way data are processed (application of the mixed assessor model), and the influence of training on PSP results and adding verbatim to the maps. There are still a lot of things to study.

Nevertheless, PSP can be applied with confidence to any product space were its use seems to be relevant. PSP has already successfully been used with cosmetics (Chréa et al. 2011), aromas, and beverages. PSP can also be adapted to consumer topics such as behavior and emotions (poles can be pictures, ambiances, or whatever you want!).

In conclusion, you can feel free to use PSP the way you want.

10.A APPENDIX

Two data sets, dealing with continuous-scale and triadic PSP, are available for download from the CRC Web site: http://www.crcpress.com/product/isbn/9781466566293.

10.A.1 Script for Analyzing Data from Continuous-Scale PSP Using MFA in R

```
# MFA is performed in the package FactoMineR, which
# has to be loaded:

library(FactoMineR)

# The first step is to load data into R from the
# clipboard. The data matrix shown in Figure 10.4
# should be copied from Excel in the clipboard and
# imported into R using the following command:

Dataset <- read.table("clipboard", header = TRUE,
  sep = "t", na.strings = "NA", dec = ".", strip.
  white = TRUE)

# The following command indicates that the name of the
# samples is included in the column identified as
# Samples:

row.names(Dataset) <- as.character(Dataset$Samples)

Dataset$Samples <- NULL

# The next step is performing MFA on the imported
# dataset. It is important to take into account that
# the MFA function needs the group structure of the data,
# as well as the type of data included in the analysis.
# Assuming that 20 assessors evaluated the samples and
# that 3 poles were used and that they evaluated samples
# using scales, the command is the following:

MFAresob<-MFA(Dataset, group = rep(3,20),
  type = rep("c",20))

# The following commands provide the eigenvalues of
# the MFA and the coordinates of the samples in the
# first five dimensions:

MFAresob$eig

MFAresob$ind$coord
```

10.A.2 Script for Analyzing Data from Triadic PSP Using CA in R

```
# CA is performed in the package FactoMineR, which has
# to be loaded:

library(FactoMineR)
```

```
# The first step is to load data into R from the
# clipboard. The data matrix shown in Table 10.1 should
# be copied from Excel in the clipboard and imported
# into R using the following command:

Dataset <- read.table("clipboard", header = TRUE,
  sep = "t", na.strings = "NA", dec = ".", strip.
  white = TRUE)

# The following command indicates that the name of the
# samples is included in the column identified as
# Samples:

row.names(Dataset) <- as.character(Dataset$Samples)

Dataset$Samples <- NULL

# The next step is performing CA on the imported
# dataset, which can be done with the following
# command:

CAresob<-CA(Dataset)

# The following commands provide the eigenvalues of
# the CA and the coordinates of the samples in the
# first five dimensions:

CAresob$eig

CAresob$row$coord
```

REFERENCES

Brockhoff, P. B., Schlich, P., and Skovgaard, I. 2011. Accounting for scaling differences in sensory profile data: Improved mixed model analysis of variance. Presented at the *Ninth Pangborn Sensory Science Symposium*, Toronto, Ontario, Canada, 2011.

Busing, F. M. T. A., Groenen, P. J. F., and Heiser, W. J. 2005. Avoiding degeneracy in multidimensional unfolding by penalizing on the coefficient of variation. *Psychometrika* 70: 49–76.

Chréa, C., Teillet, E., Navarro, S., and Mougin, D. 2011. Application of the polarized sensory positioning in the cosmetic area. Presented at the *Ninth Pangborn Sensory Science Symposium*, Toronto, Ontario, Canada, 2011.

Efron, B. 1979. Bootstrap methods: Another look at the jackknife. *The Annals of Statistics* 7: 1–26.

Green, P. E., Carmone, F. J., and Smith, S. M. 1989. *Multidimensional Scaling: Concepts and Applications*. Boston, MA: Allyn & Bacon.

MacRae, A. W., Howgate, P., and Geelhoed, E. N. 1990. Assessing the similarity of odours by sorting and by triadic comparison. *Chemical Senses* 15: 661–699.

Teillet, E. 2009. Perception, préférence et comportement des consommateurs vis-à-vis d'eaux embouteillées et d'eaux du robinet. Thèse de 3ème cycle. Dijon, France: Université de Bourgogne.

Teillet, E., Schlich, P., Urbano, C., Cordelle, S., and Guichard, E. 2010b. Sensory methodologies and the taste of water. *Food Quality and Preference* 21: 967–976.

Teillet, E., Urbano, C., Cordelle, S., and Schlich, P. 2010a. Consumer perception and preference of bottled and tap waters. *Journal of Sensory Studies* 25: 463–480.

11 Check-All-That-Apply Questions

Michael Meyners and John C. Castura

CONTENTS

11.1 ABOUT CATA QUESTIONS

11.1.1 INTRODUCTION TO CATA

A check-all-that-apply (CATA, often also known as choose-all-that-apply) question is a question format that has been used in recent years to obtain rapid product profiles from consumers. Consumers are presented with a list of attributes and asked to indicate which words or phrases appropriately describe their experience with the sample being evaluated. The terms might include sensory attributes, as well as hedonic responses, emotional responses, purchase intentions, potential applications, product positioning, or other terms that the consumer might associate with the sample.

Figure 11.1 shows an example text for a possible CATA question: "From the following list, check the attributes that describe the strawberry you just tasted (choose all that apply)." Emphasis can be added to key words, such as "attributes," "describe," "product," and "all" to provide clarity. The CATA attributes would appear underneath this question in one or several columns. The respondent checks the attributes that are appropriate for characterizing the sample being evaluated.

Adams et al. (2007) had consumers evaluate food products; they collected consumers' sensory responses using CATA questions and hedonic responses using *liking* questions. Their presentation sparked considerable interest in using CATA questions to obtain a rapid profile.

> From the following list, check the *attributes* that *describe* the *strawberry* you just tasted (choose *all* that apply):
>
> ☐ Sweet   ☐ Small
> ☐ Sour   ☐ Big
> ☐ Strawberry flavor   ☐ Firm
> ☐ Strawberry odor   ☐ Hard
> ☐ Flavorsome   ☐ Soft
> ☐ Tasteless   ☐ Juicy
> ☐ Red color   ☐ Dry
> ☐ Irregular shape   ☐ None of these
> ☐ Regular shape   ☐ Other: _____

FIGURE 11.1 Example CATA question.

In order to relate CATA results to consumer acceptance, CATA studies are often accompanied with *liking* questions and/or might include the evaluation of a (hypothetical) ideal product. CATA questions might be further combined with demographic and consumer psychographic questions, for example, to provide a so-called all-in-one test (Giacalone et al. 2013).

11.1.2 INTERPRETATION OF CATA DATA

CATA data are multivariate binary, but the meaning of an individual observation is ambiguous. Each consumer makes a subjective decision regarding the applicability of each term for each sample being evaluated. Checked terms are clearly considered by the consumer to be perceived as appropriate for describing the sample. In contrast, unchecked terms might indicate that the attribute was not perceived or might indicate that the attribute is perceived, but the term is nonetheless considered by the consumer an inappropriate term to characterize the sample. An unchecked term might also suggest that the consumer is uncertain or undecided about the term's applicability or has not given the term full consideration.

Adams et al. (2007) indicate that questions in the CATA format allow the consumers to provide sensory information using a shared set of terms without affecting hedonic responses, an assertion that merits greater scrutiny. The task is relatively simple, and this might avoid leading consumers to think in an uncharacteristically analytical manner that might alter their hedonic assessments.

When an attribute is selected for both a real and an ideal product, it does not necessarily indicate that the real product has an ideal intensity but only that the consumer thought that the attribute applied to both the real and an ideal product. The intensity of the real product could be just right or could be too high or too low.

11.1.3 RELATIONSHIP TO OTHER METHODS

The CATA question format has been used to indicate presence/absence of sensory attributes, especially to characterize aromas in wines (McCloskey et al. 1996; Le Fur et al. 2003; Campo et al. 2008, 2010). Findlay et al. (2006) use CATA and comment questions in a descriptive sensory study to provide additional attributes that were not among the attributes evaluated using line scales. These additions to the ballot are intended to alleviate so-called dumping effects, which occur when an assessor perceives an attribute not included in the questionnaire and incorporates this response into the response for another attribute (Clark and Lawless 1994). The use of the CATA methodology as proposed by Adams et al. (2007) is novel in that the evaluation is performed by the same untrained consumers that provide hedonic and other data about the samples.

CATA differs from free listing or open-ended comments in that the terms are suggested to, rather than by, the respondent. Free listing requires the respondent to write terms associated with the sample (Hough and Ferraris 2010), which the researcher must subsequently classify prior to analysis. Antmann et al. (2011) propose using a free listing task with one group of consumers to determine the consumer-relevant terms for subsequent CATA studies. If the list of CATA terms is incomplete, it may be desirable to allow the consumers to add one or more terms that they perceive are missing, rendering an intermediate approach that falls between CATA and free listing. Ultraflash profiling data (Perrin et al. 2008) is another name for free listing data, especially when these data are used to supplement projective mapping data. After arranging products on the surface in the projective mapping task, consumers indicate terms that they associate with each product, in a manner that they find most appropriate. As with free listing, word modifiers such as "low," "high," and "very" can be used. Variants of ultraflash profiling exist where consumers are provided with a lexicon that provides a common starting point but can add their own terms as they wish.

In free choice profiling (Williams and Langron 1984), untrained consumers generate and then scale attributes to characterize the samples being evaluated. Attributes can be highly personal, thus difficult to interpret (Lawless and Heymann 2010). Given the training required to enable panelists to use scales in a consistent manner (Meilgaard et al. 2006), it seems obvious that the consumers will use the scale differently. By contrast, CATA involves no scales. Consumers use the same terms to describe the samples but might interpret the terms differently. The CATA task is simple and seems not to force the consumer to evaluate samples in an analytical fashion but at the cost of foregoing intensity ratings.

Temporal dominance of sensations (TDS) data record the dominant attribute at each time point (Pineau et al. 2009). TDS can be considered as a variation of CATA. There is a relatively short list of attributes (Pineau et al. 2012). During the evaluation, each respondent continually updates the attribute perceived to be dominant. Recording whether an attribute has been checked during the evaluation period converts TDS data into CATA data, allowing the methods described herein to be applied. Note that the interpretation is slightly different: if the data arise from a TDS study, an attribute being checked indicates that it has been dominant at some point in time. In contrast, if not checked, it just means that the attribute has never been dominant, which does not imply that it does not describe a relevant characteristic of the product. As with descriptive sensory panelists, TDS panelists are trained to have a common understanding of the attributes. By contrast, consumers who answer CATA questions are untrained; their interpretation of an attribute might differ from the understanding of other consumers or of trained panelists. Differences in understanding might also be culturally dependent (Chung et al. 2012).

CATA questions can be presented to consumers after a sensory discrimination test. Cowden et al. (2012) provide guidance for decision making based on discrimination test results and CATA data. Where significant differences are found in the discrimination test and CATA results provide clear guidance to the product developer, reformulation can proceed without further diagnostic studies. Even if the discrimination test is not significant, trends identified in the CATA data can instigate further diagnostic work.

11.2   DESIGN OF CATA EXPERIMENTS

11.2.1   CONSUMERS

As with all experiments, the population of interest should be determined, along with consumer selection criteria. Quotas can be set to obtain a stratified sample where appropriate. Consumers should be recruited and selected for the panel or quota group in a random fashion, avoiding convenience samples that are unrepresentative of the target population.

Sample sizes used in previous studies are quite variable. Both Parente et al. (2010), who compare evaluations of cosmetics on a line scale against evaluations using CATA, and Ares et al. (2010), who studied milk desserts, involve 50 consumers per experimental group. Dooley et al. (2010) use 80 consumers to evaluate ice cream using CATA. Plaehn (2012) uses more than 100 consumers to evaluate citrus-flavored sodas.

There has not been much investigation to justify an appropriate sample size, but 50 consumers may be low. A study with business relevance should probably consider using 100 consumers or more. An appropriate approach is to consider the effect size of interest of all important variables and set the sample size such that the smallest effect size of interest can be detected with reasonable power. As we will discuss, some analysis methods only rely on data from consumers that showed some variation in the attribute under investigation (i.e., checked for a subset of products); in particular, this might substantially reduce the effective sample size for pairwise comparisons and attributes that are either rarely or almost always checked.

11.2.2   STUDY DESIGN AND PRESENTATION ORDER

The product design used may vary depending on the objectives of the test. Typically, the study is conducted as a full crossover where each assessor evaluates each sample in a sequential monadic presentation. To avoid systematic bias, it is highly recommended to balance the presentation order of products across assessors and to randomize the assignment of presentation order to the assessors. A Latin square or similar design often provides a good option. Other designs such as incomplete designs or repeated evaluations are possible but rarely used; they might be chosen to address specific project needs. Note that some of the analyses proposed might require appropriate modifications if deviating from a full crossover design.

The order that CATA and hedonic questions appear in the same ballot is an area that requires additional study. Ares and Jaeger (2013) did not observe differences in *overall liking* whether the *liking* question was preceded by a sensory CATA question or an emotion CATA question, implying that the priming effect of the type of CATA question does not have a strong impact on the reported hedonic response. King et al. (2013) reported that placing an emotion CATA question decreased average overall acceptance (OA) of respondents' recollection of "vanilla ice cream" from 7.4 to 7.2, whereas placing a rating question before the OA question led to a decrease in average OA from 7.4 to 6.8. Both decreases are statistically significant but the latter is apparently larger in size. These findings provide preliminary evidence that support the assertion by Adams et al. (2007) that CATA questions do not introduce the same biases as scale evaluations impose on the *liking* question. We do not have a strong recommendation for the reader on this matter, but if practical, we would tend to place the *liking* question before any CATA questions. If an ideal product is to be evaluated according to the same CATA question(s), the ideal product should always appear at the end of the study, after all the real samples have been presented.

11.2.3 CATA QUESTIONS AND ATTRIBUTE LISTS

A single CATA question can include a list of different types of terms. Dooley et al. (2010) present 13 taste, flavor, and texture terms in a single CATA question when collecting consumer data on commercial ice creams. Ares et al. (2011) present 19 hedonic and sensory terms in a single CATA question when collecting consumer data on powdered drinks. In a consumer test on cosmetic emulsions, Parente et al. (2010) merge attributes related to the sensory characteristics of the emulsions, sensory experience after application on skin, and skinfeel into a single CATA question with 20 attributes. In a subsequent consumer test involving antiaging creams, Parente et al. (2011) present consumers with a single question that includes 42 terms related to sensory properties of the creams, emotions, skin feel, applications, and product positioning.

Other researchers opt to split terms of different types into different CATA questions. For example, Meullenet et al. (2009) have consumers evaluate commercial orange juices. They arrange 41 sensory attributes into three separate CATA questions: one question each with 5 appearance attributes, 28 flavor attributes, and 8 texture attributes, respectively. Plaehn (2012) describes a CATA study on citrus-flavored sodas in which all CATA terms represented positive and negative emotions experienced during product consumption. Generally, emotion words are regularly presented in CATA format (King and Meiselman 2010).

The impact of attribute list size on the number of selections in CATA studies is still to be investigated. Some of the previously mentioned studies successfully include more than 40 attributes simultaneously. However, primacy bias and satisficing strategies (discussed in the following), as well as fatigue effects and the duration of product evaluation, should be considered: If assessors can only take a small sip from a beverage, their evaluation of 40 attributes will rely heavily on memory with all its limitations. For a chewing gum, it might be easier to evaluate a long list of CATA attributes while consuming the product.

The order in which sensory attributes are perceived must also be considered when designing the questionnaire. One approach is to group CATA terms that will be perceived by a consumer at a particular time during the evaluation. If attributes are not experienced roughly simultaneously, these can be separated into different CATA questions. For example, words that describe the appearance and aroma of a sample might be separated from words that describe flavor, taste, and mouthfeel, which might ensure that appropriate words are not skipped simply because they no longer apply. Another alternative is to provide only one CATA question but order attributes such that they are listed in the order that they are likely to be perceived or to provide specific instructions to consumers to recall and check all attributes that apply throughout the evaluation (Ares and Jaeger 2013).

The position of the attribute within the CATA list can bias the consumers' responses. Appropriate balancing and randomization of the attributes within the CATA list is recommended to avoid these effects. In a list consisting of many attributes of different types, the researcher might want to maintain the grouping of attributes of a certain type and balance and randomize the order of the groups first and only then the order of the attributes within each group. Assigning each consumer the same order across all products is one strategy that avoids different position effects within an assessor, which might bias the analysis.

11.2.4 FURTHER DESIGN ISSUES FOR CATA STUDIES

CATA studies have a lot of similarity to multiple choice questions, in which the respondent is limited to one choice. It has long been appreciated that the arrangement of choices in a multiple choice question affects selection (e.g., Matthews 1927; Payne 1951). Optimal CATA responses are characterized by thoroughness in the following aspects: comprehension of the question, retrieval of any relevant information from memory, integration of information retrieved from memory into a well-considered response, and precision in response selection (Krosnick 1999). When faced with fatigue or lack of motivation to complete long or demanding questionnaires, Krosnick (1991) postulates that respondents would increasingly make use of a suboptimal response strategy called *satisficing*. "Weak satisficing" is a suboptimal execution of the aforementioned aspects leading to a response that is nevertheless believed to be adequate. By contrast, "strong satisficing" is an arbitrary response or a response due to a superficial reading of the question without reference to memory, yet nevertheless believed to be adequate (Krosnick 1999). Primacy effects are a bias for selecting choices at the beginning of a list, while recency effects are a bias for selecting choices near the end of a list. When weak satisficing occurs in visually presented questionnaires, primacy effects are more prominent (Krosnick 1999). Additionally, choices that either are more plausible or seem more plausible when contrasted with adjacent choices might be selected more frequently (Krosnick 1999). In research by Matthews (1927), position effects in multiple choice tests increased with fatigue and difficulty.

It has been shown in web surveys without financial compensation that multiple choice and CATA questions can be subject to these kinds of position effects (Israel and Taylor 1990; Smyth et al. 2004). In a sensory test on commercial orange juice in which consumers were compensated financially, Castura (2009) observed position biases in CATA data as well. In a one-column layout, an attribute in the first position tended to be selected more often. In a three-column CATA layout, attributes in the leftmost column tended to be selected more often than attributes in other columns. Clear position biases were observed for CATA terms at top and left positions, consistent with survey response literature. Meullenet et al. (2009) found that consumers responded to a CATA question more quickly and with more selections if terms were organized alphabetically rather than in a random order. Ares and Jaeger (2013) revealed biases associated with position to be subtle and noted the importance of questionnaire development taking into account various concerns, including position bias and ease of response.

It would be possible to replace CATA questions by a series of forced-choice yes/no questions as recommended by Rasinski et al. (1994). If one question per attribute is used, the increased time required to complete the longer questionnaire is a key drawback. The compact format suggested by Bradburn et al. (2004) uses a very similar format to CATA with the only difference that for each attribute, the respondents are forced to check yes or no, thereby addressing the time deficiency. Further research is needed to determine whether this format has a stronger impact on consumers' evaluation of *liking* than does CATA. Reinbach et al. (2014) presented a yes/no–CATA hybrid: following a yes response that an attribute group (e.g., *floral*) is present, the respondents answer a sensory CATA question in which more specific descriptors are indicated (e.g., *elderflower, chamomile*).

Studies of web-based survey questionnaires have shown relationships between the manner in which questions were presented and subsequent responses. Survey researchers have found that if CATA terms were presented in a series of forced-choice yes/no questions, responses would contain more yes responses than those obtained from the same question posed in a CATA format (Rasinski et al. 1994; Smyth et al. 2006). At the same time, Rasinski et al. (1994) did not observe differences between the study formats with regard to primacy effects, suggesting that sequential yes/no questions might reduce the impact of satisficing but not primacy. However, in case of a long attribute list, respondents might find the yes/no format tedious.

A series of yes/no questions might also address the ambiguity of nonselection of a CATA term. Nonselection does not necessarily indicate that the respondent had fully considered and then rejected the applicability of a term; rather, the respondent might have determined that their response for the whole list of attributes was adequate to move on, perhaps due to satisficing, as discussed previously. Meullenet et al. (2009) used a "none-of-these-apply" option; however, this can only avoid an ambiguous nonresponse at the overall CATA question level across all attributes (which occurs when no CATA terms are selected at all).

Eye-tracking technology could provide insights into why some CATA terms are not selected. We are unaware of any such study involving CATA, but results on multiple choice questions in the survey literature show that respondents tend to spend more time looking at choices presented at the top of a list than those on the bottom and many respondents will skim lists rather than read every item completely (Galesic et al. 2008). Eye-tracking research suggests that many online readers follow an *F*-shaped reading pattern (Nielsen and Pernice 2009).

11.3 IMPLEMENTATION AND DATA COLLECTION

The consumer questionnaire should be kept as short and simple as possible and to provide clear instructions to ensure testing procedures are followed. In many cases, it will be beneficial to use a specialized computer system to manage the presentation of samples and CATA choices according to an experimental design, ensuring that responses are captured and tabulated correctly, without transcription or other errors that might occur with a manual procedure. If data collection is performed on paper, additional effort must be invested to assure data integrity.

During data collection, appropriate procedures should be followed to eliminate or reduce nonsample sources of variation (Lawless and Heymann 2010). Reasonable waiting periods should be incorporated to ensure that potential carryover effects are minimized.

11.4 ANALYSIS OF CATA DATA

In this section, we discuss some ways to analyze CATA data, both graphically and by means of statistical tests. We omit the technical details but provide some R code with the example in the next section.

Table 11.1 gives an overview of the analysis methods for the CATA data discussed in this chapter, depending on which data are recorded in addition to the CATA questions. Note that it is possible to consider subsets of the data; for example, *liking* and ideal product data might be available, but this does not preclude analyzing CATA data on its own.

11.4.1 CONTINGENCY TABLE

Typically, the first approach for summarizing CATA data is to create a contingency table. The table contains counts of the number of assessors that checked each respective attribute for each product. The counts are then merged and displayed in table format and optionally displayed as percentages, which might be particularly useful if comparing results across multiple studies with substantial differences in sample sizes. The contingency

---

TABLE 11.1
Analysis Options for CATA Studies, as Covered in This Chapter

Type of Data Available	Analysis Methods
CATA data on real products only	• Contingency tables
	• Bar charts
	• Significance testing
	• Correspondence analysis (CA)[a]
	• MFA
	• MDA
	• Correlation of attributes (visualized via MDS)
CATA data + *liking* (or related)	• Penalty-lift analysis
CATA data + (hypothetical) ideal	• Comparison of elicitation rates for real and ideal product with confidence intervals
CATA data + ideal + *liking*	• Penalty analysis

[a] Metric can be based on the χ^2-distance or the Hellinger distance.

---

table (or parts thereof) is often displayed visually using bar charts (see Section 11.4.3.1) to facilitate comparisons between products and attributes.

11.4.2 STATISTICAL TEST STRATEGY FOR CATA

As CATA data are often used to gain a better understanding of the products, statistical approaches are exploratory and descriptive in nature. Nonetheless, it is useful to determine whether the differences suggested by a visual display are real or whether they might be due to chance alone. The aim of this section is to suggest some statistical tests for CATA data. At the end of this section, we suggest an order for applying the tests proposed. Following the proposed strategy allows all aspects of product differences—across products and attributes as well as between subsets of products and/or individual attributes—to be evaluated, thereby identifying which of the effects are likely to be worthy of further investigation. If the data set includes evaluations of an ideal product, depending on the purpose of the investigation, this ideal product is treated as all other products in what follows, or it is omitted from the analysis.

From now on, we assume that the study design is a full crossover (i.e., each panelist evaluates all products). For incomplete designs (each assessor only sees a subset of all products) or for parallel studies (each assessor only evaluates a single product), the tests discussed (in particular the parametric approximations) might not apply; generalization to these settings is straightforward but implementation might be more or less laborious. As incomplete and parallel CATA studies are rarely conducted, we will not go into details here.

For further usage, we define the effective sample size as the number of assessors that show some variation across products, that is, excluding those with no variability across products as they checked the respective attribute for all or for none of the products.

11.4.2.1 Cochran's Q

Cochran's Q test (Cochran 1950) is a statistical test to investigate whether there are significant differences between products in a study with related samples. Samples are related in crossover studies because each assessor evaluates all products. This test is widely used in a CATA context for statistical inference of product differences by attribute. Under the null hypothesis of no product differences, Cochran's Q statistic is asymptotically χ^2-distributed with $(n_k - 1)$ degrees of freedom, where n_k is the number of products.

Tate and Brown (1970) investigated the required sample size to warrant the χ^2-approximation to hold; they suggested that the effective sample size times the number of products should be at least 24. Cochran (1950) argued that the F-test on binary data (treating frequency data as if it were continuous data) might give very similar results in certain situations compared to the χ^2-approximation. However, he and Tate and Brown (1970) pointed out that the F-test depends on assessors reporting only 1s or only 0s, which is unlike Cochran's Q; therefore, we prefer the χ^2-approximation, if any will be used.

Cochran (1950) also pointed out that any subject's contribution to the discrimination ability among products would depend on its probability of choosing the respective attribute. This might provide the option to consider some weighting of assessors according to their probability to check a given attribute, although this approach is exceedingly rare and thus not pursued further.

McNemar's test (McNemar 1947) and Cochran's Q test are equivalent for the two-product case, and both provide a χ^2-approximation for conducting pairwise comparisons between products. A better alternative still is given by the well-known sign test (Arbuthnott 1710), which provides an exact version of McNemar's test and is therefore preferable, especially for small sample sizes (again after omitting those assessors that check the attribute for both products or none of them). For simplicity, the sign test is recommended for all pairwise comparisons.

11.4.2.2 Overall Test

Before drawing any conclusions from CATA data, we want to be sure that there are some real differences between the products under investigation. An omnibus test should be used to protect against inflated experiment-wise error rates that occur when conducting many individual tests without controlling for multiplicity. An example of an omnibus test is the F-test that is conventionally performed on continuous data prior to conducting any post hoc tests.

The classical Pearson's χ^2-test does not provide a valid omnibus test for CATA data because the assumption of independence is violated twice: each assessor evaluates multiple products, and for each product, they assess multiple attributes. An approximate global test for CATA data is proposed by Meyners et al. (2013) based on the sum of Cochran's Q statistics across attributes. It relies on the assumption of independence in evaluations of different attributes, that is, that a consumer's evaluation of any attribute is independent of his or her evaluations of all other attributes.

Unfortunately, the assumption that attributes are evaluated independently is unlikely to hold in most applications; if a consumer checks *sour*, he or she might be more unlikely to check *sweet* as well, and vice versa, indicating that the attributes are dependent, as are the Q statistics in that case. The relationships between attributes depend on the CATA attributes included, as well as on the products in the study. Also, it is unclear what minimum sample sizes are required to conduct such a test. Therefore, a randomization test provides a viable alternative. The approach is conceptually similar to what Wakeling et al. (1992) propose for the consensus derived from generalized procrustes analysis (GPA) and to what Meyners and Pineau (2010) propose for detecting product differences in TDS studies. For brevity, we refrain from discussing any details here; for the concept of randomization tests and a proof of its validity, the interested reader is referred to the textbook by Edgington and Onghena (2007). Meyners and Pineau (2010) also provide some detailed explanation on this matter. We apply the randomization test to the sum of Cochran's Q statistics as the natural generalization from Cochran's Q test by attribute. Other test statistics could also be used; for example, Meyners and Hartwig (2009) use Pearson's χ^2-statistic.

11.4.2.3 Further Tests by Attributes or Subsets of Products

If needed, the test can be broken down for subsets of products to further investigate which products differ from each other. Typically, we would apply the approach to pairs of products without correction for multiplicity. Though not formally correct for more than three products in the study and in particular for increasing numbers of products, the global test performed at the beginning provides at least some protection against multiplicity issues. As the purpose of CATA studies is usually product understanding rather than a more formal investigation such as providing claim support, this compromise is often acceptable.

The same approach can be taken for the comparison of products within a single attribute. As Cochran's Q test provides only an approximate solution for more than two products, the randomization test can and should be used especially in the case of small effective sample sizes. The smaller the subset of products, the smaller the effective sample size, which itself might become very small even in larger studies whenever some of the attributes are hardly elicited at all (consider *salty* for chocolate) or checked almost always (consider *sweet* in the same context). For pairwise comparisons of products, the exact sign test should be employed.

11.4.2.4 Test Strategy

We strongly suggest starting with a global test for product differences in nearly all cases; if the global test is not statistically significant, any further testing would inflate the type I error rate and should therefore be avoided. As the primary aim of CATA studies is typically data exploration, we usually won't apply any further correction for multiplicity; if required, Bonferroni's correction or the more powerful Bonferroni–Holm approach (Holm 1979) might be applied. We have also used Hommel's adjustment (Hommel 1988), which is even more powerful, but less known. It is available in R using the p.adjust function in the stats package (R Development Core Team 2013).

In some studies, we might encounter attributes that do not discriminate between products at the predetermined threshold (typically 5%, but as CATA is used in a descriptive and exploratory way, 10% might be reasonable as well). To increase legibility of the figures, we usually suggest omitting nonsignificant attributes from further exploration and visualization; these attributes potentially represent random noise only. Note that some multivariate analyses are sensitive to the inclusion of these attributes.

11.4.2.5 Missing Values

If missing values occur, they will not affect the validity of the inference using randomization tests if they are missing completely at random. In contrast, they might compromise the validity of the asymptotic alternatives. If values are missing due to censoring or other systematic causes, they are likely to bias any analysis.

11.4.3 GRAPHICAL ANALYSIS

11.4.3.1 Bar Charts

A simple graphical display of the contingency table is provided by bar charts. For those, the percentage (or absolute number) of assessors checking the attribute for any product is displayed as a bar. Products can be easily compared if each attribute contains product bars next to each other. In contrast, the profile of a product is best assessed if the bars for all attributes are grouped together for this product. The researcher can choose which setup best meets his or her needs.

11.4.3.2 Correspondence Analysis

This widely used tool, which permits the visualization of a contingency table, can be thought of as a generalization of principal component analysis (PCA) for ordinal and nominal data. CA determines a projection of the data into orthogonal dimensions such that they sequentially represent as much of the variation in the data as possible. While there are multiple outcomes and key measures in CA, usually we primarily look at the (2D) plot of the first two components; if the variation explained is insufficient, we might consider further dimensions as well. The interpretation of the CA map is the same as for biplots (Gabriel 1971), a common application of which is a (linear or phase 4) preference mapping. Details about CA are beyond our scope; the interested reader is referred to the textbook by Greenacre (2007).

Classical CA is based on the so-called χ^2-distances. In reviewing earlier criticisms of the application of this metric for species abundance data, Legendre and Gallagher (2001) note that the χ^2-distance will be much more influenced by the inclusion of a few very rare species than by very abundant species. Rather than removing rare species prior to conducting CA, they propose using the so-called Hellinger distance (Hellinger 1909) as an alternative. Popper et al. (2011) made similar observations regarding the inclusion of rarely selected CATA terms on CA: distances between products in the final CA plot depend on what other products are included in the analysis, and attributes that are rarely checked overall will have an undue high impact on the final results. They apply CA using the Hellinger distance to CATA data to overcome these issues. Williams et al. (2011) note that CA is best used when there are five or more products evaluated, such as in a category review.

It is possible to conduct CA in R using various packages. The package ca (Nenadic and Greenacre 2007) is dedicated primarily to CA, while ExPosition (Beaton et al. 2013), with graphical output generated with prettyGraphs (Beaton 2013), provides CA as well as functions for other multivariate methods. As far as we are aware, ExPosition is the only R package that currently has a direct implementation for CA using the Hellinger distance. The package FactoMineR (Husson et al. 2013) also offers multivariate methods, including CA.

We provide some code with the examples in Section 11.5 of this chapter. It should be noted that the standard figures from many of these packages do not necessarily maintain the aspect ratio of the plot, which means that the standard figures from many of these packages do not respect the actual distances of the products; instead, the variation explained needs to be taken into account for the interpretation; we recommend readjustment such that visual distances are proportional to the differences between products.

11.4.3.3   Multiple Factor Analysis

This alternative tool to derive a perceptual map can be considered as a generalization of PCA for continuous data and of multiple correspondence analysis (MCA) for categorical data in that it weights (groups of) variables to balance their impact on the perceptual map (Escofier and Pagès 2008). For example, if a group of taste attributes has a large contribution while another group of texture attributes has a comparably small contribution, multiple factor analysis (MFA) will weigh the taste attributes downward (each taste attribute by the same amount) and the texture attributes upward (again each by the same amount for the whole group), such that all aspects are accounted for in the results. The final map therefore depends on how the variables have been grouped together. MFA is implemented in the R package FactoMineR.

11.4.3.4   Other Perceptual Maps

There are other ways to derive a perceptual map from CATA data, for example, by means of partial least squares (PLS), MCA, or joint correspondence analysis (JCA). All these approaches use different criteria to derive a 2D graphical arrangement of the products and attributes, with different advantages and drawbacks. As most applications use CA, we will refer the interested reader to Plaehn (2012) for an application of PLS to emotions CATA data and to Greenacre (2007) for practical discussions on the strengths and weaknesses of MCA and JCA.

11.4.3.5   Relationship between Attributes and Products

The relationship between attributes and products is only partly visible in a CA map; only two dimensions (rarely three) are displayed simultaneously. It might occur that any product is more closely related to an attribute than it appears from the display of the two CA dimensions; in contrast, an attribute might be less related to a product than the plot suggests. Mathematically, attributes and products in the perceptual map derived from CA or any other appropriate method can be considered as vectors in a multidimensional space. The angle between these vectors provides us with information about the relation between the product and the respective attribute that goes beyond the 2D graphical display. Carr et al. (2009) refer to this approach as multidimensional alignment (MDA). They suggest displaying the cosines of the angles in a bar chart. Thereby, it is worth noting that absolute cosines below $\cos(45°) = 0.707$ indicate hardly any relationship at all. The cutoff is therefore much larger than typically used for correlations. Alternatively, we suggest displaying the angles themselves on a reversed scale from $180°$ (π radians), which indicates perfect negative correlation, to $0°$ (0 radians), which indicates perfect positive correlation. In between, $90°$ ($\pi/2$ radians) indicates no correlation. Alternatively to displaying the angles as values, the attributes could also be displayed in a (semi)circle, visually displaying the angle formed by the respective attributes with the product in the multidimensional space in two dimensions only (Meyners et al. 2013). The same approach could be applied to study the relationship between pairs of attributes.

Note that angles in MDA relate to the products only; we cannot interpret the relationship of two or more angles of attributes with one product in order to compare them. Two attributes might be reasonably well correlated with a product, yet be orthogonal in the multidimensional space.

11.4.3.6   Relationship between Attributes

We might be interested in investigating which attributes are typically checked together and which are chosen independently. The φ-coefficient is a measure of correlation of two binary variables. Based on all observations (across assessors), Meyners et al. (2013) suggest to use $1 - \varphi$ to measure distances between attributes and apply a multidimensional scaling (MDS) to the respective matrix. The results of MDS can then be visualized. Additional insight might be gained from the same analysis on relevant subsets of the data (e.g., observations above the mean *liking* or subsets according to demographic variables).

11.4.4 COMPLEMENTARY ANALYSES BASED ON ADDITIONAL DATA

11.4.4.1 Penalty-Lift Analysis

If *liking* for the products is collected along with the CATA data, we can average *liking* across all observations (i.e., across assessors and products) for which the attribute under consideration was elicited and across all observations for which it was not elicited. The difference between these two mean values is an estimate of how much *liking* changes when an attribute applies compared to when it doesn't apply. Williams et al. (2011) call this approach penalty-lift analysis. *Liking* might increase due to the presence of a positive attribute or the absence of a negative attribute. Conversely, *liking* might decrease due to the presence of a negative attribute or the absence of a positive attribute. Therefore, the outcome of the analysis is a change in *liking*. Positive values represent an increase in *liking* due to the attribute and negative values a decrease in *liking* due to the attribute. Results are easily visualized in a bar chart and can be used to prioritize attribute importance. Lack of lift might be an indicator of consumer segmentation (Williams et al. 2011). Note that correlations between attributes among products might cause some attributes to spuriously appear to drive *liking*. For example, if *sweetness* and *darkness* are coincident in a set of chocolate products being evaluated, the latter might appear to be an important hedonic driver even though increased *liking* is solely driven by the *sweetness* level. Product selection can have a substantial impact on the interpretation and needs to be taken into account carefully.

11.4.4.2 Comparison of Products with Ideal

A study might include CATA evaluations for both real products and a (hypothetical) ideal product. Cowden et al. (2009) compare the proportion of elicitations for the real products with the ideal product. The approach involves using a confidence interval for the ideal but not the real products, which ignores the statistical uncertainty about the ideal product. Meyners et al. (2013) build on this approach, taking the difference between the elicitations for the ideal and the real products and a confidence interval for the difference. Effective sample size has to be taken into account: Only assessors that discriminate between the products provide pertinent information; thus, confidence intervals are of differing widths. Visual display of these data reveals the attributes for which the real products differ from the ideal.

Additionally, Cowden et al. (2009) propose to ask consumers to rank the three attributes they consider to be most important in their ideal product. The percentage of respondents that ranked an attribute first (called *strength*) is then plotted against percentages selected in CATA (called *interest*), enabling a prioritization of high-strength, high-interest attributes to be identified and refined in the subsequent reformulation process.

11.4.4.3 Penalty Analysis

In case the CATA study included an ideal product as well as *liking* scores for the real products, a penalty analysis can be performed. Unlike the well-known penalty analysis used for just-about-right (JAR) questions, analysis is based on the gaps between the real products and the ideal and the impact on *liking* scores.

Ares et al. (2012) suggest looking at incongruence between elicitations of attributes for the ideal and selected real products, where incongruence is defined as an attribute applying to the ideal or the real product only, but not both. In contrast, congruence is found if the attribute applies to both (neither) the ideal and (nor) the real product. *Liking* ratings are then averaged across consumers with incongruent and congruent elicitations, respectively. The difference between the *liking* rating for congruent and incongruent elicitations indicates how much *liking* changes when the product does not match the ideal product. The change in *liking* is called the *liking* drop because it is rarely positive. This approach is easily generalized to an overall evaluation by averaging across all products instead of a by-product evaluation only.

A limitation of this approach lies in treating both types of incongruence the same way: the contingency table is needed to assess whether the impact of a given attribute on liking is positive or negative. It often makes a difference whether an attribute is checked for an ideal but not the real product, or vice versa. Therefore, Meyners et al. (2013) extend the proposal by Ares et al. (2012) by looking at the different combinations in which the endorsement for the ideal product remains constant but endorsement of the real product changes. By doing so, it is possible to distinguish "must have" attributes from "nice to have" or "to be avoided" attributes. Plotting the observed differences in *liking* from either of the analyses against the percentage of consumers for which incongruence occurred helps to interpret the data: the higher this rate, the more important the attribute to consumers. Apparently, the percentages of the second approach are smaller than those from the approach of Ares et al. (2012): On average, they should be half as large, but the percentage can vary from attribute to attribute and is loosely related to the number of consumers checking an attribute for the ideal and the number of consumers that leave the attribute unchecked.

11.5 EXAMPLE

To illustrate the different approaches to analyze and interpret CATA data, we consider the analysis of a real data set. Some (sometimes simplified) code is provided that allows these analyses to be run easily using the software R (R Development Core Team 2013; version R 2.15.1). The data come from a study on six different types of strawberry cultivars. The study was carried out in a sensory laboratory with 116 consumers recruited at a supermarket in Montevideo, Uruguay, when buying fruits and vegetables. The presentation of the cultivars followed a balanced rotation. Consumers were asked to taste the strawberries, rate their *liking* on a 9-point scale, and answer a CATA question with 16 attributes. Four of the cultivars (Yurí, Yvahé, Guenoa, and Festival) were commercially available in Uruguay, while the two other cultivars (L20.1 and K31.5) were new cultivars developed by the Instituto Nacional de Investigación Agropecuaria (INIA, Uruguay). The data set is arranged as illustrated in Table 11.2 and is available for download from the CRC Web site: http://www.crcpress.com/product/isbn/9781466566293.

TABLE 11.2
Strawberry Data Set (Excerpt)

Consumer	Sample	Liking	Sweet	Sour	Strawberry Flavor	Strawberry Odor
1	Festival	9	1	0	0	1
1	Yvahé	5	0	0	0	1
1	Yurí	7	1	0	0	0
1	Guenoa	2	0	0	0	0
1	L20.1	8	1	0	0	0
1	K31.5	1	0	0	0	0
2	Festival	7	1	0	1	1
2	Yurí	4	0	0	0	0
2	L20.1	8	1	0	1	1
2	K31.5	2	0	0	0	0
2	Yvahé	3	0	1	0	0
2	Guenoa	6	1	1	1	1
3	K31.5	9	1	0	0	0

Note: Sample identifies the six cultivars.

The first step of our analysis is to summarize the data in a contingency table and to visualize the results by product and by attribute, respectively:

```
contingency.table <-
  aggregate(dataset[,variables],
            list(product=dataset[,product]), sum,
            na.rm=TRUE)
```

where `variables` is a character vector with the names of the variables to be used and `product` identifies the column in `dataset` identifying the product evaluated in each observation.

The contingency table is shown in Table 11.3. It should be mentioned that the data set contains a small number of missing values relative to the number of observations. There are 20 missing values for *liking* and two missing values for the attribute *dry*. For this discussion, all missing data are assumed to be missing completely at random. For the purpose of providing an illustrative example, the value of 0 is substituted for all missing values for the attribute *dry* (i.e., the attribute was considered not to have been chosen).

The following R code permits visualization of selected attributes from the contingency table in a bar chart:

```
plot.variables <- variables[1:5]
barplot(as.matrix(contingency.table[,plot.variables]),
        beside=T, col=1:nprod,
        legend.text=contingency.table[,"product"],
        args.legend=list(horiz=TRUE, x="topleft"))
```

Figure 11.2 visualizes the data for selected attributes, grouped by attribute. It can be seen that most consumers characterized product L20.1 as *sweet*, Yvahé as *small*, and Yurí as *hard*. Yurí, Yvahé, and K31.5 were characterized more often as *tasteless* compared to the other strawberries. Yvahé, Yurí, and Festival were more often attributed as having an *irregular shape* than the other cultivars; however, average elicitation rate for this attribute is only about 17%.

Applying the test strategy outlined previously allows investigation into which differences, if any, can be assumed real rather than due to chance. We do not apply any corrections for multiplicity in the analyses that follow. Table 11.4 gives the results from the overall analysis, both based on Cochran's Q and on the respective randomization approach using 1000 randomizations. The results are in excellent agreement, indicating the acceptability of the χ^2-approximation to Cochran's Q in this data set. Effective sample sizes (i.e., number of consumers showing at least some variation between products with regard to the respective attribute) varied from 67 (*small*) to 104 (*red color*), supporting this finding. The products

TABLE 11.3

Contingency Table of the Strawberry Data Set with 6 Products and 16 CATA Attributes

Cultivar	Sweet	Sour	Strawberry Flavor	Strawberry Odor	Flavorsome	Tasteless	Red Color	Irregular Shape	Regular Shape	Small	Big	Firm	Hard	Soft	Juicy	Dry	Total
Yurí	29	16	19	19	25	43	50	25	36	21	52	55	37	14	52	23	516
Yvahé	39	14	21	21	38	40	40	30	35	39	28	49	13	32	74	13	526
Guenoa	35	16	22	25	45	32	64	17	44	18	54	44	5	49	76	13	559
Festival	37	30	21	32	50	23	54	25	31	14	47	70	22	22	62	19	559
L20.1	52	10	24	31	59	22	62	13	43	11	71	53	10	39	63	23	586
K31.5	14	38	18	16	21	41	45	11	63	16	48	57	21	18	62	24	513
Total	206	124	125	144	238	201	315	121	252	119	300	328	108	174	389	115	

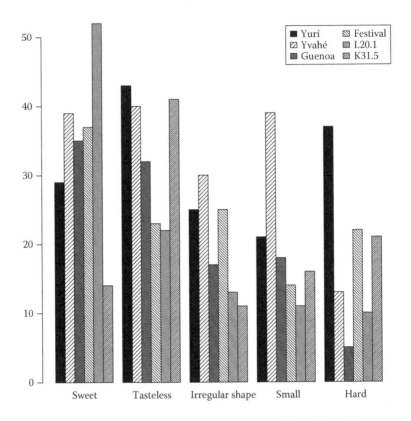

FIGURE 11.2 Bar chart of the number of elicitations for each product, grouped by attribute.

differed on almost all attributes; the only nondiscriminating attributes were *strawberry flavor* and *dry*, consistent with the contingency table displayed in Table 11.3. It might be surprising that in this test of strawberries, *strawberry flavor* was one of the least elicited attributes! We might want to inquire how the consumers interpreted this variable but will not pursue this here as this question is beyond the scope of this chapter. As the number of attributes that do not discriminate is small, we retain all attributes in the analysis despite the fact that two of these attributes might contain just random noise; it should not impact our interpretation either way.

The next step is to explore which pairs of products differ and on which attributes. Applying an overall test for each pair of products reveals significant differences for all pairs of products across attributes (details not shown). The pairwise comparisons allow further investigation of the source of differences detected by the global tests. Table 11.5 shows the results for the comparison between products Yurí and Yvahé. It is seen that these two products did not differ much regarding the taste or shape attributes but apparently differed in size (Yvahé was perceived as *big* by

TABLE 11.4

Uncorrected p Values from Statistical Testing for Overall Product Differences and Effective Sample Sizes

Attribute	p Value (Randomizations)	p Value (Cochran's Q)	Effective Sample Size
Sweet	**0.001**	**<0.001**	94
Sour	**0.001**	**<0.001**	76
Strawberry flavor	0.907	0.903	71
Strawberry odor	**0.031**	**0.023**	73
Flavorsome	**0.001**	**<0.001**	94
Tasteless	**0.001**	**0.001**	94
Red color	**0.005**	**0.005**	104
Irregular shape	**0.005**	**0.003**	71
Regular shape	**0.001**	**<0.001**	82
Small	**0.001**	**<0.001**	67
Big	**0.001**	**<0.001**	102
Firm	**0.007**	**0.005**	99
Hard	**0.001**	**<0.001**	69
Soft	**0.001**	**<0.001**	84
Juicy	**0.012**	**0.008**	101
Dry	0.126	0.115	68
Overall	**0.001**	**<0.001**	

Note: Significant p values at level 5% are set in bold.

more consumers) and texture (Yurí was perceived as *hard* and *dry* by more consumers). Here, the effective sample sizes vary from 22 (*sour*) to 55 (*tasteless*). Conclusions from Figure 11.2 are confirmed in other pairwise comparisons, which are omitted for brevity. Effective sample sizes range from 15 to 69, indicating that the overall sample size of 116 might have been relatively low in order to be able to potentially discriminate on all pairs of products and attributes.

Based on the contingency table shown in Table 11.3, CA was performed both using χ^2-distance and the Hellinger distance. The results and figure for the χ^2-distance were obtained, including the plot shown in Figure 11.3, using R code similar to the following:

```
library(ExPosition)
ca.ex <- epCA(contingency.table[,-1], graphs=FALSE)
epGraphs(ca.ex, biplots=T)
```

The CA using the Hellinger distance (not shown) is very similar and yields the same interpretation. The similarity might be ascribed to the fact that

TABLE 11.5

Summary Measures and Uncorrected p Values from Statistical Testing for Product Differences between Cultivars Yurí and Yvahé

	Assessors Endorsements				
	Yurí		Yvahé	p Value	Effective
Attribute	Only	Both	Only	(Sign Test)	Sample Size
Sweet	19	68	29	0.193	48
Sour	12	94	10	0.832	22
Strawberry flavor	13	88	15	0.851	28
Strawberry odor	13	88	15	0.851	28
Flavorsome	14	75	27	*0.060*	41
Tasteless	29	61	26	0.788	55
Red color	30	66	20	0.203	50
Irregular shape	17	77	22	0.522	39
Regular shape	23	71	22	1.000	45
Small	10	78	28	**0.005**	38
Big	35	70	11	**0.001**	46
Firm	22	78	16	0.418	38
Hard	29	82	5	**<0.001**	34
Soft	10	78	28	**0.005**	38
Juicy	14	66	36	**0.003**	50
Dry	19	88	9	*0.087*	28

Note: Attribute p values below 5% are set in bold; attribute p values between 5% and 10% are set in italics.

the number of attributes selected across assessors was very similar for all products and that the expected value for any of the cells is sufficiently high, based on the marginals shown in Table 11.3. As expected, contrasting attributes nicely plot in opposite directions (*small* vs. *big*, *regular* vs. *irregular shape*, *hard* vs. *soft*, etc.), such that the overall display seems reasonable. In line with previous interpretations, we conclude that consumers associated Yvahé with attributes *small* and *irregular shape*. By contrast, L20.1 and Guenoa were characterized as *flavorsome*, *big*, *sweet*, and *soft*. Guenoa is not well represented in Figure 11.3, as is indicated by the smaller symbol used for its display. K31.5 and Yurí were attributed as *sour*, *hard*, and *tasteless*. Overall, the properties of Yvahé, L20.1, and K31.5 are relatively well explained, while Festival is poorly explained in these figures. The contingency table in Table 11.3 shows that the latter product is not really polarizing in any of the attributes, explaining this observation.

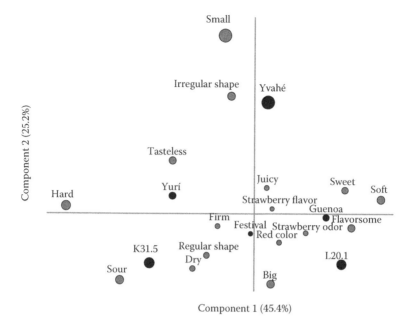

FIGURE 11.3   CA for the strawberry data using the χ^2-distance.

Based on the outcomes of the CA, we determined the multidimensional relationship between products and attributes using MDA. The relationship between Yurí and all attributes is shown in Figure 11.4. Again, previous interpretations are supported. In addition, we see that Yurí and K31.5 are rather negatively correlated with the flavor attributes *flavorsome*, *strawberry flavor*, and *odor*. The reversed scale on the horizontal axis in Figure 11.7 warrants the usual interpretation of values to the right of the center indicating positive correlation, and those to the left indicate negative correlation.

In order to further investigate the relationship between attributes, we determined the φ-coefficient for all pairs of attributes based on the full data set, translated these into a distance measure, and applied a classical MDS to the respective values (cf. Meyners et al. 2013). The following R code might be used to accomplish this. Alternatively, the package ExPosition also offers algorithms for MDS:

```
mds <- cmdscale(1-cor(dataset[,-c(1:3,20)],
  use="complete.obs"))
plot(mds, type="n", asp = 1)
text(mds, rownames(mds))
```

Figure 11.5 shows the resulting MDS. We conclude that attributes like *strawberry flavor*, *strawberry odor*, and *sweet* are frequently coelicited,

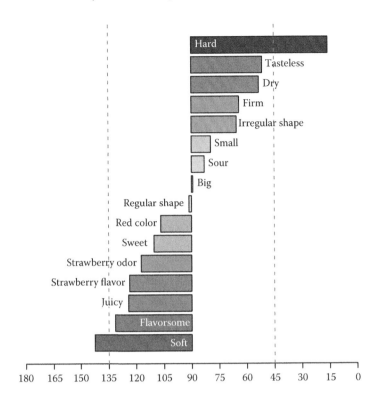

FIGURE 11.4 Bar chart of MDA results (angles) between product Yurí and all attributes. Vertical dashed lines indicate the cutoff of 45° and 135°, respectively. Gray scales are used to further emphasize the strength of the relation.

as are *dry* and *hard*. On the other hand, *firm* and *soft* seem to be rarely selected simultaneously by consumers. Attributes in the center of the plot are not well represented in these two dimensions. For example, we cannot conclude that *small* and *big* are frequently coelicited; rather, they share a certain independence from most other attributes.

As an alternative to CA, MFA was applied to this data set in order to derive a perceptual map. For this purpose, we grouped the attributes into three groups, namely, taste/flavor (*sweet, sour, strawberry flavor, strawberry odor, flavorsome,* and *tasteless*), appearance (*red color, irregular shape, regular shape, small, big*), and texture (*firm, hard, soft, juicy, dry*). The results resemble those of the CA and are therefore omitted for brevity.

Next, penalty-lift analysis is performed to investigate which attributes have the highest impact on *liking*. Figure 11.6 displays the outcomes visually and suggests that *sweet* was the main driver for *liking*, closely followed by *flavorsome, strawberry odor,* and *strawberry flavor. Juiciness* also seemed to increase *liking*. In turn, strawberries characterized as *tasteless*,

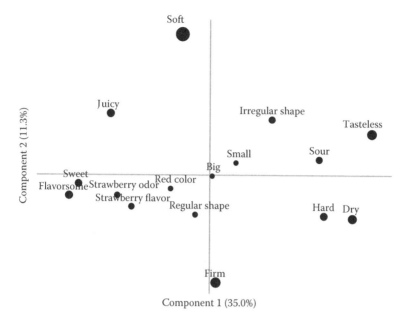

FIGURE 11.5   MDS on distances between attributes based on the φ-coefficient.

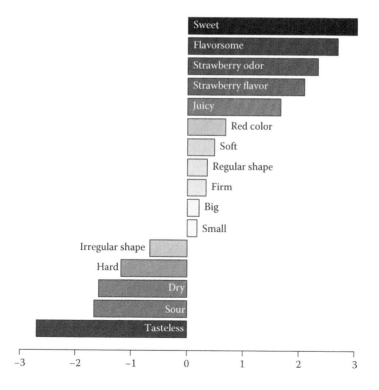

FIGURE 11.6   Visualization of the results from the penalty-lift analysis.

sour, dry, and *hard* were disliked. These findings are hardly surprising but illustrate the potential of CATA studies in general and this analysis in particular to reveal such hedonic drivers.

Unfortunately, in this study, no ideal product has been evaluated. To be able to nevertheless illustrate the methods requiring ideal data described previously, we simulated elicitations of the attributes for an ideal product. Without going into details here, we took the average number of elicitations per attribute into account and increased the likelihood for selection for positive drivers of *liking* according to the penalty-lift analysis and decreased it for negative drivers. Thus, simulations are somewhat realistic.

Figure 11.7 shows the differences between cultivar Yurí and the ideal in the proportion of elicitations along with a 95% confidence interval. In this display, for ease of interpretation, the attributes are ordered by decreasing effective sample size. The difference in elicitation rates is based on the total base size of 116. The figure indicates that for Yurí, the terms *sweet, juicy, strawberry odor, flavorsome,* and *soft* were cited significantly less frequently than the ideal, while *tasteless, hard, dry,* and *sour* were

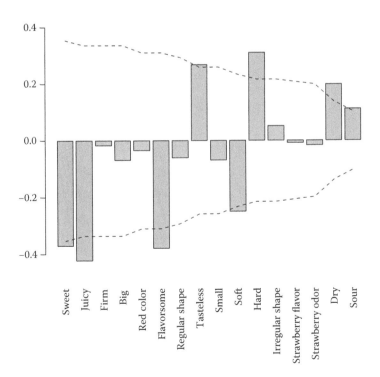

FIGURE 11.7   Differences in elicitation rates between Yurí and the ideal product including a 95% confidence interval for this difference.

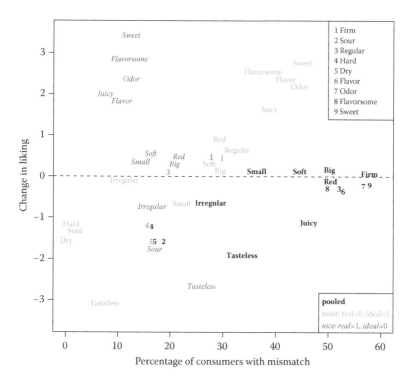

FIGURE 11.8 Visualization of the results of a penalty analysis.

cited more frequently. The analysis gives a clear picture regarding the weaknesses of the Yurí cultivar.

The penalty analysis was conducted according to both approaches described previously. Figure 11.8 shows the results of both analyses overlaid after pooling across products. It is seen that *sweetness* was indeed a driver of *liking* here; irrespective of whether it is used for the ideal product (light gray, plain font) or not (dark gray, italics), if *sweet* was elicited, the average *liking* was substantially higher than if *sweet* was not elicited. It is worth noting that in the pooled analysis (black, bold font), the impact of *sweet* on *liking* is close to zero and would not show up as a main driver, illustrating the limitation of the analysis if we only look at congruent vs. incongruent pairs. On the other hand, *tasteless* is identified as a negative driver of *liking*.

From the display proposed by Ares et al. (2012), it is not apparent that *tasteless* is a *negative* driver without reference to the contingency table. Had the attribute been *tasteful* instead of *tasteless*, it would most likely have appeared in a similar position in the display, as products endorsed for *tasteless* would not be so for *tasteful*, and vice versa, and (in)congruences with the ideal would be pretty much unchanged. From our approach, *tasteless* is clearly shown to be a negative driver of *liking*. Had the attribute been *tasteful*, it would be expected to appear in the upper portion of Figure 11.8. We refrain from detailed interpretation of the data here, as it is based on simulated data and therefore without intrinsic meaning.

11.6 DISCUSSION

CATA is one of the recently emerging rapid profiling methods employing consumers. One of the benefits of the CATA methodology is that it provides a straightforward task for consumers (Adams et al. 2007). CATA can be used to collect information about consumers' perception of products, possibly without inducing an analytical mindset that might alter hedonic assessments. Assuming that the consumers' interpretation of the terms is appropriate, CATA data can provide insights into consumers' perceptions and potentially explain why particular samples are more liked than others. In contrast to unstructured comment data, CATA data are not onerous to analyze. Speed is another advantage; Ares et al. (2010) report that consumers completed *liking* and descriptive CATA much faster and with less instruction than a projective mapping task.

CATA data might also be combined with other data to enable the investigation of relationships between sensory perception, *liking*, and consumer information, including demographics and psychographics. CATA studies that include *liking* questions have the potential of providing insights into the perception of products from the same consumers who give the hedonic scores. In case a hypothetical ideal product is evaluated, the differences between ideal and real products might yield additional insights into what drives the hedonic responses.

A drawback of CATA is that it does not give direct information regarding attribute intensities; although observed CATA frequencies correlate well with intensities in some published studies (Bruzzone et al. 2012; Reinbach et al. 2014). There is no direct information regarding whether an attribute is liked at the intensity that it exists in the product. Even if an ideal product is evaluated, the fact that an attribute is checked for both the ideal and the real product does not indicate that the intensity for the real product is optimal; it might be far too high or far too low. As in TDS, an implicit assumption sometimes made is that the true intensity will correlate with the proportion of consumers selecting the attribute; this is a strong assumption for which we have not found compelling evidence and which is likely to be wrong in many situations. Furthermore, consumers might use a CATA term differently, and many consumers might also understand a CATA attribute to have a meaning that is quite different than would be expected by a trained descriptive panel or product developers (Ares et al. 2010), which if unappreciated could result in misapplied reformulation efforts.

To summarize, CATA is another valuable method in the toolbox of sensory scientists and consumer researchers, so long as its advantages are leveraged, and its limitations considered carefully.

ACKNOWLEDGMENT

We are grateful to Gastón Ares for making the strawberry data set available to us.

REFERENCES

Adams, J., Williams, A., Lancaster, B., and Foley, M. 2007. Advantages and uses of check-all-that-apply response compared to traditional scaling of attributes. *Seventh Rose-Marie Pangborn Sensory Science Symposium.* Minneapolis, MN.

Antmann, G., Ares, G., Varela, P., Salvador, A., Coste, B., and Fiszman, S. M. 2011. Consumers' texture vocabulary: Results from a free listing study in three Spanish-speaking countries. *Food Quality and Preference* 22: 165–172.

Arbuthnott, J. 1710. An argument for divine providence, taken from the constant regularity observed in the births of both sexes. *Philosophical Transactions of the Royal Society of London* 27: 186–190.

Ares, G., Dauber, C., Fernández, E., Giménez, A., and Varela, P. 2012. Penalty analysis based on CATA questions to identify drivers of liking and directions for product reformulation. *11th Sensometrics*. Rennes, France.

Ares, G., Deliza, R., Barreiro, C., Giménez, A., and Gámbaro, A. 2010. Comparison of two sensory profiling techniques based on consumer perception. *Food Quality and Preference* 21: 417–426.

Ares, G. and Jaeger, S. R. 2013. Check-all-that-apply questions: Influence of attribute order on sensory product characterization. *Food Quality and Preference* 28: 141–153.

Ares, G., Varela, P., Rado, G., and Giménez, A. 2011. Identifying ideal products using three different consumer profiling methodologies. Comparison with external preference mapping. *Food Quality and Preference* 22: 581–591.

Beaton, D. 2013. prettyGraphs: Publication-quality graphics. R package version 2.0.3. http://CRAN.R-project.org/package=prettyGraphs (accessed September 27, 2013).

Beaton, D., Chin Fatt, C. R., and Abdi, H. 2013. ExPosition: Exploratory analysis with the singular value decomposition. R package version 2.1.3. http://CRAN.R-project.org/package=ExPosition (accessed September 27, 2013).

Bradburn, N. M., Sudman, S., and Wansink, B. 2004. *Asking Questions: The Definitive Guide to Questionnaire Design—For Market Research, Political Polls, and Social and Health Questionnaires*, Rev. edn. San Francisco, CA: Jossey-Bass.

Bruzzone, F., Ares, G., and Giménez, A. 2012, Consumers' texture perception of milk desserts. II—Comparison with trained assessors' data. *Journal of Texture Studies* 43: 214–226.

Campo, E., Ballester, J., Langlois, J., Dacremont, C., and Valentin, D. 2010. Comparison of conventional descriptive analysis and a citation frequency-based descriptive method for odor profiling: An application to Burgundy Pinot noir wines. *Food Quality and Preference* 21: 44–55.

Campo, E., Do, B. V., Ferreira, V., and Valentin, D. 2008. Aroma properties of young Spanish monovarietal white wines: A study using sorting task, list of terms and frequency of citation. *Australian Journal of Grape and Wine Research* 14: 104–115.

Carr, B. T., Dzuroska, J., Taylor, R. O., Lanza, K., and Pansini, C. 2009. Multidimensional Alignment (MDA): A simple numerical tool for assessing the degree of association between products and attributes on perceptual maps. *Eighth Rose-Marie Pangborn Sensory Science Symposium.* Florence, Italy.

Castura, J. C. 2009. Do panelists donkey vote in sensory choose-all-that-apply questions? *Eighth Rose-Marie Pangborn Sensory Science Symposium.* Florence, Italy.

Chung, L., Chung, S.-J., Kim, J.-Y., Kim, K.-O., O'Mahony, M., Vickers, Z., Cha, S.-M., Ishii, R., Baures, K., and Kim, H.-R. 2012. Comparing the liking for Korean style salad dressings and beverages between US and Korean consumers: Effects of sensory and non-sensory factors. *Food Quality and Preference* 26: 105–118.

Clark, C. C. and Lawless, H. T. 1994. Limiting response alternatives in time-intensity scaling: An examination of the halo-dumping effect. *Chemical Senses* 19: 583–594.

Cochran, W. G. 1950. The comparison of percentages in matched samples. *Biometrika* 37: 256–266.

Cowden, J., Gould, M., and Korn, B. 2012. Choose-All-That-Apply: A valuable tool for assessing differences in Triangle testing. *Third Meeting of the Society of Sensory Professionals.* Jersey City, NJ.

Cowden, J., Moore, K., and Vanleur, K. 2009. Application of check-all-that-apply response to identify and optimize attributes important to consumer's Ideal product. *Eighth Rose-Marie Pangborn Sensory Science Symposium.* Florence, Italy.

Dooley, L., Lee, Y., and Meullenet, J.-F. 2010. The application of check-all-that-apply (CATA) consumer profiling to preference mapping of vanilla ice cream and its comparison to classical external preference mapping. *Food Quality and Preference* 21: 394–401.

Edgington, E. and Onghena, P. 2007. *Randomization Tests*, 4th edn. Boca Raton, FL: Chapman & Hall/CRC.

Escofier, B. and Pagès, J. 2008. *Analyses factorielles simples et multiples; objectifs, méthodes et interprétation*, 4th edn. Paris, France: Dunod.

Findlay, C. J., Castura, J. C., Schlich, P., and Lesschaeve, I. 2006. Use of feedback calibration to reduce the training time for wine panels. *Food Quality and Preference* 17: 266–276.

Gabriel, K. R. 1971. The biplot graphic display of matrices with application to PCA. *Biometrika* 58: 453–467.

Galesic, M., Tourangeau, R., Couper, M. P., and Conrad, F. G. 2008. Eye-tracking data: New insights on response order effects and other cognitive shortcuts in survey responding. *Public Opinion Quarterly* 72: 892–913.

Giacalone, D., Bredie, W. L. P., and Frøst, M. B. 2013. "All-In-One Test" (AI1): A rapid and easily applicable approach to consumer product testing. *Food Quality and Preference* 27: 108–119.

Greenacre, M. 2007. *Correspondence Analysis in Practice*, 2nd edn. Boca Raton, FL: Chapman & Hall/CRC.

Hellinger, E. 1909. Neue Begründung der Theorie quadratischer Formen von unendlichvielen Veränderlichen. *Journal für die reine und angewandte Mathematik* 136, 210–271.

Holm, S. 1979. A simple sequentially rejective multiple test procedure. *Scandinavian Journal of Statistics* 6: 65–70.

Hommel, G. 1988. A stagewise rejective multiple test procedure based on a modified Bonferroni test. *Biometrika* 75: 383–386.

Hough, G. and Ferraris, D. 2010. Free listing: A method to gain initial insight of a food category. *Food Quality and Preference* 21: 295–301.

Husson, F., Josse, J., Lê, S., and Mazet, J. 2013. FactoMineR: Multivariate exploratory data analysis and data mining with R. R package version 1.25. http://cran.r-project.org/package=FactoMineR (accessed September 27, 2013).

Israel, G. D. and Taylor, C. L. 1990. Can response-order bias evaluations? *Evaluation and Program Planning* 13: 365–371.

King, S. C. and Meiselman, H. L. 2010. Development of a method to measure consumer emotions associated with foods. *Food Quality and Preference* 21: 168–177.

King, S. C., Meiselman, H. L., and Carr, B. T. 2013. Measuring emotions associated with foods: Important elements of questionnaire and test design. *Food Quality and Preference* 28: 8–16.

Krosnick, J. A. 1991. Response strategies for coping with the cognitive demands of attitude measures in surveys. *Applied Cognitive Psychology* 5: 213–236.

Krosnick, J. A. 1999. Survey research. *Annual Review of Psychology* 50: 537–567.

Lawless, H. T. and Heymann, H. 2010. *Sensory Evaluation of Food: Principles and Practices*, 2nd edn. New York: Springer.

Le Fur, Y., Mercurio, V., Moio, I., Blanquet, J., and Meunier, J. M. 2003. A new approach to examine the relationships between sensory and gas chromatography-olfactometry data using generalized procrustes analysis applied to Six French Chardonnay wines. *Journal of Agricultural and Food Chemistry* 51: 443–452.

Legendre, P. and Gallagher, E. 2001. Ecologically meaningful transformations for ordination of species data. *Oecologia* 129: 271–280.

Matthews, C. O. 1927. The effect of position of printed response words upon children's answers to questions in two-response types of tests. *Journal of Educational Psychology* 18: 445–457.

McCloskey, L. P., Sylvan, M., and Arrhenius, S. P. 1996. Descriptive analysis for wine quality experts determining appellations by Chardonnay wine aroma. *Journal of Sensory Studies* 11: 49–67.

McNemar, Q. 1947. Note on the sampling error of the difference between correlated proportions or percentages. *Psychometrika* 12: 153–157.

Meilgaard, M. C., Carr, B. T., and Civille, G. V. 2006. *Sensory Evaluation Techniques*, 4th edn. Boca Raton, FL: CRC Press.

Meullenet, J.-F., Findlay, C. J., Tubbs, J. K., Laird, M., Kuttappan, V. A., Tokar, T., Over, K., and Lee, Y. S. 2009. Experimental consideration for the use of check-all-that-apply (CATA) questions to describe the sensory properties of orange juices. *Eighth Rose-Marie Pangborn Sensory Science Symposium*. Florence, Italy.

Meyners, M., Castura, J. C., and Carr, B. T. 2013. Existing and new approaches for the analysis of CATA data. *Food Quality and Preference* 30: 309–319.

Meyners, M. and Hartwig, P. 2009. Consumer associations with a toddlers' product color evaluated by a choose-all-that-apply questionnaire. *Eighth Rose-Marie Pangborn Sensory Science Symposium*. Florence, Italy.

Meyners, M. and Pineau, N. 2010. Statistical inference for temporal dominance of sensations data using randomization tests. *Food Quality and Preference* 21: 805–814.

Nenadic, O., and Greenacre, M. 2007. Correspondence analysis in R, with two- and three-dimensional graphics: The ca package. *Journal of Statistical Software* 20: 1–13.

Nielsen, J. and Pernice, K. 2009. *Eyetracking Web Usability*. Berkeley, CA: New Riders Press.

Parente, M. E., Ares, G., and Manzoni, A. V. 2010. Application of two consumer profiling techniques to cosmetic emulsions. *Journal of Sensory Studies* 25: 685–705.

Parente, M. E., Manzoni, A. V., and Ares, G. 2011. External preference mapping of commercial antiaging creams based on consumers' responses to a check-all-that-apply question. *Journal of Sensory Studies* 26: 158–166.

Payne, S. L. 1951. *The Art of Asking Questions*. Princeton, NJ: Princeton University Press.

Perrin, L., Symoneaux, R., Maître, I., Asselin, C., Jourjon, F., and Pagès, J. 2008. Comparison of three sensory methods for use with the Napping® procedure: Case of ten wines from Loire valley. *Food Quality and Preference* 19: 1–11.

Pineau, N., de Bouillé, A. G., Lepage, M., Lenfant, F., Schlich, P., Martin, N., and Rytz, A. 2012. Temporal Dominance of Sensations: What is a good attribute list? *Food Quality and Preference* 26: 159–165.

Pineau, N., Schlich, P., Cordelle, S., Mathonnière, C., Issanchou, S., Imbert, A., Rogeaux, M., Etiévant, P., and Köster, E. 2009. Temporal Dominance of Sensations: Construction of the TDS curves and comparison with time–intensity. *Food Quality and Preference* 20: 450–455.

Plaehn, D. 2012. CATA penalty/reward. *Food Quality and Preference* 24: 141–152.

Popper, R., Abdi, H., Williams, A., and Kroll, B. J. 2011. Multi-block hellinger analysis for creating perceptual maps from check-all-that-apply questions. *Ninth Rose-Marie Sensory Science Symposium*. Toronto, Ontario, Canada.

R Development Core Team. 2013. *R: A Language and Environment for Statistical Computing*. Vienna, Austria: R Foundation for Statistical Computing, http://www.R-project.org/ (accessed September 27, 2013).

Rasinski, K. A., Mingay, D., and Bradburn, N. M. 1994. Do respondents really "Mark All That Apply" on self-administered questions? *Public Opinion Quarterly* 58: 400–408.

Reinbach, H. C., Giacalone, D., Ribeiro, L. M., Bredie, W. L. P., and Frøst, M. B. 2014. Comparison of three sensory profiling methods based on consumer perception: CATA, CATA with intensity and Napping®. *Food Quality and Preference* 32: 160–166.

Smyth, J. D., Dillman, D. A., Christian, L. M., and Stern, M. J. 2004. How visual grouping influences answers to internet surveys. Technical Report #04-023. Pullman, WA. Retrieved from http://survey.sesrc.wsu.edu/dillman/papers/2004/howvisualgrouping.pdf (accessed December 16, 2013).

Smyth, J. D., Dillman, D. A., Christian, L. M., and Stern, M. J. 2006. Comparing check-all and forced-choice question formats in web surveys. *Public Opinion Quarterly* 70: 66–77.

Tate, M. W. and Brown, S. M. 1970. Note on the Cochran Q Test. *Journal of the American Statistical Association* 65: 155–160.

Valentin, D., Chollet, S., Lelièvre, M., and Abdi, H. 2012. Quick and dirty but still pretty good: A review of new descriptive methods in food science. *International Journal of Food Science & Technology* 47: 1563–1578.

Wakeling, I. N., Raats, M. M., and MacFie, H. J. H. 1992. A new significance test for consensus in generalized procrustes analysis. *Journal of Sensory Studies* 7: 91–96.

Williams, A., Carr, B. T., and Popper, R. 2011. Exploring analysis options for check-all-that-apply (CATA) questions. *Ninth Rose-Marie Sensory Science Symposium.* Toronto, Ontario, Canada.

Williams, A. A. and Langron, S. P. 1984. The use of free choice profiling for the examination of commercial ports. *Journal of the Science of Food and Agriculture* 35: 558–568.

12 Open-Ended Questions

Ronan Symoneaux and Mara V. Galmarini

CONTENTS

12.1 THEORY BEHIND THE METHOD

Open-ended questions consist of asking for an opinion or a comment about something and allowing a spontaneous answer. Participants provide a free answer, without the need of adjusting to preformed phrases or a list of words. Asking for information in this way has long been used (Woodward and Franzen 1948), and it is still used in different areas, such as medical studies (Carson et al. 2011; Hanson et al. 2011), political and social sciences (Geer 1991; Jordan-Zachery and Seltzer 2012), and marketing (Heath and Chatzidakis 2012).

In the area of sensory science, it has long been questioned if consumers should be asked something other than whether they like a product or not. Some authors have argued that consumers respond to a product as a whole and that they can find it difficult to accurately explain the basis of their choice (Elmore et al. 1999; Lawless and Heymann 1998). Others have established that asking consumers to analyze and quantify their liking in terms of a set of given attributes can strongly affect their hedonic response (Prescott et al. 2011). However, recent research has proved that consumers are able to accurately describe products (Moussaoui and Varela 2010; Varela and Ares 2012) and that their comments could provide valuable information. Moreover, enabling consumers to freely describe a product, or their likes and dislikes can provide valuable insights into consumer vocabulary and also give them a chance to focus on the attributes that are important to them, without forcing their attention to attributes previously selected by the researchers.

It should be noted that the input that consumers' comments can have in sensory research will be highly dependent on the type of statistical analysis the data are subjected to. Free comments have often been considered supplementary information to other sensory methodologies (e.g., in sorting task or napping, Chollet et al. 2011; Faye et al. 2006; Nestrud and Lawless 2010). However, when treated with appropriate statistical tools, consumers' comments can be used on their own (Bécue-Bertaut et al. 2008; Perrin et al. 2008; Symoneaux et al. 2012).

Open-ended questions with subsequent comment analysis have proved to be a good methodology to obtain products' description in consumer vocabulary (Ares et al. 2010; Symoneaux et al. 2012; ten Kleij and Musters 2003). Moreover, the recent addition of the use of chi-square per cell has allowed a deeper and more statistically reliable analysis of the contingency table, providing a more accurate data interpretation that complements the representation of comments obtained with correspondence analysis (CA; Symoneaux et al. 2012).

It is the core, and the aim, of this chapter to help the reader get the most out of consumers' comments in terms of their statistical analysis. The focus will not be on the fields of linguistic, semantics, and syntax; only some practical recommendations will be given in these areas.

12.2   IMPLEMENTATION AND DATA COLLECTION

Any question that can be answered freely on a given space, without the need of adjusting to previously given words, options, or pre-elaborated sentences, becomes an open-ended question. Even if data acquisition can be considered quite simple, some considerations should be taken into account.

As stated, the only thing consumers need when answering open-ended questions is sufficient space in an answer box (Figure 12.1). In this matter, several studies performed on paper surveys (Christian and Dillman 2004;

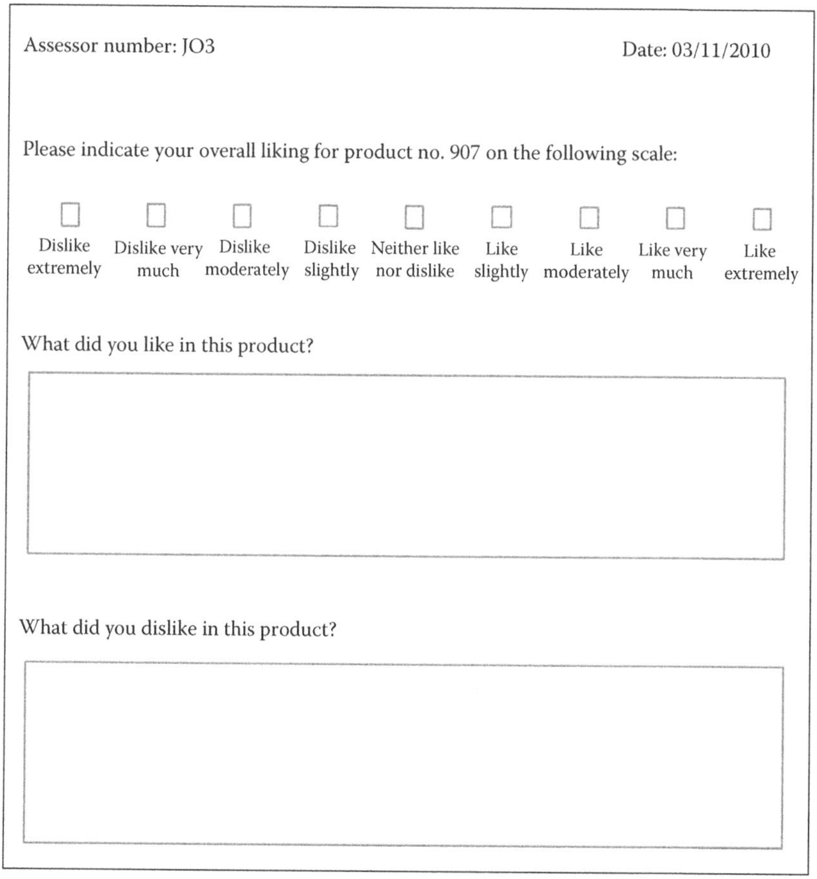

Assessor number: JO3   Date: 03/11/2010

Please indicate your overall liking for product no. 907 on the following scale:

☐   ☐   ☐   ☐   ☐   ☐   ☐   ☐   ☐
Dislike   Dislike very Dislike   Dislike Neither like   Like   Like   Like very   Like
extremely   much   moderately slightly nor dislike slightly moderately   much   extremely

What did you like in this product?

What did you dislike in this product?

FIGURE 12.1   Example of an open-ended questionnaire. In this particular case, likes and dislikes were asked separately as an extension of a hedonic test.

Israel 2006) showed that larger spaces encouraged longer answers. Given the increase of online surveys and digitalized methods for sensory data acquisition, Smyth et al. (2009) studied if this behavior was also observed when dealing with a digital support. According to these authors, although a piece of paper has real boundaries, computer-mediated space is seemingly limitless, and they found that increasing the size of the answer box had little effect on the quality of the answers obtained.

Consumers are always encouraged to answer in their own personal way, and only basic explanations should be given in order to avoid influencing their responses. In this area, Smyth et al. (2009) studied if adding a small introduction to the question could encourage higher involvement and increase the quality of the responses. These authors studied the effect of adding the statement "This question is very important to understand [the subject of study in the paper]. Please take your time for answering it" and a shortened version of each (e.g., just stating "Please take your time for answering"). Their findings showed that motivating instructions were effective for improving the quality of the answers. Using a small introduction can be helpful to emphasize the importance of responses to the research question and therefore increase the number of words and the elaboration of the answer. However, care must be taken when considering this approach since the introduction may also lead consumers to try to please the researchers, giving as a result forced answers that do not reflect their real perception.

When used for product descriptions in sensory analysis, three main approaches for data collection can be found in the literature (Varela and Ares 2012). ten Kleij and Musters (2003) allowed consumers to voluntarily write down remarks (full sentences) after their overall liking evaluations. Alternatively, Ares et al. (2010) forced consumers to provide a fixed number of words (in this particular case, up to four) to describe the samples after their overall liking evaluation. More recently, Symoneaux et al. (2012) asked consumers to state separately what they liked and what they disliked about the evaluated products. Finally, when working with consumers' expectations (instead of product descriptions), Galmarini et al. (2013) also asked consumers to state separately what they would like to find and what they would not like to find in a certain product category, being responses mandatory but without a limited amount of words. This shows that different alternatives can be used depending on the specific aim of the experiment, namely, asking for a minimum/maximum amount of words, asking separately or not for likes and dislikes, and making answering mandatory or not, in addition to the general rules for questionnaire design (Lawless and Heymann 1998).

12.3 DATA ANALYSIS

Data analysis of open-ended questions can be divided into three stages:

1. Preprocessing of the data
2. Construction of the contingency tables (*cross tabulation*)
3. Statistical analysis

12.3.1 PREPROCESSING OF THE DATA

Consumers write their comments without any guidance. Thus, they use their own writing style and sometimes make orthographic and grammatical mistakes. For these reasons, comments need to be revised and processed from this initial verbalization to get the final descriptions used in data analysis. This step is crucial to retrieve faithful results without losing important information.

Rostaing et al. (1998) mentioned the following necessary steps for text treatment: removing mistakes, elimination of connectors and auxiliary terms, location of phrases and terms that make them up, lemmatization, regrouping synonyms, and managing ambiguous words (polysemy and homographs). In the literature, many authors have used verbalization to describe products, but the way in which text was treated was rarely precise, and only word counting was mentioned without further explanations (Blancher et al. 2007; Chollet et al. 2011; Faye et al. 2004, 2006; Sinesio et al. 2010). Faye et al. (2013) stated that terms that were deemed to be synonymous or inflections of the same word should be grouped, while Perrin et al. (2008) and Perrin and Pagès (2009) explored whether or not to integrate quantifying words.

In most cases, text treatment is done manually, and at some point, arbitrary decisions are taken concerning synonymy, ambiguous terms, and/or intensities. These decisions are not always easy to make since they may lead to loss of information, but unfortunately, they are unavoidable. In Ares et al. (2010), word classification and word synonymy were done manually and individually by three different researchers. Each researcher evaluated the data individually, and they met afterward in order to check their agreement on the task. In the work done by Galmarini et al. (2013), comment analysis was done in two different countries, and data were compared. Four native Spanish speakers worked together on the Argentinean data set, and two French native speakers on the French one, and a bilingual (French–Spanish) speaker harmonized the criteria between both countries. To study variability due to coders, Niedomysl and Malmberg (2009) used five people to independently treat 500 open-ended responses from a large survey on migration motives and found that there was no variability due to coders. Therefore, it would be correct to assume that the transcribing process can be done by a unique transcoder expert in sensory sciences (Symoneaux et al. 2012).

Nonetheless, in all cases, a protocol is needed to reduce bias and error. For this purpose, some practical considerations on text processing from consumers' comments to simple terms are presented in Section 12.3.1.1. The use of certain functions from *Excel* and some *Visual Basic macros* is also described.

12.3.1.1 Practical Considerations on Text Processing

To start with the text processing, all obtained answers should be entered in a table under the following categories: (1) identification of the consumer, (2) product/question, and (3) complete comment (Table 12.1a). It should be pointed out that this information should be kept throughout the data analysis process, even if the data were simplified afterward.

Once all the answers are displayed on a table, the comments' reduction is done turning the sentences and phrases (Table 12.1a) into words and descriptors (Table 12.1b). For this purpose, the use of a particular tool in

TABLE 12.1

From Consumers' Comments to Simple Terms

Consumer	Product	Initial Comment
(a) Original data		
C1	P1	Tasty, crisp, and quite sweet
C1	P2	Distinct apple odor, somewhat sweet
C2	P1	I found this apple juicy and sugary, slightly acidic. It's crispy with a cellar odor.
C2	P2	The color yellow-green
C3	P1	Not enough taste
C3	P2	None
(b) Data split into separate words and phrases		
C1	P1	Crisp
C1	P1	Quite sweet
C1	P1	Tasty
C1	P2	Somewhat sweet
C1	P2	Distinct apple odor
C2	P1	It's crispy
C2	P1	Sugary
C2	P1	A cellar odor
C2	P1	I found this apple juicy
C2	P1	Slightly acidic
C2	P2	The color yellow-green
C3	P1	Not enough taste
C3	P2	None
(c) Final categories		
C1	P1	Crispy
C1	P1	Sweet
C1	P1	Tasty
C1	P2	Odor_Apple
C1	P2	Sweet
C2	P2	Color_Yellow_Green
C2	P1	Odor_Cellar
C2	P1	Juicy
C2	P1	Crispy
C2	P1	Sour
C2	P1	Sweet
C3	P1	No_Taste
C3	P2	None

the Microsoft Excel software called "*text to column*" (function found in *data, text to column*, Microsoft Excel v.2010), and some particular macros (developed by the authors and available under request) can be very useful.

The "*text to column*" function allows separating the content of one Excel cell into separate columns. Depending on the way the data are arranged, the cell content could be split based on a delimiter, such as a space or a character (comma, a period, or a semicolon). For example, the use of this function will move each term, separated by a comma (that would be the case of an enumeration of attributes), into different cells on the same line.

After this step, the user will have a table with one row per judge per product with all the different attributes in subsequent columns. But for the final analysis, a row per product per judge per attribute is needed, presenting each different attribute in different rows. To facilitate this process, the authors have developed a *Visual Basic macro* (available free under request) that moves the information allowing to obtain the final table. Even if more tedious, this can also be done manually.

At the end of this step, the table obtained will present consumers, products (or question in the case of a survey), and the individual terms (Table 12.1b). Other information such as preference clusters, gender, age, consuming habits, or if the answer represents a like or a dislike could be added for further analyses.

Once the data table (Table 12.1b) is obtained, the real work on the text begins. It is now that transcoders will verify typing and/or spelling mistakes. Even if this could be done before, mistakes will appear more evident now that terms are in different rows and could be ranked alphabetically. Then it would be possible to regroup terms (lemmatization) like nouns and adjectives that refer to the same characteristic (e.g., acidity, acid, and acidic could be grouped and renamed as acid) (Table 12.1c). Some arbitrary assumptions based on sensory expertise will allow grouping synonyms when they are quite evident. In case of ambiguous comments, authors recommend to take no action and keep the original term without any regrouping.

Last, but not least, it is important to take into account the way consumers describe the intensity of the different mentioned attributes, descriptors, or ideas (word combination). It is not the same to state "*I like this apple because it is acid* or *I like this apple because it is quite acid*" or "*I like this apple because it is not acid.*" Unfortunately, keeping all the different words used by consumers to describe subtleties would dilute the obtained information by decreasing the frequency of citation of words. For this reason, the words used to describe intensities can also be grouped. Table 12.2 (adapted from Galmarini et al. 2013) shows a way in which this grouping can be done, taking into account if the intensity refers to a positive or a negative answer. Even if the presented choice could be further discussed, it shows the need for standardization in the procedure.

TABLE 12.2

Example of the Transformation of Subtleties Using the Term "Sour"

	Answer Given as a Positive Characteristic[a]		Answer Given as a Negative Characteristic[b]
Original Term	**After Simplification**	**Original Term**	**After Simplification**
Slightly sour	*A little sour*	Too sour	*Too sour*
Not too sour	*Sour*	Missing a little sourness	*Not sour*
Little sour	*A little sour*	Not sour enough	*Not sour*
No sour	*Not sour*	Absence of sourness	*Not sour*
Enough sour	*Sour*	Without sourness	*Not sour*
Not sour	*Not sour*	Strong sourness	*Too sour*
Relatively sour	*Sour*	Too strong sourness	*Too sour*
Good sourness	*Sour*	Extreme sourness	*Too sour*
Rather sour	*Sour*		
Without sourness	*Not sour*		
A lot of sourness	*Too sour*		
Light sourness	*A little sour*		
Not too sour	*A little sour*		
Too sour	*Too sour*		

Source: Adapted from Galmarini, M.V. et al., *Appetite*, 63, 27, 2013.

[a] Please list all positive flavor characteristics you like finding in an apple.

[b] Please list all negative flavor characteristics you dislike finding in an apple.

12.3.2 CONSTRUCTION OF THE CONTINGENCY TABLES (CROSS TABULATION)

A contingency table is essentially a display format showing the distribution of two (or more) variables in rows and other(s) in columns, used to study the association between these variables. In the case of comment analysis, different contingency tables can be obtained from one database (Table 12.1c; Section 12.3.1), and they can be created using the function "*Pivot Table*" in Microsoft Excel v.2010 (detailed information on how to use this simple tool can be found in the help section of the mentioned software).

The amount of possible different contingency tables will obviously depend on the type of data set obtained and of the questionnaire used. In any case, the initial step would be the analysis of the contingency table

TABLE 12.3

Example of a Contingency Table Showing the Words Used to Describe the Different Products

Terms	Product 1	Product 2	Product 3	Total
Acid	56	48	64	168
Sweet	21	40	64	125
Color	73	22	35	130
Crunchy	12	90	96	198
Taste	30	34	38	102
Total	192	234	297	723

obtained using terms in rows and the products (or questions in case of a survey) in columns. In this way, the amount of times each item was used for each product (or question) will be presented (Table 12.3). It is also possible to use additional information, such as whether the comment was answered as a liked or disliked category (in the case likes and dislikes were asked separately) or preference clusters obtained after segmentation of consumers on their liking scores. In this last case, a contingency table with terms (rows) and the different groups (in columns) would allow the identification of words most often used by each cluster, or even the words used by each group per product could be identified.

12.3.3 STATISTICAL ANALYSIS

12.3.3.1 Chi-Square and Chi-Square per Cell

Contingency tables allow looking at the relationship between variables, but they are not statistically exploitable on their own since it is difficult to see what is significant within the table. For example, was one word used significantly more than another for a particular product? A simple look at the contingency table would not be enough to answer this or many other questions.

The global chi-square test is used for testing the independence between rows and columns in a contingency table. In the present case, it indicates if some words (rows) are cited significantly more (or less) often than expected for the different products (columns). When the initial chi-square is significant, it would be possible to analyze within each cell identifying the source of variation of the global chi-square (Snedecor and Cochran 1957).

In order to analyze this source of variation and to test its significance, Symoneaux et al. (2012) proposed the use of the chi-square per cell test. This statistical test allows showing for each cell of the contingency table if the observed values are significantly higher, lower, or equal to the theoretical values. This test is performed on a 2×2 table where one cell is a [i, j] cell of the original contingency table and the others contain values for the row i minus [i, j], for the column j minus [i, j], and for the rest of the table. A chi-square test is then performed for each 2×2 table. To perform this test, the authors of this chapter have developed a *Visual Basic macro* that runs in Microsoft Excel (available free under request). Thanks to this tool, the contingency table will appear presenting the result of the global chi-square, and also the contribution of each cell to the global chi-square will be indicated by means of a (+) or (−) showing if the frequency observed is higher or lower than expected. Table 12.4 shows the information presented in Table 12.3 (Section 12.3.2) after performing the chi-square per cell analysis.

Looking at the contingency table (Table 12.3), it is not possible to know if the observed number of citations is significant when compared among products and/or items. As an example, we can observe that there are more citations for "taste" for product 3, but is "taste" really more important for product 3 than for product 1? The chi-square per cell (Table 12.4) allows answering this question. It can be seen that "taste" is not significant, meaning that "taste" is not more important in the description of one product than in the other. It is more mentioned in product 3 than in the other two, but since this product also received a larger total amount of words, its contribution is not more significant than in the other products. Moreover, within the product, taste and color were mentioned a similar number of times (38 and 35, respectively). However, given that color was more frequently mentioned in product 1, this term becomes significantly important in comparison to the other products, being its mention significantly lower

TABLE 12.4

Example of a Contingency Table (Data Presented in Table 12.3) after a Chi-Square per Cell Analysis

Terms	Product 1	Product 2	Product 3	Total
Acid	56 (+)*	48	64	168
Sweet	21 (−)**	40	64 (+)*	125
Color	73 (+)***	22 (−)***	35 (−)***	130
Crunchy	12 (−)***	90 (+)***	96 (+)*	198
Taste	30	34	38	102
Total	192	234	297	723

Note: The probability (p) value of the global chi-square test was <0.00001.

***$p \leq 0.001$, **$p \leq 0.01$, and *$p \leq 0.05$; effect of the chi-square per cell.

for product 3. In this way, chi-square per cell test allows underlining the main attributes in each product when compared to the other ones.

12.3.3.2 Correspondence Analysis

In cases where the contingency table has many words and/or many products, the use of chi-square test is interesting from a statistical point of view, but it does not allow a clear visualization of all the obtained information. This is when a multidimensional approach becomes more appealing, enabling a clear visualization of the relationship between products (or questions/concepts) and the words used by consumers to answer the open-ended question.

CA creates a "map" of the data generated from a contingency table, showing rows and columns in the same geometric space. The obtained representation will be similar to those traditionally used in sensory science (e.g., PCA) depicting comments and products on a same graph: the closer a word and a product are, the more frequently consumers used this word to describe it.

In R software, this CA is easy to implement using the function CA (f.1) presented in FactoMineR package.

```
library(FactoMineR)

CA(X, ncp = 2, row.sup = NULL, col.sup = NULL,
  graph = TRUE, axes = c(1,2), row.w = NULL)
```

$$(f.1)$$

To use (f.1), the contingency table to be analyzed should be copied in R, and using the function CA, the analysis and graphical representation will be carried out. This function will create an object with all the information needed for the analysis such as eigenvalues and matrixes with results from the variables in columns and rows (e.g., coordinates, square cosine, and contribution) for the number of dimensions demanded a priori in the function (argument used: ncp, number of dimensions kept in the results).

For an even better representation of the data, R allows the representation of results taking into account the contribution of each word to the total description of the product. Bigger symbols would represent a higher importance of that particular word in the product description and vice versa. An example of this with its corresponding R script is presented in Section 12.5.

12.3.3.3 Multiple Factor Analysis for Contingency Tables

Quantitative data such as liking scores and/or sensory profiles can be crossed with qualitative data such as frequency of citation of comments (or several contingency tables). This comparison of data sets can be carried out by a multiple factor analysis for contingency tables (MFACT) (Bécue-Bertaut and Pagès 2008). This technique is based on the principle of multiple factor analysis (MFA) (Escofier and Pagès 2008), which allows to balance the influence of different sets of variables and to integrate the specificity of each type of data set. The implementation of MFA is programmed in R (f.2) in the `FactoMineR` packages, and it can be adapted to carry out an MFACT.

```
library(FactoMineR)
```

```
MFA (base, group, type)
```                                            (f.2)

The needed function is called MFA (`FactoMineR`), and it includes the following main arguments (f.2):

- Base: Data frame with n rows (products) and p columns (variables) (e.g., of different combinable variables: average intensity of sensory attributes, average hedonic scores, chemical measurements, and frequency of mention of the terms per product)
- Group: The number of variables per data set (e.g., number of sensory attributes, number of consumers, and number of terms)
- Type: The type of variable in each group (quantitative variables, quantitative variables scaled to unit variance, categorical variables, frequencies, and data from contingency tables)

In the examples section of this chapter (Section 12.5), the script used for the comparison of two data sets representing comments from two different countries on the same topic is presented.

12.4   ADVANTAGES AND DISADVANTAGES

The presented methodology has several advantages that make its use tempting as a complement of other sensory methodologies or as an independent tool (see Section 12.5). The initial writing and implementation is quite simple. It allows obtaining the opinion directly from consumers in their own words and highlighting what is really important to them, either in terms of expectations or in product description.

When implemented together with a hedonic test, liking scores and the reasons for this liking will be given by the same people and in "consumer vocabulary." This is highly relevant since, sometimes, when describing a product with a trained sensory panel, some characteristics important for consumers can be ignored. In addition, when developing communication strategies, it is necessary to describe a product with terms that are meaningful and familiar for consumers, which can be easily obtained by this method. Also, there is no need to find a functional relation between liking scores and product attributes since the relation is articulated by the consumers themselves (ten Kleij and Musters 2003). Finally, if questions are formulated to ask separately for likes and dislikes, precise descriptive information can be obtained about drivers of liking and disliking.

Moreover, the statistical methods recently applied to the analysis of open-ended questions allow finding significant differences among samples for each of the words used by consumers. This means that results are not only based on frequency of mention and tendencies, but that it can also be stated if terms are mentioned in a significantly different way when describing products in a data set.

However, even if the use of this methodology in sensory science has been improved in the late years, it still has some drawbacks. Techniques in the area of numerical data analysis have become more advanced; notwithstanding in the analysis of textual data, there is no software that can be used on its own, and the task of text processing remains time-consuming and labor-intensive. This is a disadvantage common to many qualitative research methods. Since the human brain is the only "tool" that can understand a full text together with the nuances of co-occurrent words, it is difficult to think that this stage will change radically in the next years, even with the advances of technology.

Particularly for open-ended questions, text processing is a crucial stage, and it is here that most of the subjective decisions are taken leading to loss of information and inclusion of bias. In this way, the transcoder's mediation in the deconstruction of the form leads to impoverishment of the meaning. Indeed, despite rules and a high attention during the transcription process, this one step stays arbitrary.

Moreover, asking consumers to describe a product can be considered controversial by some sensory scientists. Nonetheless, the approach of open-ended questions presents the advantage of not focusing on a particular attribute imposed in the test, as would be the case when using just-about-right scales. Even when asked to describe separately why they liked or disliked a product, consumers are not asked to precise or to quantify a particular attribute; instead, they are simply asked to give their opinion. Maybe, in this way, their liking scores would not be influenced by these supplementary questions. However, further research is needed to confirm this hypothesis.

12.5 EXAMPLES OF APPLICATIONS

12.5.1 Complementary Tool to Preference Mapping

Several authors have proved the relevance of open-ended questions as a complementary tool of traditional trained panel description (e.g., QDA). ten Kleij and Musters (2003) analyzed the comments from 165 consumers in a hedonic test on mayonnaise and compared the results to those obtained in a QDA by nine trained assessors finding that product description was similar in both cases. In this case, commenting was not mandatory, and only 72% of the consumers gave their opinion. However, text analysis of the open-ended questions enabled the authors to corroborate the quantitative results from the preference map and to obtain a description of mayonnaises in terms of consumer language. The data obtained were visualized by CA plotting word usage across products.

Another case of the use of open-ended questions to identify drivers of liking in comparison to preference mapping technique was done by Ares et al. (2010) on milk desserts. These authors tried to improve this application by (a) including a mandatory open-ended question after rating products and (b) setting a maximum number of words that consumers could provide to describe the samples. In this case, authors also found that the technique enabled the identification of liked and disliked samples, as well as the attributes that drove their preferences. Moreover, CA of consumers' descriptions provided similar information than internal and external preference mapping. In this case, MFA on consumers' descriptions, sensory data, and overall liking scores allowed the simultaneous evaluation of consumers and trained assessors' description on the sensory characteristics of a food product, as well as consumer overall liking scores.

More recently, Symoneaux et al. (2011) used this method on apple consumers with the same purpose, but asking separately to describe the characteristics that they liked and disliked. During a consumer test, 87 participants evaluated different apple batches on a hedonic scale and then answered to the nonmandatory open-ended questions, which require them to state separately what they liked and disliked from each batch. In parallel, an expert panel described the sensory profiles of the studied products. Partial results of this research are presented in the following pages (Table 12.5 and Figure 12.2) in order to provide information on how data should be treated in these cases. These data sets are available for download from the CRC Web site: http://www.crcpress.com/product/isbn/9781466566293.

Table 12.5 is an extract of the contingency table obtained when analyzing comments from these 87 consumers. The CA map obtained from this contingency table is presented in Figure 12.2 of this data set. Data were analyzed as detailed in Sections 12.3.1 through 12.3.3.2 of this chapter.

The script used for this CA of this data set is presented as follows (f.3):

```
res<-CA(Dataset[,c(1:6)],ncp=5, graph = FALSE)

# For the creation of the CA graph

rx=range(res$row$coord[,1],res$col$coord[,1])
ry =range(res$row$coord[,2],res$col$coord[,2])

plot (rx,ry,col = NA,xlab=paste("Dim 1
  (", signif(res$eig[1,2],4), "%)"), ylab=paste
  ("Dim 2 (", signif(res$eig[2,2],4), "%)"))
```

```
points (res$row$coord[,1],res$row$coord[,2],
  col = "red" ,cex = (res$row$contrib[,1]
  +res$row$contrib[,2])*0.3 )

text (res$row$coord[,1],res$row$coord[,2],
  row.names (res$row$coord), pos = 4, col = "red",
  cex = .7)

points (res$col$coord[,1],res$col$coord[,2],
  col = "blue", pch =2,cex = (res$col$contrib
  [,1]+res$col$contrib[,2])*0.05)

text (res$col$coord[,1],res$col$coord[,2], row.
  names (res$col$coord), pos = 1, col = "blue",
  cex = .7)
```

(f.3)

After the creation of an object called "*res*" with the results from the CA function of FactoMineR, the plot, text, and points are used to create a graph using the coordinates of the products (rescolcoord) and terms (resrowcoord) from *res*. The contribution results (resrowcontrib or rescolcontrib) with the argument *cex* are used to modify the size of points according to the contribution. Represented components can be changed, modifying the numbers between square brackets (resrowcoord[,1]).

The data presented here allow the identification of liked and disliked attributes of each batch and reveal the importance of the chi-square per cell analysis. For example, when looking at the CA, it appears that batches D and A were disliked because of their taste (D_Taste), tasteless (D_Tasteless), mealiness (D_Mealy), and for not being crunchy (D_Not_Crunchy). However, when looking at the chi-square per cell analysis (Table 12.5), it can be observed that the amount of times that D_Mealy and D_Taste were mentioned was significantly higher only for sample D and that D_Tasteless together with D_Not_Crunchy were significantly high for both.

Results obtained from consumers and from trained panel were compared by means of the regression vector (RV) coefficient (Perrin and Pagès 2009), calculated using the function CoeffRV from FactoMineR. In this case, the RV coefficient was found to be 0.8656 ($p=0.011$), showing that the information obtained by the comment analysis of likes and dislikes was similar to that resulting from sensory characterization performed by the trained panel. With both methods, crunchiness and sweetness appeared as main sensory key drivers of liking, while mealiness was considered as a defect. At the same time, some sensory characteristics, such as juiciness, were important for consumers but did significantly discriminated among samples for the trained panel. This result may show that the perception of juiciness was different between trained panelists and consumers,

TABLE 12.5

Contingency Table for the Main Terms Cited by Consumers

Terms	A	B	C	D	E	F
			Apple Batches			
Liking comments						
Crunchy	7 (−)***	32 (+)***	18	5 (−)***	28 (+)**	20
Sweet	17	14	14	14	26 (+)**	7 (−)**
Juicy	17	13	25 (+)*	13	15	23
Taste	4 (−)*	18 (+)***	10	4 (−)*	12	10
Firm	6	13	7	2 (−)**	16 (+)**	8
Sour	0 (−)**	9	8	3	9	7
Texture	3	3	9 (+)***	3	3	1
Color	4	2	4	7	8	3
Aspect	1	1	5	7	2	6
Soft	3	2	2	7 (+)*	3	3
Sweet/Sour	2	5	2	0	4	2
Soft_Flesh	3	1	1	5 (+)**	0	1
Disliking comments						
Mealy	10	1 (−)***	11	19 (+)***	4 (−)*	8
Tasteless	27 (+)**	9 (−)**	16	26 (+)**	14	16
Not_Crunchy	13 (+)***	1 (−)**	5	11 (+)*	2 (−)*	6
Not_Sweet	11	3 (−)**	13	10	5	13
Taste	4	2	3	12 (+)***	4	5
Sour	4	12 (+)***	7	2	2	5
Soft_Flesh	9 (+)**	1	2	8 (+)*	1	5
Thick_Skin	2	3	1	4	7	5
Hard_Skin	0 (−)*	5	6	3	6	1
Hard	0	6 (+)**	1	2	5	0
No_Odor	4	2	6	3	6	6
Firm	0	5 (+)**	1	1	5 (+)*	0
Aspect	0	1	1	0	5 (+)**	4

Notes: Number of citations per batch and results of the chi-square per cell are presented.

(+) or (−) indicate that the observed value is higher or lower than the expected theoretical value.

***p ≤ 0.001, **p ≤ 0.01, and *p ≤ 0.05; effect of the chi-square per cell.

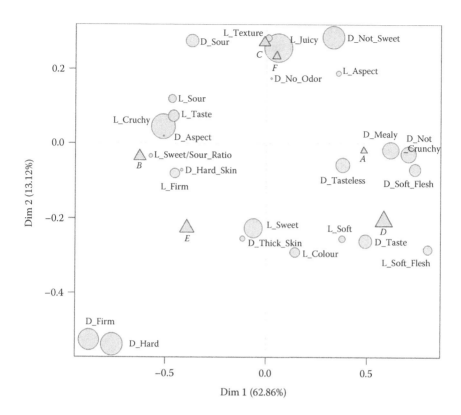

FIGURE 12.2 CA on contingency table for apple batches A–F and the terms used by consumers (D_ indicates disliking comments; L_ indicates liking comments).

emphasizing the importance of working with consumer vocabulary and getting consumer description of products.

In summary, several authors have proved that this methodology provides results comparable to those obtained by a trained panel with the advantage, as previously mentioned, of identifying the words used by consumers. This means that, in cases when a sensory profile with a trained panel cannot be done, the use of comment analysis of consumers' description can be a good alternative. However, it should be considered that the amount of consumers needed to perform a sensory characterization using open-ended questions can range from 50 to 100 (Varela and Ares 2012).

12.5.2 CONSUMERS' EXPECTATIONS AND DESCRIPTIVE VOCABULARY

Comment analysis can also be used to study consumers' expectations of a product, without the need of tasting. An example of this can be found in Galmarini et al. (2013), where comment analysis was used in an online survey to understand apple consumers' expectations in terms of likes and dislikes. In this work, comment analysis was successfully used to find out what consumers defined as quality in an apple in two different countries (France and Argentina). Moreover, authors also proved the applicability of this technique to study cultural differences regarding what consumers in different countries would and would not like to find in an apple in terms of visual characteristics, flavor, and texture, obtaining as a result a descriptive vocabulary for liked/disliked apple sensory attributes in both countries. The data set is available for download from the CRC Web site: http://www. crcpress.com/product/isbn/9781466566293.

For this purpose, the data obtained in the two countries needed to be compared. The script presented in (f.4) was used for the comparison of Argentinean and French comments about likes and dislikes in apples in an MFACT of the two contingency tables.

```
res<-MFA(Dataset[,c(1:12)], group=c(6, 6),
  type=c("f", "f"), ncp=5,

name.group=c("ARG", "FRA"), graph=FALSE)

plot.MFA(res, axes=c(1, 2), choix="freq",
  new.plot=TRUE, lab.var=FALSE,
habillage="group", lim.cos2.var=0,
  title="",invisible = "ind" )

plot.MFA(res, axes=c(1, 2), choix="group",
  new.plot=TRUE, lab.grpe=TRUE,
  title="")

plot.MFA(res, axes=c(1, 2), choix="ind",
  new.plot=TRUE, lab.ind.moy=TRUE,

lab.par=TRUE, lab.var=TRUE, habillage="group",
  partial=c("juicy", "sweet","damages", "color",
  "crunchy", "firm", "sandy", "sour", "bright",
  "mealiness", "soft", "col_red", "hard",
  "tasteless", "taste", "texture",

  "flavor", "fresh", "aroma", "smooth", "acidulé",
  "size", "dry", "shape" ,"wrinkled" ,"paposa"),
  title="")
```

$$(f.4)$$

Now a step-by-step description of each argument will be presented:

- In this case, the contingency tables were formed by questions in columns (instead of products as in other cases) and the used terms in rows. Therefore, the data set consisted of two contingency tables concatenated forming a 12-column table with the 6 questions in Argentina and the 6 questions in France (referred to as Q05_1, Q05_2, Q06_1, Q06_2, Q07_1, and Q07_2 in the example). That is to say, the database was composed of 12 columns divided into two groups of 6.

- All the data presented consisted of frequencies of mention, so the type of variable used is "f".
- The object *res* in the script (f.4) contains all the information needed to interpret MFA results.
- To make the different MFA graphs according to the stated parameters, the function *plot.MFA* (f.3) is used. By changing the *choix* argument, different representations will be obtained:
  - When selecting "*freq*" for the argument "*choix*," the factor map for the contingency tables is drawn (Figure 12.3).
  - When "*group*" is mentioned as "*choix*," a map is drawn with the representation of the different groups.
  - When "*ind*" is used for "*choix*," rows are represented on a graph. If the argument *partial* contains a list precising some or all rows, the partial points of these individuals are drawn (Figure 12.4).

Figure 12.3 shows the questions' (columns) results. Dimension 1 of the MFACT explained 24.2% of the inertia, and dimension 2 explained 23.1%. The first dimension opposed the liking terms to the disliking ones,

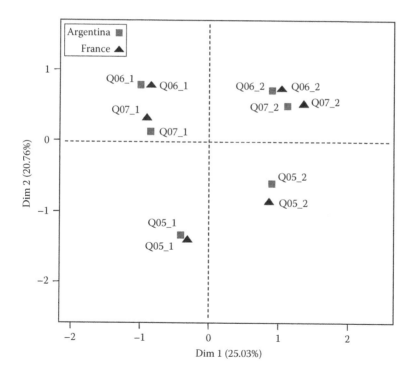

FIGURE 12.3 MFA of the contingency table of the 42 most used words to answer to 6 different questions in both countries. Representation of questions (columns).

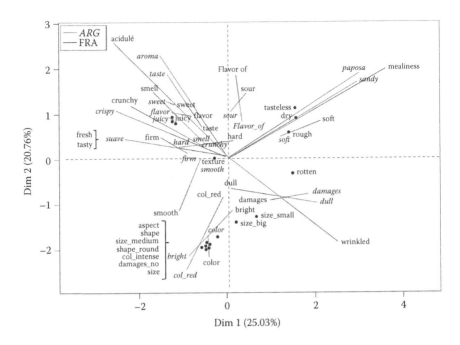

FIGURE 12.4 MFA of the contingency table of the 40 most used words to answer 6 questions in both countries. Representation of words (rows). Words not (or barely) used by one of the countries would be in the center of coordinates; they are not presented for a clearer presentation. Words used in the same amount by both countries are represented by a dot; when many words were grouped together, a brace was used.

so generally the words used for likes and dislikes were not the same. On the other hand, the words used to characterize flavor and texture (coded as Q06_1 and Q07_1 for flavor and texture likes, respectively, and as Q06_2 and Q07_2 in terms of dislikes) were the same. When this type of information is obtained, a closer look at the contingency table is recommended. Here, it was found that flavor descriptors (e.g., sweet and sour) were used only to describe flavor, while some texture attributes (namely, juicy, crunchy, and firm) were used in both categories: flavor and texture. Visual characteristics (coded as Q05_1 and Q05_2 for likes and dislikes, respectively) were clearly differentiated from the rest.

Figure 12.4 presents results from MFACT with the words used in both countries. For the purpose of clarity, all mentioned words were not included. In this case, only the 40 most cited words were selected. This is an arbitrary decision based on the data set. This representation allowed visualizing and comparing the relationship between the 40 most used terms in Argentina and in France (in addition to the chi-square per cell analysis of each question by country). In this representation, the longer the line, the larger the difference in the frequency of mention between the two countries for the given term; also the location of the word on the graph relates it to the different questions. For this, it is necessary to look at the representation shown in Figure 12.3.

It is in this way that we will be able to see that, for example, in appearance (Figure 12.4, quadrants III and IV) there were differences between the two countries. Argentineans did not like dull apples, and they did like bright red apples. On the contrary, for French consumers, red color, bright, and dull were close together in-between likes and dislikes, suggesting that there was not such a clear pattern for their preferences, while the term bright was highly mentioned in the disliked category.

In this case, the use of comment analysis allowed the identification of the terms that consumers use to describe an apple, revealing that, in general, the terms used to describe liked and disliked characteristics were not the same. Also, the influence of culture was shown by the different words used in the different countries (translation a side) and by the fact that in each country, consumers prioritized different aspects and characteristics.

12.5.3 Describing Likes and Dislikes within Consumers' Clusters

According to their preference ratings on a certain product, consumers can be grouped into clusters. Usually, when products have been described by a sensory panel, the preference pattern of each cluster can be characterized. Unfortunately, in some cases, this information is not available since product profiling can be expensive for certain companies. If open-ended questions were added—as a complement of the hedonic test—asking why consumers liked or disliked the product together with the pertinent data analysis can provide an approximate description of the clusters' preference pattern. This is briefly illustrated in the following, in an example with French cider consumers.

First, the number of cluster the consumer belongs to should be identified in the data set of the consumers' comments. In this way, contingency tables can be analyzed for each group (by chi-square and chi-square per cell) to identify the specific terms used by consumers in every cluster saying why they liked or disliked a particular product.

In our example, a preference test on 5 sensory different ciders (B1–B5) was carried out using 254 consumers. The task consisted of rating the tasted ciders and describing, separately, what they liked and what they disliked in each one. Based on the hedonic ratings, two different clusters were identified (Table 12.6).

TABLE 12.6
Mean Preference Ratings
Given by Each Cluster
(C1 and C2) for the Five
Ciders (B1–B5)

Cider	C1	C2
B1	5.0	2.5
B2	3.7	6.0
B3	1.6	3.5
B4	5.9	3.0
B5	7.6	4.9

Final comments from consumers were analyzed obtaining information on what they liked and disliked in each cluster. The 13 most often cited words to describe the group of products are presented in Table 12.7.

It is to be noted that the number of consumers in each cluster was 169 and 87, respectively, which explains the huge difference in the number of terms per product in both groups. However, thanks to the chi-square per cell, data obtained from the two clusters can be compared. It can be observed that ciders with the highest liking scores were mostly described in terms of likes, while those with the lowest scores were described in terms of dislikes. In this way, it can be observed that C1 dislikes were highly driven by bitterness, while C2 dislikes were mainly driven by light color. As for likes, they were particularly driven by sweet, taste, odor, apple aroma, and color for C1, while for C2, the most important aspects were taste, apple aroma, and sparkling. In addition, a characteristic sparkling term characterized as liking would be revealing the possibility of a subdivision within C2.

12.5.4 OTHER POTENTIAL USES

Chi-square per cell analysis can be also used with a trained panel when generating descriptive vocabulary. Also when working with a trained panel, Galmarini et al. (2011) used this technique to analyze descriptive data using a citation frequency-based method. In this work, a semiquantitative method (Quality Attribute Checklist; ASTM 1977) was used because of the limitations of intensity scoring in low-intensity complex aromas. Chi-square per cell was used to find the pertinent descriptors and then to analyze results. In the same way, the use of chi-square per cell can also be interesting for the analysis of data from check-all-that-apply (CATA) questions.

TABLE 12.7

Most Cited Words by Consumers to Describe Their Likes and Dislikes of the Tasted Ciders

	C1					C2				
	B1	B2	B3	B4	B5	B1	B2	B3	B4	B5
Likes										
Apple_Aroma	4 (−)*	7	1 (−)**	9	28 (+)***	0 (−)*	8 (+)*	1	1	8 (+)*
Aroma	15	11	2 (−)***	7 (−)*	36 (+)***	4	7	3	0	4
Color	32	34	20 (−)**	42	61 (+)**	2 (−)**	18 (+)*	10	3 (−)*	13
Mild	11	3 (−)*	0 (−)***	17 (+)*	22 (+)**	0	1	0	1	2
Sparkling	30	46	21 (−)**	43	55	4 (−)**	25 (+)*	16	9	18
Sweet	13	10 (−)*	0 (−)***	28 (+)*	46 (+)***	1	5	0 (−)*	5	6
Taste	39 (+)**	12 (−)**	1 (−)***	34	47 (+)***	1 (−)**	22 (+)**	4	1 (−)*	10
Dislikes										
Aroma	6 (−)***	15	55 (+)***	23	5 (−)***	8	2 (−)**	13	11	8
Bitter	13 (−)*	33 (+)**	53 (+)***	7 (−)***	3 (−)***	1	4	10 (+)***	1	1
Cloudy	0 (−)**	14 (+)**	22 (+)***	0 (−)**	0 (−)**	0	4 (+)***	0	0	0
Light_Color	26 (+)***	0 (−)**	3	1 (−)*	4	18 (+)***	2 (−)**	9	3	6
Sparkling	7 (−)*	43 (+)***	15	8 (−)*	7 (−)**	2	13 (+)***	8	0 (−)*	2
Taste	7 (−)***	34 (+)***	50 (+)***	11 (−)*	5 (−)***	8	2 (−)**	13	11	8

Note: ***p ≤ 0.001, **p ≤ 0.01, and *p ≤ 0.05; effect of the chi-square per cell.

REFERENCES

Ares, G., Giménez, A., Barreiro, C., and Gámbaro, A. 2010. Use of an open-ended question to identify drivers of liking of milk desserts. Comparison with preference mapping techniques. *Food Quality and Preference 21*: 286–294.

ASTM. 1977. *Manual on Sensory Testing Methods*, Vol. SPT 434. Philadelphia, PA: American Society for Testing and Materials.

Bécue-Bertaut, M., Álvarez-Esteban, R., and Pagès, J. 2008. Rating of products through scores and free-text assertions: Comparing and combining both. *Food Quality and Preference 19*: 122–134.

Bécue-Bertaut, M. and Pagès, J. 2008. Multiple factor analysis and clustering of a mixture of quantitative, categorical and frequency data. *Computational Statistics & Data Analysis 52*: 3255–3268.

Blancher, G., Chollet, S., Kesteloot, R., Hoang, D. N., Cuvelier, G., and Sieffermann, J. M. 2007. French and Vietnamese: How do they describe texture characteristics of the same food? A case study with jellies. *Food Quality and Preference 18*: 560–575.

Carson, R. T., Lewis, B. E., Shiff, S. J., Johnston, J. M., Dennee-Sommers, B., Lasch, K. E., Hwang, S., and Marquis, P. 2011. Pgi20 ibs-c patient symptom reports: Analysis of exploratory open-ended questions. *Value in Health 14*: A183–A184.

Chollet, S., Lelièvre, M., Abdi, H., and Valentin, D. 2011. Sort and beer: Everything you wanted to know about the sorting task but did not dare to ask. *Food Quality and Preference 22*: 507–520.

Christian, L. M. and Dillman, D. A. 2004. The influence of graphical and symbolic language manipulations on responses to self-administered questions. *Public Opinion Quarterly 68*: 57–80.

Elmore, J. R., Heymann, H., Johnson, J., and Hewett, J. E. 1999. Preference mapping: Relating acceptance of "creaminess" to a descriptive sensory map of a semi-solid. *Food Quality and Preference 10*: 465–475.

Escofier, B. and Pagès, J. 2008. *Analyses factorielles simples et multiples: Objectifs, méthodes et interprétation*, 4ème edn. Paris, France: Dunod.

Faye, P., Brémaud, D., Durand Daubin, M., Courcoux, P., Giboreau, A., and Nicod, H. 2004. Perceptive free sorting and verbalization tasks with naive subjects: An alternative to descriptive mappings. *Food Quality and Preference 15*: 781–791.

Faye, P., Brémaud, D., Teillet, E., Courcoux, P., Giboreau, A., and Nicod, H. 2006. An alternative to external preference mapping based on consumer perceptive mapping. *Food Quality and Preference 17*: 604–614.

Faye, P., Courcoux, P., Giboreau, A., and Qannari, E. M. 2013. Assessing and taking into account the subjects' experience and knowledge in consumer studies. Application to the free sorting of wine glasses. *Food Quality and Preference 28*: 317–327.

Galmarini, M. V., Symoneaux, R., Chollet, S., and Zamora, M. C. 2013. Understanding apple consumers' expectations in terms of likes and dislikes: Use of comment analysis in a cross-cultural study. *Appetite 63*: 27–36.

Galmarini, M. V., van Baren, C., Zamora, M. C., Chirife, J., Di Leo Lira, P., and Bandoni, A. 2011. Impact of trehalose, sucrose and/or maltodextrin addition on aroma retention in freeze dried strawberry puree. *International Journal of Food Science & Technology 46*: 1337–1345.

Geer, J. G. 1991. Do open-ended questions measure "salient" issues? *Public Opinion Quarterly 55*: 360–370.

Hanson, J. L., Balmer, D. F., and Giardino, A. P. 2011. Qualitative research methods for medical educators. *Academic Pediatrics 11*: 375–386.

Heath, T. P. and Chatzidakis, A. 2012. The transformative potential of marketing from the consumers' point of view. *Journal of Consumer Behaviour 11*: 283–291.

Israel, G. 2006. Visual cues and response format effects in mail surveys. In *Annual Meeting of the Southern Rural Sociological Association*, Orlando, FL.

Jordan-Zachery, J. S. and Seltzer, R. 2012. Responses to affirmative action: Is there a question order affect? *The Social Science Journal 49*: 119–126.

Lawless, H. T. and Heymann, H. 1998. *Sensory Evaluation of Food: Principles and Practices*. New York: Chapmann & Hall.

Moussaoui, K. A. and Varela, P. 2010. Exploring consumer product profiling techniques and their linkage to a quantitative descriptive analysis. *Food Quality and Preference 21*: 1088–1099.

Nestrud, M. A. and Lawless, H. T. 2010. Perceptual mapping apples and cheeses using projective mapping and sorting. *Journal of Sensory Studies 25*: 390–405.

Niedomysl, T. and Malmberg, B. 2009. Do open-ended survey questions on migration motives create coder variability problems? *Population, Space and Place 15*: 79–87.

Perrin, L. and Pagès, J. 2009. Construction of a product space from the ultra-flash profiling method: Application to 10 red wines from the Loire valley. *Journal of Sensory Studies 24*: 372–395.

Perrin, L., Symoneaux, R., Maître, I., Asselin, C., Jourjon, F., and Pagès, J. 2008. Comparison of three sensory methods for use with the Napping® procedure: Case of ten wines from Loire valley. *Food Quality and Preference 19*: 1–11.

Prescott, J., Lee, S. M., and Kim, K. 2011. Analytic approaches to evaluation modify hedonic responses. *Food Quality and Preference 22*: 391–393.

Rostaing, H., Ziegelbaum, H., Boutin, E., and Rogeaux, M. 1998. Analyse de commentaires libres par la techniques des réseaux de segments. In *Fourth International Conference on the Statistical Analysis of Textual Data, JADT'98*.

Sinesio, F., Peparaio, M., Moneta, E., and Comendador, F. J. 2010. Perceptive maps of dishes varying in glutamate content with professional and naive subjects. *Food Quality and Preference 21*: 1034–1041.

Smyth, J. D., Dillman, D. A., Christian, L. M., and Mcbride, M. 2009. Open-ended questions in web surveys: Can increasing the size of answer boxes and providing extra verbal instructions improve response quality? *Public Opinion Quarterly 73*: 325–337.

Snedecor, G. W. and Cochran, W. G. 1957. *Statistical Methods*, 6th edn. Ames, IA: The Iowa State University Press.

Symoneaux, R., Galmarini, M. V., and Mehinagic, E. 2012. Comment analysis of consumer's likes and dislikes as an alternative tool to preference mapping. A case study on apples. *Food Quality and Preference 24*: 59–66.

ten Kleij, F. and Musters, P. A. D. 2003. Text analysis of open-ended survey responses: A complementary method to preference mapping. *Food Quality and Preference 14*: 43–52.

Varela, P. and Ares, G. 2012. Sensory profiling, the blurred line between sensory and consumer science. A review of novel methods for product characterization. *Food Research International 48*: 893–908.

Woodward, J. L. and Franzen, R. 1948. A study of coding reliability. *Public Opinion Quarterly 12*: 253–257.

13 Dynamic Sensory Descriptive Methodologies

Time–Intensity and Temporal Dominance of Sensations

Rafael Silva Cadena, Leticia Vidal,
Gastón Ares, and Paula Varela

CONTENTS

13.1 INTRODUCTION

It is a well-known fact that perception of aroma, taste, flavor, and texture of food and beverages is a dynamic phenomenon (Lawless and Heymann 2010). The perceived intensity of the different sensory attributes changes along with the in-mouth transformation of food (Sudre et al. 2012). The different processes involved in food breakdown, such as chewing, salivation, tongue movements, and swallowing, are deeply related to the dynamic nature of food sensations (Lawless and Heymann 2010). Together with changes in texture, flavor perception varies as different taste and olfactory compounds are released during food breakdown (Sudre et al. 2012).

The most common methods for sensory profiling do not consider the temporal aspect of sensory attributes, as assessors are instructed to rate the perceived intensity of each attribute only once. Thus, ratings correspond either to the peak or to a "time-averaged" intensity. However, products with similar time-averaged profiles may differ in the way that the different sensory attributes evolve during consumption. Therefore, traditional approaches may miss crucial information as sensory appeal of many food products can be influenced by their temporal profile (Lawless and Heymann 2010).

Although the temporal aspect of sensory perception was first approached in the 1950s, it started to be seriously considered in the 1970s with the advances in the development of the time–intensity (TI) methodology (Sudre et al. 2012). This method records how the perceived intensity of a given attribute evolves over time (Lee and Pangborn 1986). Even though TI has been increasingly used in the last decades, it has some limitations that led to the development of an alternative methodology to assess the dynamics of perception: temporal dominance of sensations (TDS) (Pineau et al. 2009). In this methodology, assessors have to assess the temporal aspects of a product, evaluating simultaneously all the sensations perceived (Bruzzone et al. 2013; Pineau et al. 2009). The method consists of presenting a list of attributes to the panelists, who are asked to select which attribute is perceived as dominant and to rate its intensity. Along the evaluation, each time the dominant attribute changes, the panelists have to select the new dominant sensation and score it. This methodology has a great potential to evaluate dynamic aspects of sensory perception and has been used to evaluate a wide range of products since its development (Albert et al. 2012; Bruzzone et al. 2013; Dinnella et al. 2012; Labbe et al. 2009; Laguna et al. 2013; Meillon et al. 2009; Ng et al. 2012; Pineau et al. 2009).

In this chapter, both methodologies to assess the dynamic aspects of sensory perception, TI and TDS, are described.

13.2 TIME–INTENSITY METHODOLOGY

The evaluation of temporal aspects of sensory perception is not a new concept in sensory science. TI methodology basically consists of asking assessors to continuously evaluate the intensity of a sensory attribute over a period of time. Sjöström (1954) was the first to evaluate the intensity of a sensory attribute as a function of time. In this study, the bitterness of beer was evaluated using trained assessors, who were instructed to rate the intensity of the stimulus at 1 s intervals on a scorecard, using a clock to indicate time. A TI curve was constructed by graphing average bitterness intensity as a function of time. Larson-Powers and Pangborn (1978) improved the methodology by using a moving chart recorder equipped with a foot pedal for evaluating the evolution in time of sweetness and sourness in beverages and gelatin. However, this methodology also had some great disadvantages as assessors needed considerable training and the data recovery was labor intensive. Although many alternatives to carry out TI method were developed, they all presented great difficulties, especially to characterize samples using curve parameters (Birch and Munton 1981; Lawless and Skinner 1979; Schmitt et al. 1984).

The obstacles for the development and use of TI were overcome with the advent and dissemination of personal computers in the 1980s. Several approaches for collecting and analyzing TI data were developed (Barylko-Pikielna et al. 1990; Cliff 1987; Guinard et al. 1985; Lee 1985; Rine 1987; Takagaki and Asakura 1984; Yoshida 1986). Together with the development of systems for TI data acquisition, the advance in data analysis of summary curves enabled to obtain more parameters from the TI curves, which contributed to a better understanding of the underlying processes of temporal perception and encouraged the use of TI by both the academia and the industry (see Cliff and Heymann 1993, for more historical details). In 30 years, from Sjöström (1954) to Takagaki and Asakura (1984), 19 studies were carried out using TI analysis. However, after the development of personal computers and computerized systems for collecting TI data, less than 10 years were necessary to achieve the same number of studies (Cliff and Heymann 1993; Wendin et al. 2003).

13.2.1 TRAINING AND DATA COLLECTION

13.2.1.1 Assessor Selection and Training

Since its development, training has been one of the biggest difficulties for the application of TI methodology. Recruitment, selection, and training of assessors are essential steps to get a consistent panel and thus reliable results.

Recruitment and selection of assessors follow the same principles of traditional sensory profile methods, that is, assessors need to have normal taste and smell capabilities, motivation, availability, and willingness to participate, and they must be able to concentrate on the task. Training of the panel for TI is similar to descriptive evaluation, being characterized by long and exhaustive sessions. Training should address both the perception of the target stimulus as well as the evaluation of changes over time for getting a TI curve. The panel needs to be familiarized with the dynamic of the test for evaluating the products and with the system used for data acquisition. According to van Buuren (1992), the TI curves of assessors are different and considered as an individual "signature." Only effective training sessions can minimize differences among assessors and improve the quality of the results. Considering that general guidelines for training a sensory panel are available (Civille and Szczesniak 1973; ISO 1994, 2008), Peyvieux and Dijksterhuis (2001) proposed a three-step guideline for training assessors for TI studies:

1. Introducing the method to the assessors
2. Familiarization of the assessors with the task and the system used for data acquisition using basic taste solutions
3. Training the assessors using real products
   a. Sensory profiling
   b. Pilot TI study

Each of the previously mentioned steps is explained in detail by Peyvieux and Dijksterhuis (2001). Using basic taste solutions (sweet, salty, bitter, and sour) of concentrations higher than every assessor's threshold consists of an interesting alternative for the assessors to learn the TI methodology using pure and clearly distinguishable stimuli (Peyvieux and Dijksterhuis 2001). According to Peyvieux and Dijksterhuis (2001), after the familiarization with the methodology and the data acquisition system using basic taste solution, most assessors showed an increase in the quality of their curves and were more consistent over replications in the evaluation of real products.

13.2.1.2 Design of the Study

The aim of the study and the characteristics of the product mainly determine the sensory attributes to be evaluated by the panel in the TI methodology.

The selection of the attributes can be done during the training sessions and/or using previous results from a descriptive analysis with a trained panel. Considering that the application of a TI study is time consuming and expensive, it is essential to select the most relevant attributes. Together with the selection of attributes, assessors need to the define samples with maximum intensity of each attribute, to be considered as references in the evaluations. During the training sessions with the products, it is important to accurately define the maximum time to be considered when evaluating the samples. Maximum time has to be optimized to assure that it is longer than the time that the stimulus lasts and to prevent the evaluations for being excessively tedious. The evaluation of flavor and texture attributes requires differences in the experimental design and how to administrate the time. In a flavor evaluation, the stimulus is perceived while the product is in the mouth, and after it has been swallowed, while in a texture evaluation, perception of the stimulus ends when the sample is swallowed. Besides, some products have particularities that should be taken into account before the test. For example, when performing a TI study to evaluate sweetness in chewing gum samples, Rocha-Selmi et al. (2012) explicitly indicated that samples cannot be swallowed.

Product texture is an important feature that has to be taken into account when determining the maximum time needed for a TI evaluation. When dealing with solid products, the time needed to chew and swallow the sample has to be considered in the design. For example, assessors may need more time to chew and swallow meat compared to French fries. The time needed for chewing and swallowing the sample is defined during the training sessions. On the other hand, oral processing of liquid products is usually shorter and has short in-mouth residence times (5–10 s). Special characteristics, such as high oiliness (or unctuous) or high adhesiveness, should be considered.

Tasting protocols of different complexity have been used to collect TI data. In general, TI studies use simple protocols that involve a single sip or bite to taste the products. For example, when evaluating sweetness, bitterness, and cream flavor in ice cream samples, Cadena and Bolini (2011) asked assessors to place a predetermined amount of ice cream sample in their mouth, to swallow it after 5 s, and to evaluate the attributes for 60 s. However, more complex protocols can be used. For example, Courregelongue et al. (1999) used a complex design to evaluate the effect of sweetness, viscosity, and oiliness on temporal perception of soymilk astringency. To reach the objective of the study, assessors were asked to try samples and expectorate them four times at predetermined time intervals while evaluating astringency. After the last expectoration, they had to continue rating the astringency for another 30 s. The amount of sample to be tasted by assessors should also be defined during the training sessions since it strongly affects the total time of perception and in some cases, the intensity of the stimulus. For example, if a panelist tastes a cookie of 30 g, the time to chew and swallow the sample, as well as the total duration of the stimulus, will be different than if he or she tastes a cookie of 15 g. Thus, all assessors have to evaluate the same amount of sample that has to enable the evaluation of all the attributes while considering oral comfort. In practice, assessors receive the sample and are instructed to take the full amount of the sample in their mouth.

13.2.2 Data Analysis

TI data are usually represented using TI curves, which represent the intensity of the attribute as a function of time (Figure 13.1). A TI curve can be constructed for each assessor, from which it is possible to extract several parameters and, as already mentioned, determine the assessor "signature" (van Buuren 1992). Anatomical and physiological differences are responsible for differences among assessors. Some of the variables that have been associated to these natural individual differences are salivary factors (Fischer et al. 1994), different types of oral manipulation and chewing efficiency (Brown et al. 1994; Zimoch and Gullet 1997), and individual habits of scaling (Lawless and Heymann 2010). Panel training has a great influence in determining the shape of individual TI curves and, if well done, can prevent "nonnatural" differences.

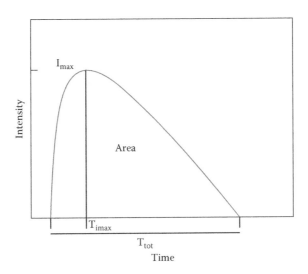

FIGURE 13.1 Example of an average TI curve showing the most commonly used parameters for data analysis. *Note:* I_{max}, Maximum intensity; T_{imax}, time to maximum intensity; T_{tot}, total time; Area, area under the curve.

From each individual curve, several parameters can be calculated (Figure 13.1). The complexity of TI data analysis has increased with computerization (Cliff and Heymann 1993). Initially, parameters that were easily calculated, such as maximum intensity, time to maximum intensity, and total time, were calculated, while other complex parameters, such as area under the curve, have only been started to be frequently used after the development of software for acquisition of sensory data (Cliff and Heymann 1993; Lawless and Heymann 2010). The initial time of the stimulus ($T_{initial}$), time of duration of the maximum intensity (Plateau or T_{plat}), and time corresponding to the point when the maximum intensity begins to decrease (T_d) can also be calculated and analyzed in TI studies. The four most commonly used parameters are described as follows:

- Maximum intensity (I_{max})—maximum perceived intensity during the test
- Time to maximum intensity (T_{imax})—time when the maximum intensity of the stimulus is perceived
- Total time (T_{tot})—total duration of the stimulus, also called persistence, finish time, or extinction time
- Area under the curve (Area)—total amplitude or total gustatory response

Intensity data from TI studies are usually analyzed using the same approaches considered for traditional descriptive analysis procedures. TI parameters (T_{plat}, T_d, I_{max}, T_{imax}, T_{tot}, Area) extracted from each individual curves can be analyzed by analysis of variance (ANOVA) or other statistical tests for comparing average values. Significant differences between each pair of samples can be determined using postcomparison tests, such as Tukey's test. However, it is important to take into account that the "signatures" of the assessors' individual curves are ignored, which can hinder the detection of an atypical response (Lawless and Heymann 2010). Individual intensity data are usually averaged over the assessors to construct a consensus TI curve, which is helpful for interpretation and presentation of the results (Dijksterhuis et al. 1994; Garrido et al. 2001; MacFie and Liu 1992). However, one of the most challenging aspects of the TI method is the analysis of these curves (McGowan and Lee 2006). MacFie and Liu (1992) proposed a statistical approach to construct average TI curves. They first averaged individual values of TI parameters, normalized the individual curves in the intensity direction and then in the time direction. This approach allows capturing some of the individual variation in the pattern of the time records (Lawless and Heymann 2010). However, the average method smoothes jagged curves, and it does not account for the different shapes of individual TI curves (Dijksterhuis and Piggot 2001; Garrido et al. 2001).

A semiparametric model was designed by Dijksterhuis and Eilers (1997) to account for the individual "signatures" of panelists (Eilers and Dijksterhuis 2004; Garrido et al. 2001; Wendin et al. 2003). A single or a set of equations are used to fit each individual TI curve. However, a curve with a good approximation to one assessor may not be good to another assessor or attribute that presents a distinct curve profile as multiple intensity peaks (Lawless and Heymann 2010; McGowan and Lee 2006). McGowan and Lee (2006) compared two methods to analyze TI curves in a corn zein chewing gum study. The first method was grouping assessors with similar individual curves, and the other approach was the enhanced method proposed by Liu and MacFie (1990). The grouping was performed by visually examining the individual curves. Nine assessors and two replications were considered in this study, so the authors analyzed a total of 18 curves for each attribute and sample. In studies involving more assessors, samples, and replications to evaluate more than one attribute, this step may be tedious and time consuming. After grouping similar curves, a representative curve for each specific group was created by averaging the individual curves. The method showed some advantages and provided an accurate and complete depiction of the individual curves and seemed to be more representative of the total data, if not all panelists returned to zero intensity (McGowan and Lee 2006). However, more studies need to be performed in order to validate this type of approach.

Principal component analysis (PCA) is an alternative method that can be used to take into account differences among the individual curves (van Buuren 1992). According to van Buuren (1992), PCA shows additional advantages for the interpretation of TI curves. PCA constructs a consensus or "principal" TI curve, which is constructed as a weighted average of the individual curves that provides a better representation of the individual curves than an average curve (Dijksterhuis and Piggot 2001). PCA gives large weights to similar curves, while curves that largely deviate from the rest receive low weights and therefore will not largely affect the resulting "principal curve." Dijksterhuis et al. (1994) compared three variants of PCA for the analysis of TI curves: centering, noncentering, and standardizing. According to these authors, when PCA was applied on the raw data matrix (noncentered PCA), all the information about differences among the individual curves was retained, while data standardizing was the least recommended approach (Dijksterhuis and Piggot 2001; Dijksterhuis et al. 1994). PCA also provides the possibility of assigning weights for each of the assessors, which indicates the degree to which they contribute to the principal curve. This approach has been regarded as an interesting tool for identifying outliers in the data or panelists with markedly different TI signatures (Peyvieux and Dijksterhuis 2001).

Multiple TI (MTI) analysis is a distinct graphical method that enables to visualize dynamic profiles of two or more sensory attributes of a single sample (Palazzo and Bolini 2009). The MTI makes the visualization of the dynamics of all the sensory attributes easier, even if they were not evaluated in the same session (Cadena and Bolini 2011). When average TI curves are constructed, it is usual to make a single graph by attribute that includes all the evaluated samples. The idea of MTI is to plot in the same graph all the attributes for each sample and to represent the TI profile. To apply MTI, it is essential the data collection for all the sensory attributes is performed in a rigorously standardized way (Cadena and Bolini 2011). All variables that may directly affect the stimulus perception must be standardized in all attribute evaluations.

Ovejero-Lopez et al. (2005) compared six statistical methods for analyzing TI data. Simple product average curves, noncentered PCA weighted average curves and normalized curves for intensity and time, ad hoc curve parameter retrieval, PLSR, dual PCA, and PARAFAC2 were evaluated. The authors concluded that all the evaluated methods provided the same information, but the amount of information gained with multivariate methods was distinct. The authors recommended that the statistical method should be chosen based on the objective of the study. Besides, more than one method can be used to provide a robust conclusion. However, more studies to explore the applicability of multivariate techniques to TI data are necessary.

13.2.3 RECENT STUDIES AND POTENTIALITIES

Since its development, TI has been widely applied and has become a standard methodology in sensory science. A quick search in Scopus with the term "time–intensity," limited to the subject area "Agricultural and Biological Sciences," retrieved 105 articles that studied, applied, or revised TI analysis in food products. Since 2000, 85 articles have been published in journal within *Food Science and Technology*, with an average of approximately 6 articles per year. *Food Quality and Preference* (29) and *Journal of Agricultural and Food Chemistry* (18) are the journals with the highest number of articles presenting TI analysis research. The most cited article, with more than 100 citations, reports the application of TI studies' flavor release and perception of flavored whey protein gels with different gel hardness and water-holding capacity (Weel et al. 2002). Meanwhile, the paper written by Liu and MacFie (1990) is the most cited article reviewing methodological aspects of TI and proposing an alternative method for averaging TI curves. The most recently published articles (2009–2013) report the application of TI for the evaluation of widely different food products, including dry-cured ham (Fuentes et al. 2013), fruit jam (De Souza et al. 2013), gluten-free bread (De Morais et al. 2013), chocolate and coffee (Chung and Lee 2012), diet chocolate (Palazzo et al. 2011), ice cream (Cadena and Bolini 2011), sausages (Ventana et al. 2010), and extra-virgin olive oils (Esti et al. 2009).

Among several recent articles, some are noteworthy and represent potential areas of research. As previously mentioned, data acquisition is one of the most challenging aspects of the methodology. Almost all systems for acquiring TI data are not free, and it is necessary for companies or research institutes to buy licenses for specific software or to develop their own software. Pinheiro et al. (2013) recently developed a free software, SensoMaker, which enables data acquisition and analysis of TI data. This software can help to disseminate the use of TI analysis among small companies and research institutes, contributing to the development of the methodology.

The combination of TI with gas chromatography (GC) also offers opportunities for studying the dynamics of sensory perception. This approach was described by Garruti et al. (2003) and recently applied by Sampaio et al. (2013) and Murat et al. (2013). Although the use of TI and GC presents some difficulties, the relationship between volatiles detected by GC and trained panelists at different moments in time has great potential, particularly when working with complex products. Hillmann et al. (2012) applied the combination of TI and GC to study "sensomics," that is, the identification of key compounds responsible for the typical taste of foods. The authors applied this approach for studying traditional balsamic vinegar and reported the identification of a new sweet taste modulator.

Pleasantness is a temporal phenomenon that can be measured using a temporal approach (Lee and Pangborn 1986). Although Taylor and Pangborn (1990) were the first to apply a temporal evaluation of liking, this type of research has been gaining attention in the last 10 years. Sudre et al. (2012) used two approaches to measure the dynamics of liking during a one-bite consumption event. In the first approach, denominated "four-step method," consumers rated their liking of each product at four specific times of their mastication process (T1, at the beginning of the mastication [after approximately three cycles of mastication]; T2, in the middle of the mastication [self-determined time]; T3, just before swallowing; and T4, just after swallowing). In the other approach, named "continuous liking method," consumers were asked to report any change in their liking over the mastication period, from the first bite to swallowing. The use of methods to measure liking over time still needs to be validated and applied in different types of products (Delarue and Loescher 2004) in order to get a deeper understanding of the determinants of dynamics of overall liking (Sudre et al. 2012).

Finally, the simultaneous evaluation of more than one attribute is noteworthy. Duizer et al. (1997) proposed a dual-attribute TI (DATI) analysis, but only one study has applied this technique (Zimoch and Findlay 1998). The multi-attribute TI (MATI) method was proposed by Kuesten et al. (2013) to minimize one of the biggest disadvantages of traditional TI analysis, the evaluation of only one attribute by session, enabling a rapid collection of data from multiple attributes. Although MATI has been applied with consumers and a small number of children, the practical implementation of MATI by trained assessors and, especially, consumers is one of the biggest obstacles of the method. The eye–hand coordination and the use of computer were difficult for some adults, while children considered that the test was easy and enjoyable (Kuesten et al. 2013). In this study, the authors used taffy products as samples and asked assessors to evaluate four attributes based on dissolution time (60–70 s). They reported that MATI was useful, required minimal training to be used with trained assessors and adult and children consumers. However, the application of MATI is too recent and further research is needed to determine its validity. Kuesten et al. (2013) listed some points to future research: investigation of the influence of tasting protocol on results, comparison of MATI with other temporal and nontemporal sensory methods, and development of statistical tools for data analysis.

13.2.4 Advantages and Disadvantages

TI analysis is a dynamic sensory descriptive method that allows the evaluation of the intensity of a predetermined attribute over time. The use of TI analysis allows the researcher to understand the dynamics of the sensory profile of a stimulus over consumption, which cannot be easily determined using traditional descriptive methods as quantitative descriptive analysis (QDA®). As already mentioned, when compared with traditional sensory profile methods, TI allows a different view of the stimulus, and the construction of TI curves enables to visualize changes of the intensity over time. This consists of an advantage over TDS that only evaluates the dominance of the attributes. Further, the fact that TI requires focusing the attention on mainly one attribute at a time over evaluation (Kuesten et al. 2013) can be seen as an advantage and a disadvantage depending on the specific aim of the research.

TI analysis requires an extensive training to provide valid and reproducible data from a consensual trained panel. The guideline for training proposed by Peyvieux and Dijksterhuis (2001) is less extensive than training for a traditional sensory descriptive technique, involving other aspects and less attributes. However, during TI training, special attention should be paid to the eye–hand coordination and the use of computer. This step is extremely important, and if not executed properly, individual TI curves will present very distinct profiles among assessors. Warm-up sessions with the product and/or basic taste solutions are a good strategy for making the assessors familiar with the TI task and the data acquisition system. Although TI is already an established sensory method, data analysis is still a deficiency. The particularity of the data, characterized by large individual differences on the TI curves, and the many statistical approaches available hinder the establishment of a unique method for analyzing TI data. The limitation of the evaluation of only one attribute, which leads to possible subsequent dumping effects (Clark and Lawless 1994), and the extensive time needed make this methodology difficult to apply, particularly as a routine analysis in the food industry. Alternative methods such as MATI are very recent, and much effort is still necessary to study and to improve them before they get established as reliable alternatives for studying the dynamic aspects of sensory perception. However, if this is done, MATI may complement TI when the evaluation of more than a single attribute is necessary.

13.3 TEMPORAL DOMINANCE OF SENSATIONS

TDS is a relatively new method, which was developed in the late 1990s at the "Centre Européen des Sciences du Goût" in order to overcome some of the disadvantages of the TI method, such as the duration of the experiments and the halo-dumping effect (Pineau et al. 2009). This methodology enables the evaluation of several attributes simultaneously and to study the sequence of dominant sensations of a product during a certain time period (Meyners 2011; Meyners and Pineau 2010).

Basically, TDS consists of presenting assessors an attribute list on a computer screen and asking them to determine which sensation is dominant. Some authors also ask the panelists to simultaneously rate attribute dominance and intensity. Assessors are asked to select a new attribute when the dominant sensation changes, until the perception is over (Pineau et al. 2009).

In a comparative study, Labbe et al. (2009) showed that although TDS provided product spaces close to the ones obtained with traditional sensory profiling, it also uncovered additional information about the dynamics of complex and long-lasting sensations, which enabled to discriminate among products. In other studies comprising samples with subtle differences, TDS provided valuable additional information that was not available with conventional sensory profiling (Dinnella et al. 2012; Meillon et al. 2009, 2010). However, as TDS focuses only on dominant sensations, it does not provide detailed information of all product attributes. Thus, TDS is not meant to replace sensory profiling but to be used as a complementary tool, adding information about the temporality of the main attributes (Meillon et al. 2009; Ng et al. 2012).

This method has been increasingly used in the last few years to study the dynamic perception of different products, such as wine (Meillon et al.

2009, 2010; Sokolowsky and Fischer 2012), blackcurrant squashes (Ng et al. 2012), coffee (Dinnella et al. 2013), water (Teillet et al. 2010), gels (Labbe et al. 2009), dairy products (Bruzzone et al. 2013; Pineau et al. 2009), wheat flakes (Lenfant et al. 2009), biscuits (Laguna et al. 2013), fish sticks (Albert et al. 2012), extra-virgin olive oil added to vegetables (Dinnella et al. 2012), and salmon–sauce combinations (Paulsen et al. 2013). Although TDS has gained a lot of popularity, it is still evolving, and thus, it is important to stress that experimental procedures are not standard and differ among researchers (Lawless and Heymann 2010).

13.3.1 Implementation and Data Collection

In TDS, assessors have to continuously characterize the dynamics of product perception by simultaneously evaluating all the sensations perceived (Pineau et al. 2009). During the evaluation, they have to indicate the attribute they perceive as dominant at each time, until the perception ends. Each time the assessors consider that the dominant attribute has changed, they have to select a new dominant attribute.

The definition of "dominant" is not standard and differs among studies. "Dominant" has been defined by Labbe et al. (2009) as "the most intense sensation," whereas Pineau et al. (2009) described it as "the most striking sensation," "the new sensation popping up, and not necessarily the most intense." However, in the majority of the studies conducted in the last few years, the definition proposed by Pineau et al. (2009) has been used. At the last Pangborn conference, "dominant" was defined as the sensation that catches the attention at that time, a mixture of intensity and rising/new (Pineau 2013).

Data collection is performed using computerized systems such as FIZZ® (Biosystemes, Couternon, France), Compusense® (Compusense Inc., Guelph, Ontario, Canada), or the freely available SensoMaker (Pinheiro et al. 2013).

13.3.1.1 Number of Assessors

The number of assessors used in TDS differs among studies but ranges from 9 (Albert et al. 2012) to 16 (Meillon et al. 2009; Pineau et al. 2009; Teillet et al. 2010). The most usual number of trained assessors is 10–13 (Bruzzone et al. 2013; Dinnella et al. 2013; Ng et al. 2012), which is similar to the number of trained assessors usually recommended for descriptive analysis (Lawless and Heymann 2010). Evaluations are usually performed in duplicate or triplicate (Albert et al. 2012; Bruzzone et al. 2013; Meillon et al. 2009; Ng et al. 2012; Pineau et al. 2009). However, research is still needed to determine the optimum number of assessors and replicates needed for the implementation of TDS studies.

13.3.1.2   Tasting Protocol

Assessors are presented with a list of attributes on a computer screen. They are asked to put the sample in their mouth and simultaneously start the software by clicking on the "start" button. Immediately after the test starts, assessors have to determine which of the attributes from the list is dominant by clicking the attribute on the screen. An example of a TDS evaluation using SensoMaker is shown in Figure 13.2. In the figure, the assessor considers that the attribute sweet is dominant 3 s after the test started.

Each time assessors feel that the perception has changed (either in intensity or in quality) and that a new attribute is dominant, the button corresponding to that attribute has to be clicked. Assessors are told that they do not have to use all the attributes in the list and that they are free to select an attribute several times, but they have to take into account that only one attribute can be selected at a time. Besides, assessors are usually not forced to choose an attribute as dominant. The evaluation continues until assessors no longer perceive sensory sensations and they click the "stop" button (Albert et al. 2012; Bruzzone et al. 2013). Alternatively, the test can continue until a predetermined time is reached or until the sample is completely swallowed (Laguna et al. 2013). It is important to take into account that the specific tasting protocol differs depending on the characteristics of the products and the aim of the study.

In addition to asking assessors to indicate the dominant attribute, several authors have asked them to also rate the intensity of the dominant attribute at each time of the evaluation. Unstructured line scales anchored with "not at all intense" and "very intense" or "weak" and "intense" are usually used for this purpose (Albert et al. 2012; Ng et al. 2012; Teillet et al. 2010). However, some authors have reported that asking assessors to rate the intensity of the dominant attributes increases the difficulty of the

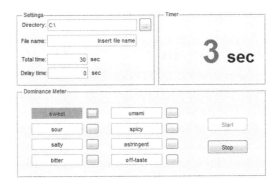

FIGURE 13.2   Example of a TDS evaluation using the freely available software SensoMaker. (Adapted from Pinheiro, A.C.M. et al., *Ciência e Agrotecnologia*, 37, 199, 2013.)

task and may lead to a delay in the peaks of the dominance curves as assessors spend more time evaluating the dominant attribute and take longer to switch to the next dominant sensation (Dinnella et al. 2013; Paulsen et al. 2013). As much of the valuable information gathered with TDS does not involve intensity measures, it is currently recommended not to ask assessors to rate intensity (Paulsen et al. 2013; Pineau 2013; Schlich 2013).

13.3.1.3 Selection of the List of Attributes

The selection of the list of attributes is one of the key steps of the implementation of TDS. The list of attributes can include different sensory modalities (flavor and texture attributes) (Albert et al. 2012) or being restricted to a single modality (e.g., texture attributes, Bruzzone et al. 2013). According to Pineau et al. (2012), assessors are able to simultaneously evaluate several sensory modalities, and including attributes within different modalities has no impact on the number of attributes selected during the evaluation.

Different approaches can be considered for selecting the list of attributes for TDS. Teillet et al. (2010) performed a TDS study of mineral waters considering the same nine attributes evaluated in descriptive analysis. Similarly, Bruzzone et al. (2013) held an open session in which trained assessors selected five of the eight texture attributes evaluated in descriptive analysis for studying the dynamics of yogurt texture perception using TDS. Assessors excluded homogeneity and smoothness since they were evaluated over the whole mastication period, not presenting dominance at any moment of the evaluation. Besides, ropiness was not considered since it is evaluated before consumption of the product.

The list of attributes can also be selected considering results from preliminary studies with the trained panel. Albert et al. (2012) asked assessors to try fish nuggets and to describe all the in-mouth sensations they felt. The seven most mentioned sensations were retained for TDS. Similarly, Meillon et al. (2009) asked the assessors to evaluate wine samples using a detailed tasting protocol and to note on a paper all the in-mouth sensations they felt during the tasting. They kept the 10 more mentioned terms for the TDS evaluations.

Pineau et al. (2012) reported that assessors tend to select a small number of attributes during TDS evaluations (two to six), which are independent of the total number of attributes included in the list. These authors recommended to use lists with a maximum of 10 attributes.

Primacy bias has been reported when assessors perform TDS tasks (Pineau et al. 2012). Assessors tend to more frequently and earlier select as dominant the attributes located at the top of the list, compared to those located at the bottom. Although the influence of attribute position is smaller than attribute effect, it is recommended that each assessor evaluates the products using a list that includes the attributes in a different order to reduce the potential influence of primacy bias on the results (Pineau et al. 2012). Thus, it is recommended to balance the order in which the attributes are included in the list between assessors following a Williams Latin square design. However, in order to facilitate the evaluation, attribute order is usually kept constant for each assessor.

13.3.1.4 Panel Training

Most of TDS applications have been carried out with trained assessors. Assessor selection and training has remained very similar to training for descriptive analysis, focusing on assessors' ability to recognize and, if needed, rate the intensity of the different product attributes. In fact, in many studies in which both TDS and descriptive analysis were applied, the same panel performed both tasks (Bruzzone et al. 2013; Ng et al. 2012; Paulsen et al. 2013). However, given the fact that TDS deals with dynamic perception in a very specific way, some additional training must be considered in order to introduce the assessors with the notion of TDS as well as to allow familiarization of the assessors with the software used for data collection (Dinnella et al. 2013; Labbe et al. 2009; Laguna et al. 2013; Meillon et al. 2009, 2010; Pineau et al. 2009). Ng et al. (2012) introduced the concept of temporality of sensations to assessors using the analogy of an orchestra playing music.

TDS has also been carried out with untrained assessors. Albert et al. (2012) used assessors who were familiar with sensory evaluation but untrained in the specific product type. They attended only two preliminary training sessions: the first one to be introduced to the notion of temporality and to the TDS methodology and to generate the attributes, and the second one to get used to the methodology and the software used in the study. Albert et al. (2012) showed that TDS can provide valuable information even without the need of a time-consuming training period. In fact, in the 10th Pangborn Sensory Science Symposium, some studies with untrained consumers were presented, both for studies in a sensory laboratory and at home (Schlich 2013). It was stated that TDS can be performed by consumers provided the researcher makes sure that the task is understood by the assessors, no intensity ratings are asked, and time for each consumers is standardized for data analysis.

13.3.2 DATA ANALYSIS

13.3.2.1 TDS Curves

An example of TDS raw data is shown in Table 13.1. For each assessor, information about the periods in which each attribute was chosen as dominant is collected. In the example, Assessor 1 has considered that sweet was the dominant attribute from time 0.5 to 3 s, while creamy was dominant from time 3.5 to 6 s.

TABLE 13.1

Example of the Raw Data Obtained from a TDS Study

Assessor	Sample	Time	Sweet	Sour	Bitter	...	Creamy
1	A	0.5	1	0	0	...	0
1	A	1	1	0	0	...	0
1	A	1.5	1	0	0	...	0
1	A	2	1	0	0	...	0
1	A	2.5	1	0	0	...	0
1	A	3	1	0	0	...	0
1	A	3.5	0	0	0	...	1
1	A	4	0	0	0	...	1
1	A	4.5	0	0	0	...	1
1	A	5	0	0	0	...	1
1	A	5.5	0	0	0	...	1
1	A	6	0	0	0	...	1
1	A	6.5	0	0	0	...	0
...		...	...	...	...	...	...
n	X	...	...	...	...	...	...

TDS data are usually represented by TDS curves showing the dominance rate of each of the sensations at each time for each sample (Pineau et al. 2009). Dominance rate is calculated as the proportion (or percentage) of citations of an attribute across the panel, that is, by dividing the number of selections of an attribute (across replications) at each time by the number of assessors and the number of replications. The higher the dominance rate for the attribute, the higher the proportion of assessors who considered it as dominant, and therefore, the higher its dominance at the panel level. Figure 13.3 shows an example of the dominance rate (expressed as percentage) as a function of time. It can be seen that, across replicates, the attribute was regarded as dominant by 29.4% of the assessors from the period elapsed between 2 and 2.5 s, while after 9.5 s, none of the assessors regarded it as dominant.

The dominance rate curve for each attribute is then smoothed using spline-type polynomial and plotted against time for each sample to obtain TDS curves (Bruzzone et al. 2013; Pineau et al. 2009). The pspline package of R language (R Development Core Team 2007) can be used for this purpose. Figure 13.4 shows the smoothed TDS curve from data shown in Figure 13.3.

Considering that the time elapsed from the start of the mastication to swallowing can differ among assessors, time scales of sensory perception can also differ (Lenfant et al. 2009). Thus, TDS data from each assessor are usually normalized according to the individual duration of the test, so

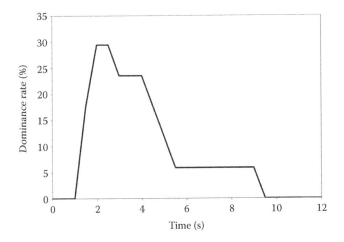

FIGURE 13.3   Example of the dominance rate of an attribute (expressed as percentage) as a function of evaluation time.

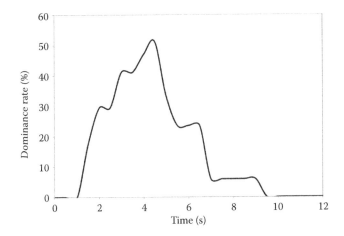

FIGURE 13.4   Smoothed TDS curve from data shown in Figure 13.3.

that time data from each assessor are expressed from $x = 0$ (start of the test) to $x = 100$ (end of the test), which enables a standardized comparison of the individual curves (Albert et al. 2012).

According to Labbe et al. (2009) and Pineau et al. (2009), TDS curves should be interpreted considering two parameters: chance level and significance level. Chance level (P_0) is the dominance rate that an attribute can obtain by chance, considering all the evaluated attributes. When assessors are allowed not to select any attribute as dominant, the value P_0 is equal to the inverse of the total number of attributes plus one, as suggested by Labbe et al. (2009). Alternatively, if assessors have to select a dominant attribute at each time, P_0 is equal to the inverse of the total number of attributes. Attributes with dominant rates below P_0 are not considered as dominant.

Significance level (P_s) is the minimum value of dominance rate an attribute has to obtain to be significantly higher than P_0. Attributes with dominance rates higher than this value are considered significantly dominant.

Significance level can be calculated using the confidence interval of a binomial distribution based on a normal approximation (Pineau et al. 2009), as shown in the following equation:

$$P_s = P_0 + 1.645 \sqrt{\frac{P_0 (1 - P_0)}{n}}$$

where

P_0 is the chance level

n is the number of judgments (assessors × replicates)

1.645 is the z value for a one-tailed normal distribution considering a significance level of 5%

This approximation is when $n \cdot P_0 (1 - P_0) > 5$ (Rosner 1995), which is usually achieved when the number of judgments (assessors × replicates) is high. Alternatively, when the number of judgments in the experiment is low, significance level can be calculated considering an exact test using binomial test, as recommended by Pineau et al. (2009). The smallest value for which the cumulative binomial distribution is greater than the chance level for a 95% confidence level can be calculated using the CRITBINOM function of Excel. For example, for an experiment in which 10 assessors evaluate samples in triplicate ($n = 10 \times 3 = 30$) and $P_0 = 0.091$, CRITBINOM (30, 0.091, 0.95) returns 6, which is the smallest value for which the cumulative binominal distribution is greater than chance level for a confidence level of 95%. Considering that 1 should be added to the result (Pineau et al. 2009), significance level is determined as $(6 + 1)/30 = 0.33$.

Chance and significance levels are plotted as horizontal lines on the plot of TDS curves to show when attributes are dominant and significantly dominant. Figure 13.5 shows an example of smoothed TDS curves for texture attributes of yogurt. As shown, the dominance rate of gelatinousness was lower than chance level during the whole evaluation, suggesting that it was not dominant. The dominance rate of thickness was larger than chance level but lower than significance level from 2 to 3.5 s, suggesting that this attribute was dominant but not significant. Creaminess and melting were both significantly dominant. Creaminess was the first significantly dominant attribute, from the time period elapsed between 2 and 7.5 s, while melting was significantly dominant during the period elapsed between 6.5 and 11s.

TDS curves also allow identifying the sequence in which the attributes are perceived as dominant. In the example shown in Figure 13.5, creaminess is

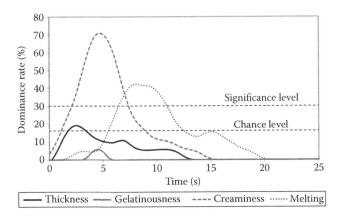

FIGURE 13.5   Example of smoothed TDS curves for four texture attributes. Chance and significance levels are included to aid interpretation.

the first attribute to catch assessors' attention, while after approximately 7 s, melting becomes the dominant attribute until the perception ends.

13.3.2.2   Evaluation of Differences among Products

Comparison of products allows determining differences in the dynamic profile of the products and the exact moments in which the differences are perceived.

The dominance rates of each pair of samples for each attribute can be compared using TDS difference curves. These curves are constructed by subtracting the dominance rate of each pair of samples at each time. Dominance rate differences are considered significant when they are significantly different from zero according to a classical test of comparison of binomial proportions (Pineau et al. 2009). The least significance difference in attribute dominance between two samples at time t ($P_{diff,t}$) for a 95% confidence level can be calculated as follows:

$$P_{diff,t} = 1.96 \sqrt{\left(\frac{1}{n_1} + \frac{1}{n_2} \right) P_{average,t} \left(1 - P_{average,t} \right)}$$

$$P_{average,t} = \frac{P_{1,t} n_1 + P_{2,t} n_2}{n_1 + n_2}$$

where
n_1 is the number of judgments (assessors × replicates) for sample 1
n_2 is the number of judgments (assessors × replicates) for sample 2
$P_{1,t}$ is the dominance rate of sample 1 at time t
$P_{2,t}$ is the dominance rate of sample 2 at time t

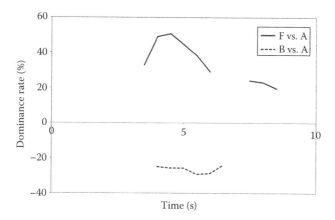

FIGURE 13.6   Example of difference TDS curves for two pairs of samples.

Figure 13.6 shows an example of difference curves for two pairs of samples. The dominance rate of the evaluated attribute was significantly higher for sample F than for sample A in two periods of the evaluation, from 3.5 to 6 s and from 7.5 to 8.5 s. Meanwhile, the dominance of the evaluated attribute was significantly higher for sample F than for sample B between 4 and 6.5 s. For the rest of the evaluation, there were no significant differences between samples F and A and between samples B and F.

Alternatively, Meyners and Pineau (2010) introduced a randomization test based on distances between matrices to compare a set of products globally or in pairwise comparisons. These authors proposed to unfold TDS data into a series of matrices, in which each row represents an attribute and each column represents a point of time, for each assessor, sample, and repetition. In each of these matrices, 1 indicates that the attribute was selected by the assessor at a certain point in time, while 0 indicates that the attribute was not selected. The distance between each pair of matrices is calculated using squared Euclidean distances. Distances are then used as a test statistic for determining global significant differences among products, for pairwise comparisons, and for making inferences by attribute or time point, based on rerandomizations. Following this approach and in order to simplify the calculation and to be able to use standard statistical methods, the use of aggregated data in time intervals was proposed (Pineau et al. 2011). For each time period, the binomial data are summarized and the frequencies of attribute dominance by subject are computed. This approach that summarizes subject responses as frequency values in a given number of time periods was further validated and extended for estimating differences in attribute dominance among products, by investigation of the residuals of the ANOVA models (Dinnella et al. 2013).

13.3.2.3   Panel and Assessor Monitoring

Meyners (2011) proposed methods for testing panel and assessor agreement based on randomization tests. This author proposed to measure the disagreement as the interaction between assessor and product, that is, that one assessor systematically considers that an attribute is dominant for sample 1 and not for sample 2, while the rest of the assessors consider that the attribute is dominant for sample 2 and not for sample 1 or choose both or none of them. The interaction is estimated using randomization tests for estimating the differences between pair of samples. It is important to take into account that this approach has been claimed to have some computational burdens (Meyners 2011), suggesting that there is still a need to develop tools for assessing panel performance in TDS studies.

13.3.2.4   Other Data Analyses

Pineau et al. (2009) characterized TDS curves using maximum dominance rate, time to reach the maximum dominance from the beginning of tasting, and the period of time during which dominance was at least 90% of the maximum dominance value.

In order to simultaneously take into account the duration and dominance of each attribute and to get a measure of total dominance of an attribute during evaluation, Bruzzone et al. (2013) calculated the area under the TDS curves and above the significance level for each attribute and sample. This approach is similar to that used in TI methods for calculating the total intensity of an attribute during evaluation (Cliff and Heymann 1993). Bruzzone et al. (2013) used PCA on the correlation matrix of the areas under TDS curves to get a sample map based on attribute dominance.

When attribute intensity is measured, TDS scores have been calculated as a measure of the intensity of dominant attributes. For each attribute, TDS scores are calculated as follows (Labbe et al. 2009):

$$TDS\,score = \frac{\sum_{scoring} intensity \times duration}{\sum_{scoring} duration}$$

On the calculated scores, ANOVA, PCA, or canonical variate analysis (CVA) can be used to study differences among samples (Albert et al. 2012; Labbe et al. 2009).

13.3.3   Example of Application

Texture perception of semisolid foods depends on the product's initial rheological properties and the mechanical and structural changes that occur during oral manipulation (Van Vliet 2002). Considering that these processes are dynamic and occur simultaneously during consumption and that texture is a multiparameter sensory property (Szczesniak 2002), TDS seems an interesting tool for studying the temporal aspects of texture perception and to better understand this sensory property.

Bruzzone et al. (2013) used TDS to characterize the texture of eight yogurts, formulated with different fat, modified starch, and gelatin concentration. A sensory panel, composed of 10 trained assessors, evaluated the samples in duplicate using TDS. A list of five attributes was considered during the evaluations: thickness, gelatinousness, creaminess, melting, and mouthcoating.

Figure 13.7 shows TDS curves for two clearly different samples (1 and 2). The samples clearly differed in their first in-mouth impact. The first dominant attribute for sample 1 was creaminess, while for sample 2, the first dominant attribute was thickness. Besides, sample 2 was less complex regarding its dominance profile than sample 1, since it only showed thickness as the significantly dominant attribute during consumption. Differences between samples were explained by their formulation. Sample 2 was formulated with whole pasteurized milk, 1% modified starch, and 0.5% gelatin, which explains its high thickness dominance. On the other hand, sample 1 was formulated with skimmed pasteurized milk, 1% modified starch, and without gelatin. Compared to sample 2, the fact that sample 1 was characterized by the high dominance rate of creaminess can be explained by the lack of gelatin. This ingredient is usually associated with a gel-like structure due to its interaction with the casein matrix of yogurt (Fiszman et al. 1999).

When comparing TDS with descriptive analysis for characterizing the texture of yogurt samples, Bruzzone et al. (2013) reported that although both methodologies provided similar information regarding similarities and differences among samples, some of the information provided by TDS was not available in the analysis of data of descriptive analysis. According to these authors, the largest differences between descriptive analysis and TDS were found for complex attributes that are perceived throughout consumption, such as creaminess and mouthcoating. For example, the addition of starch (in a concentration of 1%) significantly increased creaminess dominance rate but not its intensity. This suggests that the extent to which creaminess caught assessors' attention was larger in samples formulated with modified starch, which can have a positive influence on consumer's hedonic perception.

13.3.4 POTENTIALITIES AND CHALLENGES

TDS is a relatively novel methodology for assessing the dynamics of sensory perception. It evaluates samples based on attribute dominance,

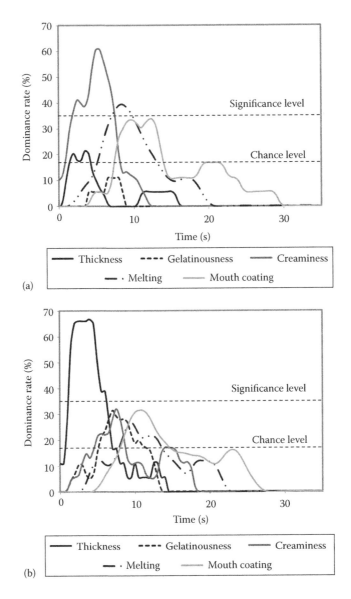

FIGURE 13.7   TDS curves for the evaluation of yogurt texture of (a) sample 1 and (b) sample 2.

which is a new approach within sensory and consumer science. As several studies have shown, the attribute that catches assessors' attention at a given time is not necessarily the most intense (Bruzzone et al. 2013; Labbe et al. 2009; Meillon et al. 2009), which suggests that information provided by TDS may not be obtained using classical descriptive analysis or TI methods, particularly when dealing with products with complex and dynamic sensory characteristics. Therefore, TDS may provide information about aspects of sensory perception that has not been extensively studied. Considering that TDS is focused on dominant attributes of the perception instead of quantifying attribute intensity, results from this methodology could better explain consumers' perception or the sensations that determine their hedonic perception. Studies should be carried out to investigate this hypothesis in different product categories.

Research is still necessary to develop methodological guidelines for the implementation of TDS, particularly regarding the minimum number of assessors and replications for obtaining reliable data, as well as the development of standard and simple approaches for data analysis.

One very promising area of research is the dynamics of consumer liking, as oral process has been identified as a very important factor in consumer appreciation. Some recent works have tried to relate liking to dynamic sensory perception. Laguna et al. (2013) studied the oral trajectory of biscuits with a trained panel via TDS and related it to consumers' penalties using JAR scales with different attributes driving liking. Similarly, Varela et al. (2013) combined dynamic techniques with consumer sensory description by check-all-that-apply (CATA) and liking scores to have a better understanding of which particularities of the oral process drove consumer liking during ice cream consumption. Further into this, TDS has been recently shown to provide additional insight about drivers of liking in a preference mapping context (Paulsen et al. 2013). Through this technique, it was possible to identify the time of dominance of certain attributes and the number of perceived sensations and blends in specific time periods as potential drivers of liking and disliking. They also showed that TDS has a potential to provide complementary information to classical descriptive techniques about the dynamic nature of sensory interactions in food–food combinations. At the recent *10th Pangborn Symposium*, Pineau (2013) proposed that the application of multiple-bite or multiple-sip TDS had a great potential to better understand the dynamics of sensory perception. In this sense, the application of this approach with consumers opens new challenges for studying the dynamics of preferences and how it relates with the dynamics of sensory perception.

13.4 CONCLUSIONS

When facing any research problem, the first step is to select the best method to approach it (Lawless and Heymann 2010). So, the first question to answer is whether the product set under study is likely to have a dynamic profile and which is the best methodology to identify the temporal profile of the products. In this sense, it is important to take into account that TI and TDS have different principles and therefore are appropriate for fulfilling different objectives.

When comparing TI and TDS methodologies, Pineau et al. (2009) reported that both methods provided similar information regarding differences among products, attributes, and evolution over time but also showed that TI and TDS are suitable for different needs. TDS is better when several attributes need to be compared and provides information about the sequence in which the attributes are perceived, as well as about the interaction among attributes. However, it is not well suited to obtain a TI profile for each attribute over time. When the kinetics of specific attributes are of interest, TI seems to be the best option (Pineau et al. 2009).

ACKNOWLEDGMENTS

The authors are grateful to CAPES-Brazil for the scholarship granted to Rafael Silva Cadena and to the Spanish Ministry of Science and Innovation for the financial support (AGL2009-12785-C02-01) and for the contract awarded to author Paula Varela (Juan de la Cierva Programme). They would also like to thank the Comisión Sectorial de Investigación Científica (CSIC, Universidad de la República).

REFERENCES

Albert, A., A. Salvador, P. Schlich, S. Fiszman. 2012. Comparison between temporal dominance of sensations (TDS) and key-attribute sensory profiling for evaluating solid food with contrasting textural layers: Fish sticks. *Food Quality and Preference* 24:111–118.

Barylko-Pikielna, N., I. Matuszewska, U. Hellemann. 1990. Effect of salt on time-intensity characteristics of bread. *LWT—Food Science and Technology* 23:422–426.

Birch, G. G., S. L. Munton. 1981. Use of 'SMURF' in taste analysis. *Chemical Senses* 6:45–52.

Brown, W. E., K. R. Landgley, A. Martin, H. J. H. MacFie. 1994. Characterisation of patterns of chewing behavior in human subjects and their influence on texture perception. *Journal of Texture Studies* 15:33–48.

Bruzzone, F., G. Ares, A. Giménez. 2013. Temporal aspects of yoghurt texture perception. *International Dairy Journal* 29:124–134.

Cadena, R. S., H. M. A. Bolini. 2011. Time–intensity analysis and acceptance test for traditional and light vanilla ice cream. *Food Research International* 44:677–683.

Chung, S., S.-Y. Lee. 2012. Modification of ginseng flavors by bitter compounds found in chocolate and coffee. *Journal of Food Science* 77:S202–S210.

Civille, G. V., A. S. Szczesniak. 1973. Guidelines to training a texture profile panel. *Journal of Texture Studies* 4:204–223.

Clark, C. C., H. T. Lawless. 1994. Limiting response alternatives in time–intensity scaling: An examination of the halo-dumping effect. *Chemical Senses* 19:583–594.

Cliff, M. 1987. Temporal perception of sweetness and fruitiness and their interaction in model system. MS thesis, University of California, Davis, CA.

Cliff, M., H. Heymann. 1993. Development and use of time-intensity methodology for sensory evaluation: A review. *Food Research International* 26:375–385.

Courregelongue, S., P. Schlich, A. C. Noble. 1999. Using repeated ingestion to determine the effect of sweetness, viscosity and oiliness on temporal perception of soymilk astringency. *Food Quality and Preference* 10:273–279.

De Morais, E. C., A. G. Cruz, H. M. A. Bolini. 2013. Gluten-free bread: Multiple time-intensity analysis, physical characterisation and acceptance test. *International Journal of Food Science and Technology* 48:2176–2184.

De Souza, V. R., P. A. P. Pereira, A. C. M. Pinheiro, H. M. A. Bolini, S. V. Borges, F. Queiroz. 2013. Analysis of various sweeteners in low-sugar mixed fruit jam: Equivalent sweetness, time-intensity analysis and acceptance test. *International Journal of Food Science and Technology* 48:1541–1548.

Delarue, J., E. Loescher. 2004. Dynamics of food preferences: A case study with chewing gums. *Food Quality and Preference* 15:771–779.

Dijksterhuis, G., P. Eilers. 1997. Modeling time–intensity curves using prototype curves. *Food Quality and Preference* 8:131–140.

Dijksterhuis, G., M. Flipsen, P. Punter. 1994. Principal component analysis of TI-curves: Three methods compared. *Food Quality and Preference* 5:121–127.

Dijksterhuis, G., J. R. Piggott. 2001. Dynamic methods of sensory analysis. *Trends in Food Science & Technology* 11:284–290.

Dinnella, C., C. Masi, T. Naes, E. Monteleone. 2013. A new approach in TDS data analysis: A case study on sweetened coffee. *Food Quality and Preference* 30:33–46.

Dinnella, C., C. Masi, G. Zoboli, E. Monteleone. 2012. Sensory functionality of extra-virgin olive oil in vegetable foods assessed by temporal dominance of sensations and descriptive analysis. *Food Quality and Preference* 26:141–150.

Duizer, L. M., K. Bloom, C. J. Findlay. 1997. Dual-attribute time–intensity sensory evaluation: A new method for temporal measurement of sensory perceptions. *Food Quality and Preference* 8:261–269.

Eilers, P. H. C., G. B. Dijksterhuis. 2004. A parametric model for time–intensity curves. *Food Quality and Preference* 15:239–245.

Esti, M., M. Contini, E. Moneta, F. Sinesio. 2009. Phenolics compounds and temporal perception of bitterness and pungency in extra-virgin olive oils: Changes occurring throughout storage. *Food Chemistry* 113: 1095–1100.

Fischer, U., R. B. Boulton, A. C. Noble. 1994. Physiological factors contributing to the variability of sensory assessments: Relationship between salivary flow rate and temporal perception of gustatory stimuli. *Food Quality and Preference* 5:55–64.

Fiszman, S. M., M.A. Lluch, A. Salvador. 1999. Effect of addition of gelatin on microstructure of acidic milk gels and yoghurt and their rheological properties. *International Dairy Journal* 9:895–901.

Fuentes, V., J. Ventanas, D. Morcuende, S. Ventanas. 2013. Effect of intramuscular fat content and serving temperature on temporal sensory perception of sliced and vacuum packaged dry-cured ham. *Meat Science* 93:621–629.

Garrido, D., A. Calviño, G. Hough. 2001. A parametric model to average time-intensity data. *Food Quality and Preference* 12:1–8.

Garruti, D. S., M. R. B. Franco, M. A. A. P. Silva, N. S. Janzantii, G. L. Alves. 2003. Evaluation of volatile flavour compounds from cashew apple (*Anacardium occidentale* L.) juice by the Osme gas chromatography/olfactometry technique. *Journal of the Science of Food and Agriculture* 83:1455–1462.

Guinard, J. X., R.-M. Pangborn, C. F. Shoemaker. 1985. Computerized procedure for the time-intensity sensory measurement. *Journal of Food Science* 50:543–546.

Hillmann, H., J. Mattes, A. Brockhoff, A. Dunkel, W. Meyerhof, T. Hofmann. 2012. Sensomics analysis of taste compounds in balsamic vinegar and discovery of 5-acetoxymethyl-2-furaldehyde as a novel sweet taste modulator. *Journal of Agricultural and Food Chemistry* 60:9974–9990.

ISO. 1994. *Sensory Analysis—Methodology—Texture Profile, ISO 11036*. Geneva, Switzerland: International Organization for Standardization.

ISO. 2008. *Sensory Analysis—Vocabulary, ISO 5492*. Geneva, Switzerland: International Organization for Standardization.

Kuesten, C., B. Jian, F. Yaohua. 2013. Exploring taffy product consumption experiences using a multi-attribute time–intensity (MATI) method. *Food Quality and Preference* 30:260–273.

Labbe, D., P. Schlich, N. Pineau, F. Gilbert, N. Martin. 2009. Temporal dominance of sensations and sensory profiling: A comparative study. *Food Quality and Preference* 20:216–221.

Laguna, L., P. Varela, A. Salvador, S. Fiszvman. 2013. A new sensory tool to analyse the oral trajectory of biscuits with different fat and fibre contents. *Food Research International* 51:544–553.

Larson-Powers, N. L., R.-M. Pangborn. 1978. Paired comparison and time-intensity measurements of the sensory properties of beverages and gelatins containing sucrose or synthetic sweeteners. *Journal of Food Science* 43:41–46.

Lawless, H. T., H. Heymann. 2010. *Sensory Evaluation of Food: Principles and Practices*. New York: Springer.

Lawless, H. T., E. Z. Skinner. 1979. The duration and perceived intensity of sucrose taste. *Perception & Psychophysics* 3:180–184.

Lee, W. E. III. 1985. Evaluation of time-intensity sensory response using personal computer. *Journal of Food Science* 50:1750–1753.

Lee, W. E. III., R. M. Pangborn. 1986. Time–intensity: The temporal aspects of sensory perception. *Food Technology* 40:71–78, 82.

Lenfant, F., C. Loret, N. Pineau, C. Hartmann, N. Martin. 2009. Perception of oral food breakdown. The concept of sensory trajectory. *Appetite* 52:659–667.

Liu, Y. H., H. J. H. MacFie. 1990. Methods for averaging time–intensity curves. *Chemical Senses* 15:471–484.

MacFie, H. J. H., Y. H. Liu. 1992. Developments in the analysis of time–intensity curves. *Food Technology* 46:92–97.

McGowan, B. A., S.-Y. Lee. 2006. Comparison of methods to analyze time–intensity curves in a corn zein chewing gum study. *Food Quality and Preference* 17:296–306.

Meillon, S., C. Urbano, P. Schlich. 2009. Contribution of the Temporal Dominance of Sensations (TDS) method to the sensory description of subtle differences in partially dealcoholized red wines. *Food Quality and Preference* 20:490–499.

Meillon, S., D. Viala, M. Medel, C. Urbano, G. Guillot, P. Schlich. 2010. Impact of partial alcohol reduction in Syrah wine on perceived complexity and temporality of sensations and link with preference. *Food Quality and Preference* 21:732–740.

Meyners, M. 2011. Panel and panelist agreement for product comparisons in studies of temporal dominance of sensations. *Food Quality and Preference* 22:365–370.

Meyners, M., N. Pineau. 2010. Statistical inference for temporal dominance of sensations data using randomization tests. *Food Quality and Preference* 21:805–814.

Murat, C., M.-H. Bard, C. Dhalleine, N. Cayot. 2013. Characterisation of odour active compounds along extraction process from pea flour to pea protein extract. *Food Research International* 53:31–41.

Ng, M., J. B. Lawlor, S. Chandra, C. Chaya, L. Hewson, J. Hort. 2012. Using quantitative descriptive analysis and temporal dominance of sensations analysis as complementary methods for profiling commercial blackcurrant squashes. *Food Quality and Preference* 25:121–134.

Ovejero-López, I., R. Bro, W. L. P. Bredie. 2005. Univariate and multivariate modelling of flavour release in chewing gum using time-intensity: A comparison of data analytical methods. *Food Quality and Preference* 16:327–343.

Palazzo, A. B., H. M. A. Bolini. 2009. Multiple time-intensity and acceptance of raspberry-flavored gelatin. *Journal of Sensory Studies* 24:648–663.

Palazzo, A. B., M. A. R. Carvalho, P. Efraim, H. M. A. Bolini. 2011. The determination of isosweetness concentrations of sucralose, rebaudioside and neotame as sucrose substitutes in new diet chocolate formulations using the time-intensity analysis *Journal of Sensory Studies* 26:291–297.

Paulsen, M. T., T. Naes, O. Ueland, E. Rukke, M. Hersleth. 2013. Preference mapping of salmon–sauce combinations: The influence of temporal properties. *Food Quality and Preference* 27:120–127.

Peyvieux, C., G. Dijksterhuis. 2001. Training a sensory panel for TI: A case study. *Food Quality and Preference* 12:19–28.

Pineau, N. 2013. Extension of TDS to multi-bite evaluation. In *10th Pangborn Sensory Science Symposium*, August 11–15, 2013, Rio de Janeiro, Brazil.

Pineau, N., A. Goupil de Bouillé, M. Lepage et al. 2012. Temporal Dominance of Sensations: What is a good attribute list? *Food Quality and Preference* 26:159–165.

Pineau, N., T. Neville, M. Lepage. 2011. Panel performance tool for Temporal dominance of sensations studies. In *Ninth Pangborn Sensory Science Symposium*, Toronto, Ontario, Canada.

Pineau, N., P. Schlich, S. Cordelle et al. 2009. Temporal dominance of sensations: Construction of the TDS curves and comparison with time–intensity. *Food Quality and Preference* 20:450–455.

Pinheiro, A. C. M., C. A. Nunes, V. Vietoris. 2013. Sensomaker: A tool for sensorial characterization of food products. *Ciência e Agrotecnologia* 37:199–201.

R Development Core Team. 2007. *R: A Language and Environment for Statistical Computing*. Vienna, Austria: R Foundation for Statistical Computing.

Rine, S. 1987. Computerized analysis of the sensory properties of peanut butter. MS thesis, University of California, Davis, CA.

Rocha-Selmi, G. A., E. C. Morais, C. S. Fávaro-Trindade, H. M. A. Bolini. 2012. Sensory evaluation of commercial chewing gums and the relations between the consumer sensory acceptance with time-intensity parameters of sweetness perceptions. In *Proceedings of 16th World Congress of Food Science and Technology*, August 5–9, 2012, Foz do Iguaçu, Brazil.

Rosner, B. 1995. *Fundamentals of Biostatistics*, 4th edn. Belmont, CA: Duxbury Press.

Sampaio, K. L., A. C. T. Biasoto, E. J. N. Marques, E. A. C. Batista, M. A. A. P. Silva. 2013. Dynamics of the recovery of aroma volatile compounds during the concentration of cashew apple juice (*Anacardium occidentale* L.). *Food Research International* 51:335–343.

Schlich, P. 2013. TDS with untrained consumers in lab and at home. In *10th Pangborn Sensory Science Symposium*, August 11–15, 2013, Rio de Janeiro, Brazil.

Schmitt, D. J., L. J. Thompson, D. M. Malek, J. H. Munroe. 1984. An improved method for evaluating time-intensity data. *Journal of Food Science* 49:539–542.

Sjöström, L. B. 1954. The descriptive analysis of flavor. In *Food Acceptance Testing Methodology*, eds. D. Peryan, F. J. Pilgrim, and M. Peterson, pp. 25–61. Chicago, IL: Quartermaster Food & Container Studies.

Sokolowsky, M., U. Fischer. 2012. Evaluation of bitterness in white wine applying descriptive analysis, time-intensity analysis, and temporal dominance of sensations analysis. *Analytica Chimica Acta* 732:46–52.

Sudre, J., N. Pineau, C. Loret, N. Martin. 2012. Comparison of methods to monitor liking of food during consumption. *Food Quality and Preference* 24:179–189.

Szczesniak, A. S. 2002. Texture is a sensory property. *Food Quality and Preference* 13:215–225.

Takagaki, M., Y. Asakura. 1984. Taste time-intensity measurements using a new computerized system. In *Proceedings of the 18th Japanese Symposium on Taste and Smell (JASTS)*, ed. K. Ueda, November 12–13, 1984, pp. 105–108, Tokyo, Japan.

Taylor, D. E., R.-M. Pangborn. 1990. Temporal aspects of hedonic responses. *Journal of Sensory Studies* 4:241–247.

Teillet, E., P. Schlich, C. Urbano, S. Cordelle, E. Guichard. 2010. Sensory methodologies and the taste of water. *Food Quality and Preference* 21:967–976.

van Buuren, S. 1992. Analyzing time intensity responses in sensory evaluation. *Food Technology* 46:101–104.

Van Vliet, T. 2002. On the relation between texture perception and fundamental mechanical parameters for liquids and time dependent solids. *Food Quality and Preference* 13:227–236.

Varela, P., A. Pintor, S. Fiszman. 2013. How hydrocolloids affect the temporal oral perception of ice cream. *Food Hydrocolloids* 36:220–228.

Ventana, S., E. Puolanne, H. Tuorila. 2010. Temporal changes of flavour and texture in cooked bologna type sausages as affected by fat and salt content. *Meat Science* 85:410–419.

Weel, K. G. C., A. E. M. Boelrijk, A. C. Alting et al. 2002. Flavor release and perception of flavored whey protein gels: Perception is determined by texture rather than by release. *Journal of Agricultural and Food Chemistry* 50:5149–5155.

Wendin, K., H. Janestad, G. Hall. 2003. Modeling and analysis of dynamic sensory data. *Food Quality and Preference* 14:663–671.

Yoshida, M. 1986. A microcomputer (PC 980l/MS mouse) system to record and analyze time-intensity curves of sweetness. *Chemical Sense* 11:105–118.

Zimoch, J., C. J. Findlay. 1998. Effective discrimination of meat tenderness using dual attribute time intensity. *Journal of Food Science* 63:940–944.

Zimoch, J., E. A. Gullett. 1997. Temporal aspects of perception of juiciness and tenderness of beef. *Food Quality and Preference* 8:203–211.

14 Comparison of Novel Methodologies for Sensory Characterization

Gastón Ares and Paula Varela

CONTENTS

14.1 COMPARISON OF NOVEL METHODOLOGIES FOR SENSORY CHARACTERIZATION

Interest in novel methodologies for sensory characterization has remarkably increased in the last decade. Several studies have applied these methodologies with trained and untrained assessors for sensory characterization of a wide range of products with different sensory complexity, ranging from mineral water (Teillet et al. 2010) to cosmetic products (Parente et al. 2011) or complex foods such as wine (Perrin et al. 2008) or fish nuggets (Albert et al. 2011). All published studies have reported that novel methodologies are quick and simple alternatives to gather valid and reliable information about the sensory characteristics of products. In this chapter, novel methodologies for sensory characterization are compared on the basis of the following aspects:

- Cognitive processes involved in sample evaluation
- Validity of product spaces
- Reliability of product spaces
- Discriminating ability
- Vocabulary
- Practical issues
- Number of samples
- Number of assessors
- Time and resources needed for implementation

14.1.1 Cognitive Processes Involved in Sample Evaluation

Novel methodologies for sensory characterization largely differ in how assessors evaluate samples, which has a strong influence on the cognitive processes involved in sample evaluation. This leads to differences in the information they provide about products and the applications for which they are more appropriate.

According to the type of task assessors are asked to perform, methodologies can be classified in five main groups: (1) methodologies based on the evaluation of specific attributes (check-all-that-apply [CATA] questions, free-choice profiling, and flash profile); (2) methodologies based on the temporal dominance of specific sensory attributes (temporal dominance of sensations [TDS]); (3) methodologies that provide a verbal description of the products (open-ended questions); (4) holistic methodologies, based on global similarities and differences among products (sorting and projective mapping); and (5) methodologies based on the comparison of products with references (polarized sensory positioning [PSP]).

Attribute-based methodologies encourage assessors to focus their attention on multiple attributes and to analytically evaluate them (Prescott 1999; Prescott et al. 2011; Small and Prescott 2005). In this approach, the products' sensory experience is evaluated through a set of individual attributes. This type of cognitive processing is also required for performing classical descriptive analysis, in which assessors are asked to quantify the intensity of a set of sensory attributes. Therefore, attribute-based methodologies rely on similar cognitive processes than descriptive analysis. All these methodologies provide detailed information about how assessors perceive specific sensory attributes of the products and about how samples differ in those attributes. The basic difference between novel methodologies and descriptive analysis is that the former save time and resources by reducing to a different extent the steps related to the generation of the attributes and by minimizing or eliminating the training in how to evaluate them.

Although CATA questions rely on the evaluation of specific attributes, the cognitive processes involved in sample evaluation differ from those involved in other attribute-based methodologies. When completing a CATA question, assessors have to check all the terms they consider appropriate for describing the product from a list that contains sensory attributes that are both applicable and not applicable to describe it. Therefore, they do not need to focus their attention on each of the attributes, which reduces the cognitive effort needed to complete the task (Rasinski et al. 1994; Smyth et al. 2006).

An important difference between CATA and other attribute-based methods, such as free-choice profiling and flash profile, is that the sensory attributes included in the list of terms are selected by the researcher. Therefore, the evaluation of attributes that are not selected by the assessor may affect how assessors perceive the product by making them focus their attention on attributes that may not be relevant or spontaneously generated. In this sense, one of the key issues when designing CATA questions is the selection of the list of terms and assuring that consumers understand them.

TDS also requires a different cognitive processing although it is based on the evaluation of specific attributes. In TDS, assessors are asked to evaluate the dominance of the attributes while consuming the product, that is, to indicate the attribute that catches their attention in real time. Although assessors are sometimes asked to rate the intensity of the dominant attribute (Labbe et al. 2009), the cognitive process involved in the evaluation is different from that of classic descriptive analysis since they have to evaluate the most salient attribute at each moment.

Holistic methodologies are based on the perception of global similarities and differences among products and encourage the generation of a global representation of the products, which is usually inhibited when assessors are asked to focus their attention on multiple attributes (Prescott 1999; Small and Prescott 2005). Reference-based methodologies require the same type of cognitive process since consumers are asked to evaluate global differences and similarities between samples and each of the reference samples. In these approaches, assessors evaluate the sensory experience of the products as a whole, and only after the evaluations of all the samples are completed, assessors are asked to think about specific sensory characteristics.

Regarding open-ended questions, they require consumers to evaluate products and provide a verbal description; therefore, this methodology does not encourage assessors to focus their attention on predefined attributes and can enable them to gain a synthetic representation of the products.

Holistic methodologies and those that provide a verbal description of the products enable to identify the most salient sensory characteristics of the products and the main attributes responsible for differences in how samples are perceived by assessors without forcing consumers to focus on specific sensory characteristics.

14.1.2 VALIDITY OF PRODUCT SPACES

When selecting methodologies for sensory characterization, the main feature that has to be assured is that they provide reliable and valid information. Validity is related to the extent to which the methodology measures what it is intended to (Carmines and Zeller 1979). In the context of sensory characterization, it can be argued that a methodology is valid if it provides *true* information about the sensory characteristics of the evaluated products. One way of confirming the validity of a new methodology for sensory characterization is by comparing results with those provided by a reference method. Considering the widespread use of descriptive analysis with trained assessor panels in academia and industry (Lawless and Heymann 2010), it is reasonable to consider this methodology as a reference for obtaining valid and reliable sensory characterizations.

Sensory characterizations obtained with classical descriptive analysis and novel methodologies have been compared in several studies dealing with a wide range of products of different sensory complexity. Results from descriptive analysis and novel methodologies have been compared on the basis of visual comparison of product spaces (Hopfer and Heymann 2012), sample grouping from hierarchical cluster analysis (Moussaoui and Varela 2010), multiple factor analysis (Dooley et al. 2010), and RV coefficients (Dehlholm et al. 2012b). Based on these approaches, most studies have reported that classical descriptive analysis and novel methodologies provide similar sensory maps, as well as similar information regarding the main sensory characteristics responsible for differences in how products are perceived. These results have been reported for free-choice profiling (Guàrdia et al. 2010; Jack and Piggot 1991; Moussaoui and Varela 2010), flash profile (Albert et al. 2011; Dairou and Sieffermann 2001; Dehlholm et al. 2012b; Delarue and Sieffermann 2004; Moussaoui and Varela 2010), CATA questions (Ares et al. 2010a; Bruzzone et al. 2012; Dooley et al. 2010), open-ended questions (Ares et al. 2010c; Symoneaux et al. 2012; ten Kleij and Musters 2003), projective mapping (Albert et al. 2011; Dehlholm et al. 2012b; Hopfer and Heymann 2012; Moussaoui and Varela 2010;

Risvik et al. 1994, 1997), sorting (Cartier et al. 2006; Dehlholm et al. 2012b; Faye et al. 2006; Moussaoui and Varela 2010), and PSP (Teillet et al. 2010). These results support the validity of all the recently developed methodologies and suggest that they can be used for gathering information about the sensory characteristics of products with different complexity.

Systematic differences in the validity of the methodologies have not been reported. However, several studies have shown that the information provided by attribute-based methodologies, such as flash profile, is more similar to results from descriptive analysis than the information provided by holistic approaches. This can be explained by the fact that descriptive analysis and attribute-based methodologies are based on the evaluation of specific sensory characteristics, whereas holistic methodologies rely on the evaluation of global similarities and differences among samples. Blancher et al. (2007) reported that flash profile provided sample configurations more similar to those from descriptive analysis than free sorting. Similarly, Moussaoui and Varela (2010) compared descriptive analysis with a trained assessor panel with four methodologies with naïve subjects (free-choice profiling, flash profile, sorting, and projective mapping) for sensory characterization of hot beverages. These authors reported that although all the methodologies provided similar product spaces, flash profile and free-choice profiling showed the highest correlation with descriptive analysis. Furthermore, when working with fish nuggets, Albert et al. (2011) reported that the RV coefficient between the configurations provided by flash profile and descriptive analysis was higher than the RV coefficient between descriptive analysis and projective mapping. Nevertheless, Dehlholm et al. (2012b) compared descriptive analysis, global napping, partial napping, sorting, and flash profile with two trained panels for sensory characterization of liver pâté. These authors reported that the highest similarity between sample configuration obtained from descriptive analysis and novel methodologies was found for partial napping, followed by sorting.

It should be noted that holistic methodologies are based on an integrated evaluation of the sensory characteristics of the product instead of the evaluation of specific sensory characteristics. This type of evaluation can be regarded as more natural, less analytic, and more representative of consumer evaluation of products during purchase or consumption than that performed in attribute-based methodologies. For this reason, it is expected that the sensory spaces obtained by holistic methodologies slightly differ from those obtained with descriptive analysis.

14.1.3   RELIABILITY OF PRODUCT SPACES

Novel methodologies for sensory characterization also need to provide reliable (i.e., precise) data. Reliability can be regarded as the extent to which a methodology yields the same results on repeated trials (Carmines and Zeller 1979).

The reliability of sensory characterizations from descriptive analysis is usually analyzed by monitoring the global performance of the panel, as well as the performance of each individual assessor. The main characteristics used for evaluating the reliability of descriptive analysis are panel discriminating ability, panel reproducibility, individual assessor agreement with the panel as a whole, individual assessor discriminating ability, and individual assessor reproducibility (Lawless and Heymann 2010). Several tools have been used for this purpose, including graphical tools (Tomic et al. 2007), analysis of variance (Brockhoff 2003), and multivariate techniques (Kermit and Lengard 2005). Although no standard procedure is available for evaluating the reliability of sensory characterizations obtained with novel methodologies, different approaches have been used. One common approach for estimating the reliability of product spaces from novel methodologies is to evaluate the repeatability of the panel as a whole.

Chollet et al. (2011), Hopfer and Heymann (2012), Lim and Lawless (2005), Nestrud and Lawless (2010), Moussaoui and Varela (2010), and Veinand et al. (2011) evaluated the repeatability of novel methodologies for sensory characterization at the panel level by comparing the position of a blind repeated sample on product spaces. These authors reported that repeated samples were located close to each other in sample spaces and concluded that flash profile, free-choice profiling, projective mapping, and sorting were reliable. However, some differences in the repeatability of the methodologies have been found. Moussaoui and Varela (2010) reported that flash profile was more repeatable than projective mapping and sorting, while the repeatability of projective mapping was higher than that of sorting tasks. However, it is important to take into account that when working with very similar samples, blind repeated sample could be placed apart from each other in the product space, which would not necessarily imply lack of reproducibility.

Another possible approach is using a test–retest paradigm, by comparing responses from the same group of respondents to the same sample set in different sessions. If all aspects of the empirical execution of the methodology were kept constant, differences between both sessions could be attributed to the passing of time (Yu 2005). At the panel level, several authors have compared sample configurations from different sessions to evaluate the repeatability of novel methodologies. These authors have compared RV coefficient between sample configurations obtained on different sessions and concluded that, at the panel level, CATA, flash profile, sorting, and projective mapping were repeatable (Cartier et al. 2006; Dairou and Sieffermann 2002; Falahee and MacRae 1997; Hopfer and Heymann 2012; Jaeger et al. 2013; Lawless and Glatter 1990). Regarding projective mapping, sample configurations from different sessions have been reported to strongly differ, but overall similarities and differences among samples were constant over repeated sessions (Barcenas et al. 2004; Hopfer and Heymann 2012; Kennedy 2010; Risvik et al. 1994).

Repeatability has been reported to be affected by training. Chollet and Valentin (2001) reported that untrained assessors were less repeatable than trained assessors when completing a sorting task with 12 beer samples. Using procrustes distances between sample configurations, these authors reported that untrained assessors were not very repeatable when evaluating the samples in two sessions separated by 20 min. The lack of repeatability was explained by the fatigue caused by tasting a large number of beer samples. Moreover, Hopfer and Heymann (2012) reported that trained assessors were more repeatable than untrained assessors for completing a projective mapping task with wine samples.

Comparison of sample configurations obtained with different panels has also been used to evaluate the repeatability of novel methodologies for sensory characterization. Blancher et al. (2007), Lelièvre et al. (2008), and Chollet et al. (2011) calculated the RV coefficient between sample configurations from different panels to evaluate the repeatability of sorting tasks. These authors concluded that sorting tasks performed with different panels provided similar information regarding similarities and differences among samples.

Faye et al. (2006) and Blancher et al. (2012) proposed a bootstrapping resampling approach for studying the reliability of sorting maps. Blancher et al. (2012) argued that a sorting map can be considered stable if sampling repeatedly from the population of interest provides equivalent sorting maps. They calculated the correlation of different random subsets of different sizes with the reference configuration (i.e., that obtained with all the assessors) through the RV and Mantel coefficients. The authors also calculated an average RV coefficient across different subsets as an indicator of the reliability of a sorting task. Blancher et al. (2012) argued that this approach can be extended to other methodologies and could consist of a simple tool for estimating the reliability of product spaces. This same approach has been used by Ares et al. (2014) to evaluate the stability of sample and term configurations from CATA questions and to study the minimum number of consumers needed to yield stable results.

Bootstrapping has also been used to draw confidence ellipses around samples on sensory spaces (Abdi et al. 2009; Dehlholm et al. 2012a). Using this approach, the discriminating ability of the methodology can be estimated. Two samples are regarded as significantly different if their confidence ellipses do not overlap.

The individual repeatability of the assessors has been also assessed for sensory characterizations obtained with sorting and projective mapping.

In general, results have shown that individual assessors provide repeatable evaluations. Falahee and MacRae (1997) used kappa statistics to evaluate the agreement between the sorting tasks performed by 20 trained assessors over 5 replicates. These authors reported that 2 assessors showed poor repeatability, 14 showed slight repeatability, 3 showed fair repeatability, while 1 assessor showed a moderate level of repeatability. Lelièvre et al. (2009) calculated the RV coefficient between different repetitions of sorting tasks performed by trained and untrained assessors, reporting that training increased the assessors' ability to provide repeatable results. Hopfer and Heymann (2012) proposed a people performance index to evaluate assessors' repeatability in projective mapping tasks when using blind repeated samples. This index was calculated as the Euclidean distance between blinded samples, divided by the maximum distance between different samples.

Tools for checking panel homogeneity have also been developed. One of the most common approaches for this purpose is to calculate the correlation between the configurations provided by individual assessors and those obtained with the whole panel (Lelièvre et al. 2009). Besides, when analyzing data from sorting tasks using DISTATIS, an indicator of global similarity of each panelist to the group average is obtained (Abdi et al. 2007). Faye et al. (2006) used Rand indexes and cluster analysis to identify groups of assessors who used different criteria to sort the products.

Although several approaches are available for analyzing reliability, further research is necessary to compare the reliability of novel methodologies for sensory characterization and to develop standard tools for checking the reliability of their results (Blancher et al. 2012; Valentin et al. 2012).

14.1.4   DISCRIMINATING ABILITY

Several studies have compared results from novel methodologies in a wide range of products. These studies have reported that all the methodologies provide comparable information regarding the main sensory characteristics of the products, as well as regarding similarities and differences among samples.

Several authors have reported that attribute-based and holistic methodologies provide similar results and have similar discriminating ability. Blancher et al. (2007) compared flash profile and sorting for sensory characterization of visual appearance and texture of jellies. They reported a high agreement between sample spaces provided by both methodologies (RV coefficient ranging from 0.68 to 0.73), although some few noticeable differences were identified. Similarly, Albert et al. (2011) worked with fish nuggets and reported that sample spaces provided by flash profile and projective mapping were highly correlated despite the fact that sample grouping from hierarchical cluster analysis provided slightly different results.

These authors stated that assessors tended to cluster samples according to their appearance when using flash profile, while they mainly took into account texture when grouping samples in the projective mapping task.

Ares et al. (2010b) compared CATA questions and projective mapping for sensory characterization of vanilla milk desserts and reported that sample spaces provided by both methodologies were highly comparable. Meanwhile, when comparing CATA questions, projective mapping, and sorting for sensory characterization of orange-flavored powdered drinks, Ares et al. (2011) stated that the three methodologies provided similar sample spaces (RV coefficients higher than 0.73). These authors reported that results from holistic methodologies differed from those provided by CATA questions in two of the seven samples. This difference was attributed to the fact that specific sensory characteristics not included in the CATA question were taken into account by assessors when evaluating those two samples using holistic methodologies.

On the other hand, some authors have reported that holistic methodologies are less discriminating than methodologies based on the evaluation of specific sensory attributes, particularly when working with small differences among samples. Veinand et al. (2011) compared free-choice profiling, flash profile, and projective mapping, for sensory characterization of lemon iced teas using consumers. These authors concluded that although sample spaces were similar, flash profile showed the highest discriminating ability, while projective mapping showed the lowest. Moussaoui and Varela (2010) reported that flash profile and free-choice profiling provided more accurate sample maps than similarity-based methodologies such as projective mapping and free sorting when working with hot beverages. Furthermore, Dehlholm et al. (2012b) compared flash profile, partial napping, sorting, and global napping for sensory characterization of nine commercial samples of liver pâté. These authors compared the discriminating ability of the methodologies based on size of the confidence ellipses, concluding that flash profile showed the highest discriminating ability, followed by partial napping, sorting, and finally global napping.

Holistic methodologies have been reported to provide similar information regarding the sensory characteristics of samples (Ares et al. 2011; Moussaoui and Varela 2010). Nestrud and Lawless (2010) compared projective mapping and free sorting for sensory characterization of apples and cheeses. They concluded that the sensory maps provided by the methodologies were similar for both products. However, sample grouping from projective mapping was easier to interpret based on the sensory characteristics of the samples than that provided by sorting.

It is important to highlight that no generalizations can be made regarding the superiority of any methodology in terms of its discriminating ability. Further research is needed to compare results from sensory characterization studies with different novel methodologies in product categories of different complexity.

14.1.5 VOCABULARY

Novel methodologies differ in their ability to generate vocabulary to describe the sensory characteristics of the products. CATA questions and ideal profile rely on the use of predefined sensory attributes; therefore, they are not appropriate for generating consumer-based vocabulary. These methodologies require preliminary studies for selecting the attributes to be included in the study.

On the contrary, in free-choice profiling, flash profile, open-ended questions, or holistic methodologies, assessors provide sensory terms to describe samples using their own words. These methodologies usually provide a rich vocabulary to describe the sensory characteristics of the evaluated products. However, analysis and interpretation of the sensory terms provided by assessors are difficult, time-consuming, and labor-intensive. Due to the heterogeneity of individual descriptions, the large number of terms used, and the lack of definitions and evaluation procedures, the sensory terms provided by assessors could be difficult to interpret. In particular, the interpretation of sensory terms can be difficult for complex sensations in which more than one modality is involved. Moussaoui and Varela (2010) reported that, when using projective mapping, sorting, and flash profile, the interpretation of the term *creaminess* was difficult since it was not clear if assessors were referring to flavor, texture, or aroma.

Several studies have reported that trained assessors have a greater ability to verbalize their sensory perception and that their descriptions are usually more reliable, specific, and consensual than those provided by consumers (Chollet and Valentin 2001; Chollet et al. 2005, 2011; Lawless et al. 1995; Lim and Lawless 2005; Saint-Eve et al. 2004; Soufflet et al. 2004). Differences between the descriptions provided by trained assessors and consumers have been attributed to the linguistic components involved in training, particularly for odor description (Lawless and Glatter 1990). Besides, consumers many times use hedonics or benefit-related terms to describe the products (Veinand et al. 2011). Although this can be seen as a limitation, this information can be useful to relate sensory characteristics to marketable features and consumer preference.

Several studies have reported that more detailed vocabularies are elicited with flash profile and free-choice profiling compared to holistic methodologies such as projective mapping or sorting (Albert et al. 2011; Blancher et al. 2007; Moussaoui and Varela 2010; Veinand et al. 2011). This difference has been attributed to the fact that attribute-based methodologies encourage assessors to focus their attention on specific characteristics, breaking down their sensory perception into recognizable sensory attributes (Dehlholm et al. 2012b; Valentin et al. 2012). On the other hand, holistic methodologies rely on the evaluation of global similarities and differences among samples. After assessors have completed the task, they are asked to provide words to describe the sensory characteristics of the samples. These different cognitive processes are responsible for the fact that free-choice profiling and flash profile provide more detailed and precise sensory descriptions (Blancher et al. 2007; Moussaoui and Varela 2010). Furthermore, Veinand et al. (2011) reported that projective mapping provided a richer sensory vocabulary than sorting, probably due to the fact that in the latter methodology, assessors focus their attention on groups of samples instead of on describing each individual sample.

14.1.6 Practical Issues

Apart from differing from a methodological perspective, novel methodologies also differ in practical issues related to their implementation. In particular, three aspects should be taken into account when designing sensory characterization studies: number of samples, number of assessors, and time and resources needed for implementation.

14.1.6.1 Number of Samples

Novel methodologies strongly differ in the number of samples that can be evaluated in a single session. Flash profile, free-choice profiling, free sorting, and projective mapping require assessors to compare samples; therefore, all samples should be evaluated in the same session. In order to avoid fatigue and adaptation, the number of samples should be limited. This is particularly important when working with projective mapping or sorting since samples are simultaneously evaluated to determine global similarities and differences among them. For this reason, it could be difficult to use these methodologies when working with products that require careful temperature control or that have intense and persistent sensory characteristics.

Moreover, some authors have reported that assessors can find it difficult to memorize the sensory characteristics of a large number of products. When assessors complete holistic methodologies, such as sorting or projective mapping, they have to memorize the characteristics of the samples in order to compare them and determine their global similarity. Therefore, as the number of samples increases, the greater the short-term memory load is (Chollet et al. 2011). According to Patris et al. (2007), trained and untrained assessors can have difficulties to memorize samples during a sorting task. In a recent study, Chollet et al. (2011) reported that increasing the size of the sample set decreased the discriminating ability of sorting tasks. These authors reported that a maximum of 20 beer samples were efficiently sorted in a single session. This maximum number of samples might strongly depend on the sensory complexity of the product.

Flash profile, free-choice profiling, and holistic methodologies rely on multivariate procedures for data analysis. Thus, a minimum number of 6 samples are usually considered. Taking into account the previously mentioned considerations, the number of samples to be evaluated using free-choice profiling, free sorting, or projective mapping ranges from 6 to 15 (Ares et al. 2010, 2011; Dairou and Sieffermann 2002; Jack and Piggott 1991; Nestrud and Lawless 2010; Pagès 2005). However, Tarea et al. (2007) evaluated the texture of 49 commercial apple and pear purees using flash profile in a single session that lasted between 2 and 5 h.

On the other hand, methodologies that rely on sequential monadic evaluation of samples, such as CATA questions, ideal profile, PSP or open-ended questions, do not require a minimum or maximum number of samples and evaluations can be separated into different sessions.

Approaches based on the comparison with references, such as PSP, are appropriate when working with large sample sets, when comparing products over time, or when dealing with evaluations performed on different sessions. These methodologies require the use of fixed reference products, which should be stable and easily available. However, it has to be taken into account that having to compare samples with references makes it more tiresome for the panelists.

14.1.6.2 Number of Assessors

The number of assessors required for sensory characterization using novel methodologies is more dependent on their training than on the specific methodology.

When sensory characterization is performed with trained assessors, the number of participants ranges from 9 to 30, whereas the number of consumers commonly used for sensory characterization ranges from 20 to 100. However, it is important to highlight that research aiming at studying the influence of the number of assessors on the stability of sample configurations from different methodologies is still necessary.

The usual number of trained assessors for sensory characterization ranges from 9 to 15 for sorting tasks (Cartier et al. 2006; Chollet et al. 2011) and projective mapping (Perrin et al. 2008; Risvik et al. 1994) and 6–12 trained or semitrained assessors for flash profile (Albert et al. 2011; Dairou and Sieffermann 2002; Delarue and Sieffermann 2004; Moussaoui and Varela 2010; Tarea et al. 2007).

When untrained assessors or naïve consumers are considered, the number of assessors increases. Sorting tasks have been performed with 9–98 consumers (Cadoret et al. 2009; Chollet et al. 2011). However, despite the variability in the number of untrained assessors considered in sorting tasks, most studies work with 20–50 consumers (Ares et al. 2011; Cartier et al. 2006; Falahee and MacRae 1997; Moussaoui and Varela 2010). Blancher et al. (2012) recommended to perform sorting tasks with 30 assessors and to check the stability of the maps using bootstrapping resampling techniques.

Meanwhile, flash profile and projective mapping have been commonly performed with 15–50 consumers (Albert et al. 2011; Ares et al. 2010b; Moussaoui and Varela 2010; Nestrud and Lawless 2008; Veinand et al. 2011).

CATA and open-ended questions for sensory characterization are mainly used with consumers. Therefore, the number of assessors necessary ranges from 50 to 100 (Ares et al. 2010a–c; Dooley et al. 2010; Parente et al. 2011; Symoneaux et al. 2012; ten Kleij and Musters 2003). Ares et al. (2013) recommended working with at least 60–80 consumers to get stable sample and term configurations. Campo et al. (2008, 2010) used a modified version of the CATA methodology to characterize the odor profile of wines. They asked a panel of approximately 30 trained assessors to choose a limited number of terms from a long list of sensory descriptors to describe wine aroma.

14.1.6.3   Time and Resources Needed for Implementation

Holistic methodologies are usually regarded as more intuitive and less rational than other methodologies based on the evaluation of specific sensory attributes. However, several authors have reported that they can be difficult to understand and perform. Ares et al. (2011) reported that although consumers were able to understand projective mapping and sorting tasks, they found them much significantly more difficult than CATA questions or intensity scales. Similarly, Ares et al. (2010b) reported that consumers needed further explanations to understand projective mapping compared to CATA questions. Moreover, Veinand et al. (2011) reported that projective mapping was more difficult to perform with consumers than flash profile. According to these authors, consumers found it difficult to use the sheet of paper to locate samples according to the similarities and differences among them.

Novel methodologies differ in the time needed for their implementation. Free-choice profiling requires two separate sessions, one for descriptor generation and a second one for sample evaluation. On the contrary, the rest of the methodologies can be completed in a single session. Dehlholm et al. (2012a) reported that trained assessors needed 120 min to complete a flash profile of liver pâté, compared to the 20 min they required for completing a projective mapping and the 30 min they needed for completing a sorting task.

Ideal profile, CATA questions, and open-ended questions are usually much less time-consuming than sorting tasks, projective mapping, free-choice profiling, flash profile, and PSP. According to Ares et al. (2010b), consumers took between 5 and 15 min to answer a CATA question for sensory characterization of eight milk desserts, whereas they needed between 18 and 25 min to complete a projective mapping task with the same samples. Moreover, Veinand et al. (2011) reported that assessors needed 70 min to perform free-choice profiling, 50 min to complete a flash profile, and 40 min for projective mapping. These authors also stated that assessors were less saturated and tired after completing a projective mapping task than after performing a flash profile.

Therefore, holistic methodologies seem to be slightly more difficult and time-consuming for consumers than attribute-based methodologies. Taking into account that trained assessors could more easily understand these methodologies, Veinand et al. (2011) recommended performing projective mapping with expert panelists. However, it is important to highlight that these methodologies have been used by naïve assessors, providing valid and reliable results (Albert et al. 2011; Ares et al. 2011; Moussaoui and Varela 2010; Veinand et al. 2011).

Another issue to consider when using projective mapping with paper ballots is that measuring the products' coordinates in the sheet of each assessor is tedious, tiresome, and time-consuming for the panel leader, particularly when working with a large number of consumers (Veinand et al. 2011). Besides, the software available for sensory evaluation does not enable the implementation of flash profile yet.

14.2 COMPARISON OF NOVEL METHODOLOGIES AND DESCRIPTIVE ANALYSIS

Many studies have shown that the product spaces gathered with descriptive analysis and novel methodologies are similar. However, it should be highlighted that the information provided by both types of methodologies is different.

Classical descriptive analysis provides a quantitative measure of the intensity of a set of specific sensory attributes by a trained assessor panel. In this methodology, the sensory attributes are precisely defined, assessors are trained in attribute recognition and scaling, they use a common and agreed sensory language, and products are scored on repeated trials (ASTM 1992). For these reasons, descriptive analysis provides detailed and accurate information about the sensory characteristics of the products. Data from descriptive analysis are analyzed by univariate parametric statistics such as analysis of variance, which makes it possible to identify small and subtle differences among samples in each of the evaluated attributes (Lawless and Heymann 2010). On the other hand, novel methodologies commonly use untrained assessors and do not rely on the evaluation of sensory attributes that have been previously defined and agreed among panelists.

Due to the previously mentioned differences, descriptive analysis usually provides accurate and detailed information than novel methodologies, having a higher discriminating ability to detect significant differences among samples in many situations (Cartier et al. 2006; Dehlholm et al. 2012b). Moreover, the interpretation of the sensory characteristics elicited by novel methodologies is usually difficult due to the large number of elicited terms, the lack of definitions, and standard evaluation protocols. Therefore, results from descriptive analysis are usually more actionable for product developers than those from novel methodologies. In general, descriptive analysis is usually more appropriate when evaluating similar samples, comparing different sample sets, or evaluating samples at different moments in time.

Although novel methodologies for sensory characterization are usually considered complementary to descriptive analysis with trained assessor panels, they can be regarded as interesting alternatives in many specific situations. Descriptive analysis is expensive and time-consuming. The length of the training process usually ranges from 10 to 120 h, depending on the complexity of the specific product and the number and characteristics of the sensory attributes needed to characterize the product (Dairou and Sieffermann 2002; Meilgaard et al. 1999). This makes it difficult to apply this methodology in many everyday situations in the food industry where there are constraints in terms of time and resources (Delarue and Siefferman 2004; Labbe et al. 2004).

When using descriptive analysis, the vocabulary and associated panel training must be adapted to each specific type of product, which requires substantial time necessary to get reliable results. In the context of today's highly competitive markets, the time available for new product development have become shorter. Therefore, in many occasions, it is not possible to use descriptive analysis during new product development of a novel product. Moreover, novel methodologies can be a valuable alternative to gather information about the sensory characteristics of products for companies that do not have resources to select, train, and maintain sensory panels for evaluating a specific product, which is common in small companies or developing countries. In these cases, the cost and time involved in the selection and training of the assessors might be higher than those needed to perform a consumer study with 30–150 participants. In these situations, novel methodologies can be quick, reliable, and cost-effective alternatives for gathering information about the sensory characteristics of a set of products. Novel methodologies can be particularly useful when the objective is to gather information about the most salient sensory attributes and the most relevant characteristics responsible for similarities and differences among products.

Besides, interest in gathering sensory information directly from the target consumers of products instead of the more technical descriptions provided by trained assessors has remarkably increased (Varela and Ares 2012). In most common approaches for product optimization, consumers are asked to rate their liking of a large set of products and the products' sensory properties are characterized by trained assessors (van Kleef et al. 2006). However, trained assessors may describe the product differently than consumers and/or evaluate attributes that may be irrelevant for them (ten Kleij and Musters 2003). Therefore, sensory characterization of products based on consumers can have greater external validity, which has contributed to increasing the popularity of novel methodologies for sensory characterization.

Furthermore, some methodological advantages of novel methodologies can also be highlighted. The use of common sensory attributes can be regarded as a disadvantage of descriptive analysis (Delarue and Sieffermann 2004), which is overcome by novel methodologies. Considering that a single stimulus can be perceived differently by subjects (Lawless 1999), forcing assessors to use common descriptors may lead to simplified sensory descriptions (Delarue and Sieffermann 2004). Therefore, methodologies that enable assessors to use their own sensory attributes according to their sensitivity and perception may enrich the sensory description, particularly when working with complex products (Albert et al. 2011).

In some cases, novel methodologies have been reported to provide better information than descriptive analysis. For example, Delarue and Sieffermann (2004) reported that flash profile showed a higher discriminating capacity among apricot fresh cheeses than descriptive analysis. This difference was attributed to the forced use of consensual attributes that might refer to different sensory concepts for assessors. Similarly, Albert et al. (2011) reported that flash profile with semitrained assessors provided a more detailed description of the sensory characteristics of fish nuggets than descriptive analysis. Teillet et al. (2010) reported that sorting and PSP were more discriminating than descriptive analysis for sensory characterization of mineral waters. Besides, Symoneaux et al. (2012) reported that results from an open-ended question with consumers showed better discriminating power than descriptive analysis on the juiciness of apple samples.

Regarding TDS, it provides information about the evolution of sensory sensations of products that cannot be obtained when using conventional descriptive analysis. Descriptive analysis involves the evaluation of a set of attributes immediately after perception or after a predetermined period (ASTM 1992). However, sensory perception is not a single event but a dynamic process in which a series of simultaneous events occur

(Piggott 1994, 2000). TDS studies the sequence of dominant sensations of a product during a certain time period (Pineau et al. 2009). Some studies have reported that TDS provides information that is not emphasized by QDA (Bruzzone et al. 2013; Labbe et al. 2009; Meillon et al. 2009), due to the fact that the attribute that catches assessors' attention at a given time is not necessarily the most intense. Therefore, QDA and TDS provide information about different aspects of sensory perception, particularly when working with complex and dynamic sensory characteristics.

14.3 CRITERIA FOR SELECTING NOVEL METHODOLOGIES FOR SENSORY CHARACTERIZATION

As discussed in the previous sections, no clear differences exist in the validity and reliability of novel methodologies for sensory characterization. Generally speaking, they all provide similar information regarding the sensory characteristics of the products and the differences that exist among the samples. Therefore, the selection of novel methodologies basically depends on the objective of the study and practical considerations.

14.3.1 Objective of the Study

The main criterion for selecting a methodology for a particular application is the specific aim of the study. If detailed information about the sensory characteristics of the products is necessary, attribute-based methodologies are recommended. In these methodologies, assessors focus their attention on specific sensory characteristics, and therefore, they provide rich and detailed descriptions. A relevant difference exists among attribute-based methodologies. Flash profile and free-choice profiling do not require prior selection of sensory attributes since assessors generate their own vocabulary, being the most appropriate methodologies for generating consumer vocabulary to describe samples. On the other hand, the ideal profile method or CATA questions rely on the evaluation of sensory attributes that are previously selected by the researcher based on preliminary information.

CATA and open-ended questions are simple methodologies that are strongly recommended when information about the sensory characteristics of products is obtained concurrently with hedonic scores.

The analytical evaluation performed in attribute-based methodologies can be regarded as *artificial* since consumers do not evaluate a set of specific attributes when consuming a product. Therefore, methodologies based on verbal descriptions or holistic perception would better reflect consumers' synthetic evaluation of products. These methodologies provide a measure of the global similarities and differences among samples.

Besides, holistic methodologies are also recommended when evaluating complex multimodal sensory properties. According to Saint-Eve et al. (2004), complex texture–flavor interactions should be better evaluated as a unique sensory experience instead of a series of independent sensory characteristics.

It should be taken into account that the use of language can influence the perceptual representation of products (Blancher et al. 2007). Therefore, holistic methodologies could be an interesting alternative when conducting cross-cultural studies since this type of methodology does not require translating sensory terms.

PSP is the recommended approach when results from sensory characterization are going to be compared over time or over different sessions.

TDS is clearly recommended when the objective of the study is to obtain a dynamic sensory characterization, that is, to characterize samples based on the evolution of their sensory characteristics throughout consumption.

14.3.2 Type of Assessors

The selection of a novel methodology also depends on the type of assessors to be considered in the study. When working with consumers, it would be generally easier to work with simple methodologies such as CATA questions, open-ended questions, or ideal profile. On the other hand, when a trained assessor panel is available and quick information about the sensory characteristics of food products is needed, the recommended approach would be to apply flash profile, sorting, projective mapping, or PSP due to their higher complexity.

14.3.3 Practical Considerations

Another relevant issue that has to be taken into account when selecting novel methodologies is the time and resources available for their implementation. As previously discussed, the methodologies strongly differ in the time required by assessors to complete the task, as well as the tediousness of data analysis. In this sense, it is important to take into account that free-choice profiling requires two separate sessions for its implementation, whereas the time needed for analyzing data from open-ended questions can be quite long. Also, when projecting mapping is performed on paper ballots, the data measuring and data input steps are long and tedious for the panel leader.

A summary of the main characteristics of the methodologies, which can be useful for selecting the most appropriate for a particular application, is included in Table 14.1.

TABLE 14.1

Summary of the Main Characteristics of Novel Methodologies for Sensory Characterization

Methodology	Type of Methodology	Task	Minimum Number of Sessions	Usual Number of Samples	Vocabulary	Difficulty
Free-choice profiling	Attribute-based	Rating of specific sensory attributes	2	6–15	Elicited by assessors	Medium
Flash profile	Attribute-based	Ranking of specific sensory attributes	1	6–15	Elicited by assessors	Medium
Ideal profile	Attribute-based	Rating of specific sensory attributes	1	1–10	Provided by the researcher	Low
CATA questions	Attribute-based	Selection of sensory terms from a list	1	1–8	Provided by the researcher	Low
Open-ended questions	Verbal description	Verbal description of samples	1	1–8	Elicited by assessors	Low
Sorting	Holistic	Classification of differences based on their similarities and differences	1	6–15	Elicited by assessors	Medium
Projective mapping	Holistic	Locating samples on a 2D map according to their similarities and differences	1	6–15	Elicited by assessors	Medium
PSP	Comparison with references	Evaluation of global differences between samples and fixed references	1	6–15	Elicited by assessors	Medium
TDS	Temporal evaluation of attributes	Selection of the dominant attribute at each moment	1	1–8	Provided by the researcher	Medium

14.4 CONCLUSIONS

Novel methodologies are simple and quick alternatives for sensory characterization of food products with trained and untrained assessors. They have been reported to provide valid and reliable information, similar to that gathered with classical descriptive analysis performed with trained assessor panels.

However, novel methodologies cannot be regarded as a replacement for classical descriptive analysis since this methodology is always more accurate due to the fact that assessors are extensively trained in the identification and quantification of clearly defined sensory attributes.

Recently developed methodologies enable to perform sensory characterizations with consumers, which could be useful for uncovering consumer perception of food, in their own vocabulary. This information could provide valuable information during new food product development or when designing marketing or communication campaigns, which could have greater external validity than classical external preference mapping approaches.

Finally, it is important to take into account that, unlike classical descriptive analysis, most novel methodologies for sensory characterization have been used for a relatively short period of time and have been used in a limited number of applications. For this reason, further research on the applicability, reliability, and reproducibility of new approaches for sensory characterization is still strongly needed, particularly when dealing with complex products.

ACKNOWLEDGMENTS

The authors are grateful to the Spanish Ministry of Science and Innovation for financial support (AGL2009-12785-C02-01) and for the contract awarded to author P. Varela (Juan de la Cierva program). They would also like to thank the Comisión Sectorial de Investigación Científica (CSIC, Universidad de la República) for financial support.

REFERENCES

Abdi, H., Dunlop, J. P., and Williams, L. J. 2009. How to compute reliability estimates and display confidence and tolerance intervals for pattern classifiers using the Bootstrap and 3-way multidimensional scaling (DISTATIS). *NeuroImage* 45: 89–95.

Abdi, H., Valentin, D., Chollet, S., and Chrea, C. 2007. Analyzing assessors and products in sorting tasks: DISTATIS, theory and applications. *Food Quality and Preference* 18: 627–640.

Albert, A., Varela, P., Salvador, A., Hough, G., and Fiszman, S. 2011. Overcoming the issues in the sensory description of hot served food with a complex texture. Application of QDA®, flash profiling and projective mapping using panels with different degrees of training. *Food Quality and Preference* 22: 463–473.

Ares, G., Barreiro, C., Deliza, R., Giménez, A., and Gámbaro, A. 2010a. Application of a check-all-that-apply question to the development of chocolate milk desserts. *Journal of Sensory Studies* 25: 67–86.

Ares, G., Deliza, R., Barreiro, C., Giménez, A., and Gámbaro, A. 2010b. Comparison of two sensory profiling techniques based on consumer perception. *Food Quality and Preference* 21: 417–426.

Ares, G., Giménez, A., Barreiro, C., and Gámbaro, A. 2010c. Use of an open-ended question to identify drivers of liking of milk desserts. Comparison with preference mapping techniques. *Food Quality and Preference* 21: 286–294.

Ares, G., Tárrega, A., Izquierdo, L., and Jaeger, S. R. 2014. Investigation of the number of consumers necessary to obtain stable sample and descriptor configurations from check-all-that-apply (CATA) questions. *Food Quality and Preference* 31: 135–141.

Ares, G., Varela, P., Rado, G., and Gimenez, A. 2011. Are consumer profiling techniques equivalent for some product categories? The case of orange-flavoured powdered drinks. *International Journal of Food Science and Technology* 46: 1600–1608.

ASTM. 1992. *Quantitative Descriptive Analysis (QDA)*. Philadelphia, PA: ASTM Digital Library. DOI: 608 10.1520/MNL10523M.

Barcenas, P., Elortondo, F. J. P., and Albisu, M. 2004. Projective mapping in sensory analysis of ewes' milk cheeses: A study on consumers and trained panel performance. *Food Research International* 37: 723–729.

Blancher, G., Chollet, S., Kesteloot, R., Nguyen, D., Cuvelier, G., and Sieffermann, J.-M. 2007. French and Vietnamese: How do they describe texture characteristics of the same food? A case study with jellies. *Food Quality and Preference* 18: 560–575.

Blancher, G., Clavier, B., Egoroff, C., Duineveld, K., and Parcon, J. 2012. A method to investigate the stability of a sorting map. *Food Quality and Preference* 23: 36–43.

Brockhoff, P. B. 2003. Statistical testing of individual differences in sensory profiling. *Food Quality and Preference* 14: 425–434.

Bruzzone, F., Ares, G., and Giménez, A. 2012. Consumers' texture perception of milk desserts II—Comparison with trained assessors' data. *Journal of Texture Studies* 43: 214–226.

Bruzzone, F., Ares, G., and Giménez, A. 2013. Temporal aspects of yoghurt texture perception. *International Dairy Journal* 29: 124–134.

Cadoret, M., Lê, S., and Pagès, J. 2009. A factorial approach for sorting task data (FAST). *Food Quality and Preference* 20: 410–417.

Campo, E., Ballester, J., Langlois, J., Dacremont, C., and Valentin, D. 2010. Comparison of conventional descriptive analysis and a citation frequency-based descriptive method for odor profiling: An application to Burgundy Pinot noir wines. *Food Quality and Preference* 21: 44–55.

Campo, E., Do, B. V., Ferreira, V., and Valentin, D. 2008. Aroma properties of young Spanish monovarietal white wines: A study using sorting task, list of terms and frequency of citation. *Australian Journal Grape Wine Research* 14: 104–115.

Carmines, E. G. and Zeller, R. A. 1979. *Reliability and Validity Assessment.* Thousand Oaks, CA: Sage Publications.

Cartier, R., Rytz, A., Lecomte, A., Poblete, E., Krystlik, J., Belin, E. et al. 2006. Sorting procedure as an alternative to quantitative descriptive analysis to obtain a product sensory map. *Food Quality and Preference* 17: 562–571.

Chollet, S., Lelièvre, M., Abdi, H., and Valentin, D. 2011. Sort and beer: Everything you wanted to know about the sorting task but did not dare to ask. *Food Quality and Preference* 22: 507–520.

Chollet, S. and Valentin, D. 2001. Impact of training on beer flavor perception and description: Are trained and untrained subjects really different? *Journal of Sensory Studies* 16: 601–618.

Chollet, S., Valentin, D., and Abdi, H. 2005. Do trained assessors generalize their knowledge to new stimuli? *Food Quality and Preference* 16: 13–23.

Dairou, V. and Sieffermann, J.-M. 2002. A comparison of 14 jams characterized by conventional profile and a quick original method, flash profile. *Journal of Food Science* 67: 826–834.

Dehlholm, C., Brockhoff, P. B., and Bredie, W. L. P. 2012a. Confidence ellipses: A variation based on parametric bootstrapping applicable on Multiple Factor Analysis results for rapid graphical evaluation. *Food Quality and Preference* 26: 278–280.

Dehlholm, C., Brockhoff, P., Meinert, L., Aaslyng, M., and Bredie, W. 2012b. Rapid descriptive sensory methods—Comparison of Free Multiple Sorting, Partial Napping, Napping, Flash Profiling and conventional profiling. *Food Quality and Preference* 26: 267–277.

Delarue, J. and Sieffermann, J.-M. 2004. Sensory mapping using Flash profile. Comparison with a conventional descriptive method for the evaluation of the flavour of fruit dairy products. *Food Quality and Preference* 15: 383–392.

Dooley, L., Lee, Y.-S., and Meullenet, J.-F. 2010. The application of check-all-that-apply (CATA) consumer profiling to preference mapping of vanilla ice cream and its comparison to classical external preference mapping. *Food Quality and Preference* 21: 394–401.

Falahee, M. and MacRae, A. W. 1997. Perceptual variation among drinking waters: The reliability of sorting and ranking data for multidimensional scaling. *Food Quality and Preference* 8: 389–394.

Faye, P., Brémaud, D., Teillet, E., Courcoux, P., Giboreau, A., and Nicod, H. 2006. An alternative to external preference mapping based on consumer perceptive mapping. *Food Quality and Preference* 17: 604–614.

Guàrdia, M. D., Aguiar, A. P. S., Claret, A., Arnau, J., and Guerrero, L. 2010. Sensory characterization of dry-cured ham using free-choice profiling. *Food Quality and Preference* 21: 148–155.

Hopfer, H. and Heymann, H. 2012. A summary of projective mapping observations—The effect of replicates and shape, and individual performance measurements. *Food Quality and Preference* 28: 164–181.

Jack, F. R. and Piggott, J. R. 1991. Free choice profiling in consumer research. *Food Quality and Preference* 3: 129–134.

Jaeger, S. R., Chheang, S. L., Yin, L., Bava, C. M., Giménez, A., Vidal, L., and Ares, G. 2013. Check-all-that-apply (CATA) responses elicited by consumers: Within-assessor reproducibility and stability of sensory product characterizations. *Food Quality and Preference* 30: 56–67.

Kennedy, J. 2010. Evaluation of replicated projective mapping of granola bars. *Journal of Sensory Studies* 25: 672–684.

Kermit, M. and Lengard, V. 2005. Assessing the performance of a sensory panel–panellist monitoring and tracking. *Journal of Chemometrics* 19: 154–161.

Labbe, A., Schlich, P., Pineau, N., Gilbert, F., and Martin, N. 2009. Temporal dominance of sensations and sensory profiling: A comparative study. *Food Quality and Preference* 20: 216–221.

Labbe, D., Rytz, A., and Hugi, A. 2004. Training is a critical step to obtain reliable product profiles in a real food industry context. *Food Quality and Preference* 15: 341–348.

Lawless, H. T. 1999. Descriptive analysis of complex odors: Reality, model or illusion? *Food Quality and Preference* 10: 325–332.

Lawless, H. T. and Glatter, S. 1990. Consistency of multidimensional scaling models derived from odor sorting. *Journal of Sensory Studies* 5: 217–230.

Lawless, H. T. and Heymann, H. 2010. *Sensory Evaluation of Food: Principles and Practices*, 2nd edn. New York: Springer.

Lawless, H. T., Sheng, N., and Knoops, S. S. C. P. 1995. Multidimensional scaling of sorting data applied to cheese perception. *Food Quality and Preference* 6: 91–98.

Lelièvre, M., Chollet, S., Abdi, H., and Valentin, D. 2008. What is the validity of the sorting task for describing beers? A study using trained and untrained assessors. *Food Quality and Preference* 19: 697–703.

Lelièvre, M., Chollet, S., Abdi, H., and Valentin, D. 2009. Beer-trained and untrained assessors rely more on vision than on taste when they categorize beers. *Chemosensory Perception* 2: 143–153.

Lim, J. and Lawless, H. T. 2005. Qualitative differences of divalent salts: Multidimensional scaling and cluster analysis. *Chemical Senses* 30: 719–726.

Meilgaard, M. C., Civille, G. V., and Carr, B. T. 1999. *Sensory Evaluation Techniques*, 2nd edn. Boca Raton, FL: CRC Press.

Meillon, S., Urbano, C., and Schlich, P. 2009. Contribution of the temporal dominance of sensations (TDS) method to the sensory description of subtle differences in partially dealcoholized red wines. *Food Quality and Preference* 20: 490–499.

Moussaoui, K. A. and Varela, P. 2010. Exploring consumer product profiling techniques and their linkage to a quantitative descriptive analysis. *Food Quality and Preference* 21: 1088–1099.

Nestrud, M. and Lawless, H. 2008. Perceptual mapping of citrus juices using projective mapping and profiling data from culinary professionals and consumers. *Food Quality and Preference* 19: 431–438.

Nestrud, M. A. and Lawless, H. T. 2010. Perceptual mapping of apples and cheeses using projective mapping and sorting. *Journal of Sensory Studies* 25: 390–405.

Pagès, J. 2005. Collection and analysis of perceived product inter-distances using multiple factor analysis: Application to the study of 10 white wines from the Loire Valley. *Food Quality and Preference* 16: 642–649.

Parente, M. E., Manzoni, A. V., and Ares, G. 2011. External preference mapping of commercial antiaging creams based on consumers' responses to a check-all-that-apply question. *Journal of Sensory Studies* 26: 158–166.

Patris, B., Gufoni, V., Chollet, S., and Valentin, D. 2007. Impact of training on strategies to realize a beer sorting task: Behavioral and verbal assessments. In: D. Valentin, D. Z. Nguyen, and L. Pelletier (eds.), *New Trends in Sensory Evaluation of Food and Non-Food Products*, pp. 17–29. Ho Chi Minh, Vietnam: Vietnam National University-Ho chi Minh City Publishing House.

Perrin, L., Symoneaux, R., Maître, I., Asselin, C., Jourjon, F., and Pagès, J. 2008. Comparison of three sensory methods for use with the Napping® procedure: Case of ten wines from Loire Valley. *Food Quality and Preference* 19: 1–11.

Piggott, J. 1994. Understanding flavour quality: Difficult or impossible? *Food Quality and Preference* 5: 157–171.

Piggott, J. 2000. Dynamics in flavour science and sensory methodology. *Food Research International* 33: 191–197.

Pineau, N., Schlich, P., Cordelle, S., Mathonnière, C., Issanchou, S., Imbert, A. et al. 2009. Temporal dominance of sensations: Construction of the TDS curves and comparison with time-intensity. *Food Quality and Preference* 20: 450–455.

Prescott, J. 1999. Flavour as a psychological construct: Implications for perceiving and measuring the sensory qualities of foods. *Food Quality and Preference* 10: 349–356.

Prescott, J., Lee, S. M., and Kim, K. 2011. Analytic approaches to evaluation modify hedonic responses. *Food Quality and Preference* 22: 391–393.

Rasinski, K. A., Mingay, D., and Bradburn, N. M. 1994. Do respondents really "mark all that apply" on self-administered questions? *Public Opinion Quarterly* 58: 400–408.

Risvik, E., McEvan, J. A., Colwill, J. S., Rogers, R., and Lyon, D. H. 1994. Projective mapping: A tool for sensory analysis and consumer research. *Food Quality and Preference* 5: 263–269.

Risvik, E., McEwan, J. A., and Rodbotten, M. 1997. Evaluation of sensory profiling and projective mapping data. *Food Quality and Preference* 8: 63–71.

Saint-Eve, A., Paài Kora, E., and Martin, N. 2004. Impact of the olfactory quality and chemical complexity of the flavouring agent on the texture of low fat stirred yogurts assessed by three different sensory methodologies. *Food Quality and Preference* 15: 655–668.

Small, D. M. and Prescott, J. 2005. Odor/taste integration and the perception of flavour. *Experimental Brain Research* 166: 345–357.

Smyth, J. D., Dillman, D. A., Melani Christian, L., and Stern, M. J. 2006. Comparing check-all and forced-choice question formats in web surveys. *Public Opinion Quarterly* 70: 66–77.

Soufflet, I., Calonnier, M., and Dacremont, C. 2004. A comparison between industrial experts' and novices' haptic perceptual organization: A tool to identify descriptors of the handle of fabrics. *Food Quality and Preference* 15: 689–699.

Symoneaux, R., Galmarini, M. V., and Mehinagic, E. 2012. Comment analysis of consumer's likes and dislikes as an alternative tool to preference mapping. A case study on apples. *Food Quality and Preference* 24: 59–66.

Tarea, S., Cuvelier, G., and Siefffermann, J.-M. 2007. Sensory evaluation of the texture of 49 commercial apple and pear purees. *Journal of Food Quality* 30: 1121–1131.

Teillet, E., Schlich, P., Urbano, C., Cordelle, S., and Guichard, E. 2010. Sensory methodologies and the taste of water. *Food Quality and Preference* 21: 967–976.

ten Kleij, F. and Musters, P. A. D. 2003. Text analysis of open-ended survey responses: A complementary method to preference mapping. *Food Quality and Preference* 14: 43–52.

Tomic, O., Nilsen, A., Martens, M., and Næs, T. 2007. Visualization of sensory profiling data for performance monitoring. *LWT—Food Science and Technology* 40: 262–269.

Valentin, D., Chollet, S., Lelièvre, M., and Abdi, H. 2012. Quick and dirty but still pretty good: A review of new descriptive methods in food science. *International Journal of Food Science and Technology* 47: 1563–1578.

van Kleef, E., van Trijp, H. C. M., and Luning, P. 2006. Internal versus external preference analysis: An exploratory study on end-user evaluation. *Food Quality and Preference* 17: 387–399.

Varela, P. and Ares, G. 2012. Sensory profiling, the blurred line between sensory and consumer science. A review of novel methods for product characterization. *Food Research International* 48: 893–908.

Veinand, B., Godefroy, C., Adam, C., and Delarue, J. 2011. Highlight of important product characteristics for consumers. Comparison of three sensory descriptive methods performed by consumers. *Food Quality and Preference* 22: 474–485.

Yu, C. H. 2005. Test-retest reliability. In: K. Kempf-Leonard (ed.), *Encyclopedia of Social Measurement*, Vol. 3, pp. 777–784. San Diego, CA: Academic Press.

Index

For Product Safety Concerns and Information please contact our EU
representative GPSR@taylorandfrancis.com Taylor & Francis Verlag GmbH,
Kaufingerstraße 24, 80331 München, Germany

Printed and bound by CPI Group (UK) Ltd, Croydon, CR0 4YY
01/05/2025
01858478-0001